AF114674

LIFE, BRAIN AND CONSCIOUSNESS
New Perceptions through
Targeted Systems Analysis

ADVANCES IN PSYCHOLOGY

63

Editors:

G.E. STELMACH

P.A. VROON

NORTH-HOLLAND
AMSTERDAM • NEW YORK • OXFORD • TOKYO

LIFE, BRAIN AND CONSCIOUSNESS

New Perceptions through Targeted Systems Analysis

Gerd SOMMERHOFF

1990

NORTH-HOLLAND
AMSTERDAM • NEW YORK • OXFORD • TOKYO

NORTH-HOLLAND
ELSEVIER SCIENCE PUBLISHERS B.V.
Sara Burgerhartstraat 25
P.O. Box 211, 1000 AE Amsterdam, The Netherlands

Distributors for the United States and Canada:
ELSEVIER SCIENCE PUBLISHING COMPANY, INC.
655 Avenue of the Americas
New York, N.Y. 10010, U.S.A.

ISBN: 0 444 88436 x

© ELSEVIER SCIENCE PUBLISHERS B.V., 1990

All rights reserved. No part of this publication may be reproduced, stored in a retrieval system, or transmitted, in any form or by any means, electronic, mechanical, photocopying, recording or otherwise, without the prior written permission of the publisher, Elsevier Science Publishers B.V./ Physical Sciences and Engineering Division, P.O. Box 1991, 1000 BZ Amsterdam, The Netherlands.

Special regulations for readers in the U.S.A. - This publication has been registered with the Copyright Clearance Center Inc. (CCC), Salem, Massachusetts. Information can be obtained from the CCC about conditions under which photocopies of parts of this publication may be made in the U.S.A. All other copyright questions, including photocopying outside of the U.S.A., should be referred to the publisher.

No responsibility is assumed by the Publisher for any injury and/or damage to persons or property as amatter of products liability, negligence or otherwise, or from any use or operation of any methods, products, instructions or ideas contained in the material herein.

Transferred to digital printing 2007

CHAPTER 7. - SELF-AWARENESS, IMAGINATION AND MEMORY

Recapitulation and summary 171
7.1 - Self-awareness 172
7.2 - Properties of the imagination 177
7.3 - The neural correlates of the imagination 180
7.4 - Aspects of memory 185
7.5 - Episodic memory recall 190

CHAPTER 8. - LANGUAGE AND RATIONAL THOUGHT

Introduction and summary 199
8.1 - Classes and the internal representation of classes 203
8.2 - Logical relationships between classes 204
8.3 - The brain's struggle with nameless entities 207
8.4 - What the sentence structure communicates 210
8.5 - Mental images and the propositional form 214
8.6 - The foundations of rational thought 216

CHAPTER 9. - MIND AND MATTER

Summary 219
9.1 - Categories of mental events 219
9.2 - The bridge between mind and matter 221
9.3 - The unity of the self in split-brain subjects 222
9.4 - The freedom of the will 224
9.5 - Knowledge and belief 226
9.6 - Artificial Intelligence 228
9.7 - Contrasts with the approach of academic philosophy 234

Part III: CULTURE

CHAPTER 10. - DEEPER LEVELS OF CONSCIOUSNESS (1)

Introduction to Part III and summary 243
10.1 - A crucial departure from the biological norm 248
10.2 - The roots of the human psyche 251
10.3 - Frustration, conflict and anxiety 255

CHAPTER 11. - DEEPER LEVELS OF CONSCIOUSNESS (2)

11.1 - The sense of beauty 259
11.2 - The moral sense 266
11.3 - A distinctive step in the evolution of life 273
11.4 - The religious dimension 274
11.5 - Concluding remarks 282

PART IV: APPENDICES

APPENDIX A: Further applications of the concept of directive correlation 287

APPENDIX B: Some basic concepts of set theory 291

APPENDIX C: Some basic concepts of formal logic 299

APPENDIX D: Generative and transformational grammars 305

APPENDIX E: Some basic concepts of information theory 311

APPENDIX F: Physical systems and causal relationships 315

***

BIBLIOGRAPHY 319

INDEX 329

PREFACE

The natural sciences deal with the observation and explanation of the physical world. Their language is a physical language in the sense that all key-concepts are carefully defined in terms of what is observable, measurable, or otherwise operationally accessible in the physical world - or in terms of abstractions from these elements construed with equal care. And the natural sciences owe their power and success as a collective enterprise, including the verifiability of their propositions, to the semantic precision and universally standardized meanings which can be established on this basis.

Biology and physiology are part of the natural sciences, and their language, too, is a physical language in the above sense. But the faculty of consciousness has so far eluded their grasp. The primary reason for this must be seen in a failure of scientists to find ways of defining, describing and analyzing this faculty in physical terms without violating what in everyday speech is meant by words like 'conscious' or consciousness'. The attempts of behaviourists of the Watson and Skinner school to deal with consciousness in terms that satisfied the canons of the natural sciences failed precisely on this score. Two major obstacles have been the subjectivity of what we commonly mean by conscious experience, and the reflexive nature of self-consciousness when this term is understood as knowledge of what passes in one's own mind.

Yet the problem has to be solved if the study of human cognition is to be placed on a truly sound scientific footing. Moreover, without understanding the physiology of consciousness we cannot come to understand the physiology of mental events generally. The mind/body problem, too, will remain unresolved - one of the greatest remaining challenges in modern science.

In this monograph I intend to show that the various obstacles that confront those who seek to deal with consciousness in a physical language *can* be overcome if a strictly methodical approach is followed in which from the start all analytical concepts are accurately defined in physical terms. Or rather, I intend to show how very far one can get by following such an approach.

Equally important is the realization that, since the brain acts as an integral whole in the higher brain functions, the proper scientific approach here has to be a *top-down* approach. That is to say, an approach which begins by looking at the brain as a *system*. This contrasts with the *bottom-up* approach which begins at the level of individual neurons and has remained the common approach in brain research; also with the information-processing approach (another bottom-up approach), and with the biochemical approach which studies the effect on brain functions of chemical agents of one kind or another.

This volume should not therefore be read in the expectation of finding what one may be accustomed to find in other scientific monographs in the cognitive disciplines, e.g. new data, a review of existing data, or new theories to explain some narrowly defined set of observations - all formulated in the language of one or other of the many competing schools of contemporary psychological thought. Rather, it should be read as an attempt o show how far a rigorously pursued systems-approach and strict adherence to a physical language can succeed in covering the phenomenon of consciousness in all its major aspects, including its subjectivity and reflexiveness, thus placing the whole problem of human cognition and mental activity on a sounder scientific footing. New perspectives will be opened up in consequence - perspectives that will suggest new answers to some questions commonly met in the literature, while rejecting other questions on the grounds that the new analysis shows them to rest on mistaken premises. New hypotheses can be formulated - now in well-defined scientific terms - and the empirical evidence examined in their light. And research can be pointed in new directions. I would also claim that if this physical and systems-oriented approach is followed with scrupulous care, real progress can be made towards a synoptic view of mental activity in physical terms and, in consequence, towards the still very distant goal of discovering the neural correlates of consciousness and of mental events generally.

When systems thinking is applied to any system that exhibits various levels of regulation, control and feedback (as in the brain), the conceptual framework required tends to have a closer affinity to that of control engineering than that of either neurophysiology or contemporary psychology. But to reach the phenomenon of *consciousness*, that framework has to be extended by a measure of conceptual analysis which is alien to the control engineer - and which has so far remained alien also to workers in cognitive psychology (and AI). Since the aim of this conceptual analysis is to arrive at clear *scientific* statements about the nature of consciousness, it also tends to differ from the conceptual analyses most commonly found in academic philosophy. In consequence the application of systems thinking to the vexed problem of human consciousness has no natural allies in any of these camps - which may be one of the reasons why progress has been so slow.

*

I came to all this a good many years ago as an engineer and systems analyst who had acquired a lasting interest in the relation between mind and brain from an Oxford course in Modern Greats. This contained a good chunk of philosophy and brought me into contact with Gilbert Ryle, whose famous THE CONCEPT OF MIND debunked with exceptional eloquence the 'ghost in the machine' notion of the mind. I was also privileged to have been tutored by Alfred Ayer, author of the now classical

LANGUAGE, TRUTH AND LOGIC, who impressed on my mind the need to check and double-check the precise meaning of any theoretical term one might be tempted to introduce in tackling this or any other problem. Indeed, when I first entered the literature of brain research, my strongest impression was that the scientific study of the higher brain functions required a far more thorough clarification of key concepts than had hitherto been produced. The space I have given in this volume to such a clarification should not strike the reader as excessive if he will remember that even the common concept of *time* had to be clarified in operational terms before the relativity theory could see the light of day.

To the debt which I still owe to Ryle and Ayer, I must add the even greater debt I owe to the late Sir Peter Medawar - zoologist, medical scientist, polymath and Nobel laureate. It was he who, following a publication of mine on regulation and control in living systems, persuaded me to look at the brain from the same systems-oriented standpoint that I had come to adopt. He also arranged for the original funding of that work. And it was he who, as Director of the National Institute of Medical Research, later drew my attention to the special problems posed by the phenomenon of consciousness. His exceptional erudition and breadth of vision enabled him to take a truly synoptic view of the whole problem-area and he clearly saw the need for a more rigorous and systematic approach. than had hitherto been attempted.

Turning to the more recent past, I must acknowledge also my debt to the friends and rival analysts who have read parts or all of the MS: Linda Bown, Mathew Buncombe, Mark Hill, David Jones, Tim Reid, Richard Sorabji and David Stott. Since they came from different academic backgrounds, their comments helped greatly with the almost intractable problem of presentation. For, when a topic cuts across as many different academic disciplines as does the problem of human consciousness, and may interest the general reader as well, it is an unenviable task to find a mode of presentation that will explain enough of the background in each field to meet the needs of the outsider, without at the same time boring the insider or irritating him by the inevitable sketchiness of such explanatory passages or their neglect of work that he may regard as important. Peter Medawar would have made a better job of it. I can only ask for the reader's indulgence. Finally I must thank Prof. N.J. Mackintosh for the hospitality of his department during my years at this university, and Jonathan Harlow. for his patient computer support.

Throughout this book I have for simplicity adhered to the traditional expedient of treating the masculine as a neutral gender.

G.S.
Cambridge 1989.

ACKNOWLEDGMENTS

Acknowledgments are due to the following publishers for permission to print copyright material: Basil Blackwell; Cambridge University Press; Clarendon Press; W.H.Freeman & Co.; John Wiley & Sons Ltd.; Kegan Paul, Trench & Trubner; The Listener; The Scientific American; Weidenfels & Nicolson Ltd.

Chapter 1

INTRODUCTION AND OVERVIEW

1.1 - SCIENCE AND THE NATURE OF CONSCIOUSNESS

There is probably no aspect of human life less clearly understood by science than the faculty of consciousness and the nature of its embodiment in the brain/body system. Yet there are few problems more urgently in need of a scientific solution. The scientists' prospects of eventually arriving at a clear perception of the neural mechanisms involved in the higher brain functions, and thus of the relation between mind and matter, depends on nothing less. That solution, I shall argue, demands a more methodical approach to the problem and an altogether more precise level of theoretical thought than has generally been applied in the past.

Throughout this book I mean by *science* the *natural sciences,* or as they are now often called, the *physical* sciences, for the term 'physical' has now established itself as virtually synonymous with 'material'. These sciences occupy a unique position in human knowledge in that they have jointly given us a comprehensive and coherent account of the natural world which stands out by the precision and definiteness of its concepts and the unequivocal reference they have (however indirectly) to the world of what is publicly observable. This is the secret of their success. Because only through these qualities do scientific concepts become unequivocal and effectively shared concepts – thus making it possible for science to proceed as a *collective* enterprise in which any worker can at any time pick up the work of another and test it or take it a step further.

All of this applies to the problem of consciousness. No theory about the nature of the faculty of consciousness and its physical substrates can claim to have arrived until it has been formulated in terms that are congruent with the conceptual framework of the natural sciences and satisfy the same stringent semantic standards of objectivity and precision. For consciousness is the faculty of a living organism and that organism is a *physical* system. It follows that in our analysis of the nature of this faculty we must desist from introducing any theoretical concepts other than concepts we are able to define in *physical* terms. This will be the most fundamental rule for the procedure I shall follow.

Thus the call is for an account of the nature of consciousness which is given in *physical* terms, but which must yet be able to capture what is commonly meant by consciousness, including such awkward aspects as the essential *subjectivity* and *privacy* of conscious experience.This is the rock on which attempts to tackle the problem of consciousness have foundered.in the past The ultimate aim, of course, must be to arrive at a physical account of consciousness which is sufficiently detailed to tell us in *neural* terms what happens in the brain when a person has conscious sensations, feelings, desires, volitions, thoughts, imaginings, memories, etc.

It follows from the nature of consciousness that we have to begin by looking at the *organism as a whole*. In other words, we need to follow a carefully targeted *systems* approach – or *top-down* approach as it is sometimes called. A state of consciousness, as this expression is commonly understood, is the state of a *person*, i.e. of the organism as a whole.It may be controlled by local events in the brain, but it is not itself a local event. And in its contribution to this state of the organism the brain, too, acts as an integral whole, as a system.

Systems analysis looks at the formal properties and logical structure of entities consisting of numerous interacting parts. That is to say, it begins with the functional relationships that seem to characterize the organization of the system as a whole, and then selects for detailed analysis those that seem relevant to the problems in hand: particular patterns of causal relationships for example, such as those involved in feedback loops. By contrast, the *bottom-up* approach to the brain – the most common approach in brain research – begins with the properties of individual neurons and their ascertainable reactions and interactions, their biochemistry, or their information processing capabilities. It is backed by the hope that along this road one may eventually arrive at something that can be identified with one or other of the higher brain functions involved in conscious experience. It fails in the absence of a clear perception of how these higher brain functions themselves may be conceived in precise and objective terms. In the 'top-down' approach this clear perception is one of the primary objectives.

Contemporary theories about the higher brain functions as such, e.g. theories of perception, cognition, thought and memory, tend to occupy a grey area between physiology and psychology, in which physiological concepts tend to be mixed with mental concepts and others of sub-scientific standards in order to arrive at least at some kind of theoretical explanation for the experimental data obtained in the field concerned.

To break out of this area an extensive clarification of concepts is needed. Concepts which seem important, or even essential, yet lack the qualities demanded by the natural sciences, will have to be replaced by concepts that satisfy them – but can yet be shown to do the same job. That is a drudge, but one that has to be faced. Chapters 2-4 will be mainly devoted to this task, and I shall apply here some of the basic insights gained in the development of modern systems analysis.

In this volume, too, a new conceptual framework will be introduced. But those readers who fear from these remarks that they are about to be plunged into yet another pool of proprietary concepts, can be assured that I shall be introducing only one single concept that may be wholly new to them. This is the mathematical concept of *directive correlation*, to be introduced in Chapter 3 (see also Section 1.7). All the others will be familiar concepts, though many of them will be newly defined in accordance with the standards of objectivity and conceptual precision which a sound systems analysis must demand.

***

1.2 - THE IMPORTANCE OF THE PROBLEM FOR SCIENCE

Consciousness holds the key to all our mental life. The kind of events that we commonly regard as *mental* events, such as our sensations, perceptions, thoughts, fantasies, memories, are all part of what James has called our "stream of consciousness". It is true, of course, that psychologists often talk about *unconscious* and *subconscious* 'mental' events or mechanisms. But it is, I think, correct to say that these hypothetical events or mechanisms have attracted the predicate *mental* only in so far as the authors of these hypotheses have defined or described them in terms borrowed from the description of *conscious* mental events (Freud's 'unconscious *wishes*', 'unconscious *hates*' and 'repressed *desires*', for example.) Psychologists would not, for instance, classify as an unconscious *mental* event an event which is defined in physiological terms, such as a rise in blood pressure of which the subject has remained unaware. Again, if we say that even an unconscious person has a mind, we are talking about something that we take to be preserved during the state of unconsciousness, and it is always something we conceived in terms of its manifestations in consciousness when consciousness returns. It is also worth noting in this respect that French and German have no strict equivalent to our 'mental', their 'esprit' and 'Geist' denoting the spiritual rather than what we call the 'mental' dimension.

Seen in this light, then, it is idle to think that science can discover the true nature of mental phenomena and the relation of mind to matter until it has learnt to understand the nature of consciousness. Take, for example, the problem of memory. In simple terms, to remember an event is to bring that event back into consciousness. I shall stress later that point that the mechanisms of memory cannot be fully understood unless we understand what they achieve, i.e. what it is to raise something to the level of consciousness. The same applies to the study of the mechanisms of pain, subliminal perception, dreams, sleep- walking, hypnosis, blind sight, and the like. Again, until the nature of consciousness is understood, it is idle to speculate whether AI (Artificial Intelligence) can ultimately achieve genuine mind-modelling.

Thus the failure of science to tackle the problem of consciousness and of the brain mechanisms involved - indeed, its failure even to hint how the material activities of the brain can yield, or even merely sustain, such seemingly immaterial qualities as conscious experience, has left a vast area of darkness in our understanding of human nature in general and of the higher brain functions in particular. It has left as total mystery how thoughts can influence physical movements, and it has left brain research without a pattern that would enable its accumulated wealth of neuro-anatomical and neuro-physiological data gradually to fall into place. It has also hindered progress in a variety of academic disciplines other than brain research and psychology. In the science of linguistics, for example, it has obscured the link between language and reality. In philosophy it has left academics with numerous questions about mind and consciousness which science cannot help to elucidate because they are insufficiently clarified to permit an experimental or theoretical scientific approach. It has also left such questions in the air as whether there is a conceptual link between consciousness and life, or whether a split-brain patient can be said to have psychological unity. And it has denied an answer to important questions asked by the man in the street: do animals have consciousness? Do they feel pain the way we do? Do they think? When does consciousness first arise in the human embryo? Can computers or robots have consciousness, feel pain?

The tenor of this book is that there is no way of arriving at a scientific answer to all these questions - or of reformulating them in terms of sufficient clarity that a scientist can address himself to them - except by way of an analysis which looks at the body and its brain as an integral physical system, and which is throughout conducted only with the aid of concepts that can be unequivocally defined in physical terms, while at the same time keeping its sights firmly fixed on the meaning which words like 'conscious' or 'consciousness' have in everyday life. This is a laborious process, permitting no easy short-cuts. The reader should be warned.

***

1.3 - THE 'STATE OF THE ART'

At the point at which an introduction to the science of consciousness should review the state of the art, one can only report that there is as yet no art to review. Instead one has to tell a tale of neglected issues, of evasions and lack of clarity in the questions asked or answers suggested, and of investigative approaches which, to use Wittgenstein's phrase, "cause problem and method to pass one another by" (the main reason, as he saw it, of what he called the "barrenness" of psychology).

In the contemporary scientific literature consciousness tends to be mentioned mainly in a clinical, physiological, or behavioural context to distinguish a state in which the subject is in full possession of his faculties from a state of sleep or coma. Sometimes, too, the term is used merely as a bracket for functions peripheral to consciousness, such as information storage or, more relevantly, the ability to report an event. Speculations about the faculty of consciousness tend to centre on the question of its evolution or social usefulness. Nicholas Humphrey in Cambridge and John Crook in Bristol, for example, broadly equate consciousness with self-consciousness and believe this to have primarily a social function. Humphrey holds that we need to understand ourselves in order to understand other people, while Crook holds that concepts relating to ownership need concepts relating to personal identity. None of these theories address themselves to the problem of the nature of consciousness as such. What, for example, is the physiological difference between stimuli to the brain which enter consciousness and the mass of brain-events which never do so?

A systematic scientific study of the phenomenon of consciousness should be allied with the study of perception and knowledge. This points in the first instance to the field of *cognitive psychology* as the discipline most directly concerned. Yet we look in vain in the literature of cognition for even an *attempt* at a systematic scientific treatment of the faculty of consciousness as such. It is symptomatic of the hiatus this has created, that in a recent (and very scholarly) review of the state of the 'cognitive sciences' (Gardner, 1985), the word 'consciousness' does not even appear in the seven-page index; and in the 400-page text it occurs only about a dozen times. This despite the fact that the author casts his net very wide indeed: his review covers not only cognitive psychology, but also philosophy, artificial intelligence, linguistics, anthropology, and the neurosciences. Yet, the author himself is in no way to blame for this omission. The deficit is merely symptomatic of the general neglect of the problem in the period covered by his work. It is symptomatic, too, that in the encyclopaedic OXFORD COMPANION TO THE MIND (ed. Gregory, 1987), the entry on *consciousness* is written by a philosopher (D.C. Dennett) and not a scientist. Indeed, the entry is contra-scientific, for the author deplores the absence of a "pre-theoretical" definition of consciousness, whereas in science every definition is theory-dependent. In the approach to consciousness followed in this work, too, it is regarded as crucial that theory and definition should go hand in hand (see below).

As regards the neural correlates of consciousness, some titles of books or articles tend to suggest that such correlates are already known, but to that extent these titles are also misleading. For example, the symposium entitled THE CEREBRAL CORRELATES OF CONSCIOUS EXPERIENCE (Buser *et al.*, 1978) may lead one to expect what the title would seem to promise. In fact, the book contains only an assortment of articles dealing with observations made on conscious subjects, e.g. the consequences of specific brain lesions. Although each of these articles is of great interest and considerable merit in its own right (I shall cite several later in this work), none of them enlighten us about the nature of consciousness as such and its neural basis, or, for example, about the difference between neural events in the brain that amount to a conscious experience and those that do not.

Only two of these articles actually seek to describe consciousness, and neither of these descriptions can be regarded as even remotely adequate to serve as a definition. Richard Jung describes consciousness as "a selective function for limiting actual mental experience and for choosing from amongst the many potentially psychical phenomena in the unconscious", while Kornhuber sees the essence of consciousness in "information processing in an active, self- controlled and adaptive manner". Jung's description if flawed from the start from a scientific point of view because it is couched in mental terms; while Kornhuber's description does not even begin to cover what people commonly mean by consciousness. Indeed, if taken as a definition, it would compel us to attribute consciousness to any living system, also to homing missiles and other robotic devices. In the same volume, Berlucci mentions in passing that the majority of neurophysiologists stick to an operational definition of consciousness: "regarding only those experiences as conscious experiences which can be verbally reported or, if worse comes to worse, reported in writing." This is a perfectly workable criterion and in practice it has proved to be widely acceptable. But it is no more than that and tells us nothing about the nature of consciousness and self- consciousness as such. A self-report like "I am seeing a red spot" presupposes a faculty of introspective self-awareness allied to a conscious awareness of the situation in which the report is invited. The very nature of these sophisticated brain-functions is the point at issue and the criterion sheds no light on these. In the reflexive nature of consciousness, our consciousness of being conscious, of course, we are faced with one of the most vexatious aspects of the whole problem. The lack of clear perceptions here can lead to such fallacies as the assumption that the contents of self-consciousness invariably have a propositional form. The author himself seems to have fallen into that trap. For he suggests that the efferent pathways of conscious experiences are to be identified with the descending neural pathways from the frontal lobe that appear to be needed for verbal or written reports.

As I shall cite with references in Chapter 4, some writers describe or define consciousness in terms of other mental notions, which, of course, takes us no further. Thus one sees consciousness described as "a system by which individuals become aware", as "activated memory", or as "choice". This is particularly embarrassing in the case of 'awareness', since dictionaries commonly define 'awareness' in terms of 'consciousness', thus leaving us with a neat case of circularity. Other writers have tried to make consciousness amenable to scientific discourse by radically reducing the notion itself, leaving little or none of the core of its common meaning. For example, one may see consciousness described as "the summation of total brain activity since birth" and self-awareness as being merely "a receptivity for stimuli arising within the body". Then there are those who tacitly adopt Freud's stance, that consciousness is something so obvious that it does not need defining! Again others have resorted to computer analogies and, for example, described consciousness as the brain's "high level operating system" or suggested that the relation between mind and body is to be compared with that between computer software and computer hardware. Both are metaphors which I shall reject as grossly misleading.

Many cognitive psychologists have felt themselves irresistibly drawn to AI as a possible source of paradigms for the higher brain functions. But it has not proved fruitful in the matter under discussion. Until a few years ago, workers in the field of AI were themselves generally convinced that their methods could successfully model the human mind, and cognitive psychologists here formed a receptive audience. Yet in the absence of a clear perception of the nature of consciousness, these claims eventually came to be recognized even by leaders in the field to have been ill-founded (see, for example, Winograd and Flores, 1986). In consequence, the once fashionable preoccupation with 'mind-modelling' has now been largely superseded in AI by work on the design of 'expert systems', i.e. systems dedicated to tasks of specific value in industry or elsewhere.

The brain is an integral dynamic system, ceaselessly interacting with the environment in a manner that is essentially holistic and far removed from the modular design, serial processing and sequential manipulation of symbols that tends to characterize the operations of conventionally programmed computers and robots. By a 'holistic' mode of operation is here meant no more than the dynamic way in which the whole brain seems to take part in setting up or sustaining, for example, an internal representation of the total situation which the orgnism is encountering and which its actions have to match. Moreover, even where the artefacts of AI and robotics succeed in emulating some of the special competences of the human brain, as in object recognition for example, it is a fallacy to take it for granted that such artefacts will enlighten us about the mechanisms of the brain. From a similarity of *competence* one cannot in logic infer a similarity of *method* or *mechanism*. After all, hydraulic jacks move the limbs of earth-shifting machinery with the same efficiency as does human muscle move human limbs, but one would hardly advise a myologist to look to hydraulic jacks for clues about the mechanisms of human muscle.

Conscious of this holistic character of the brain's operations, some workers in AI have turned to a new approach. Their idea is to study the theoretical properties of networks of devices which mimic neurons in the sense that they are designed as multiple-input-single-output units capable of undergoing adaptive changes. Reliance here is on network theory and on simulations by way of the growing family of computers capable of parallel- as well a serial processing. It was hoped that such 'connectivist systems' and 'neural computers' might lend themselves eventually to the simulation of some of the competences of the brain in a more realistic fashion. But I shall contend that there is little chance of achieving successful 'mind-modelling' along these lines so long as the designers of these systems remain without a clear perception of the target, viz. of the nature of conscious experience.

Cognitive psychologists sometimes excuse their enchantment with AI and computing machines on the grounds that they just cannot conceive of any way in which the brain can achieve what it does achieve, e.g. in object recognition, except the way in which similar competences have been produced in AI. But this is a sad state of affairs if it means giving up trying.

At a less controversial level, many cognitive psychologists have concentrated on the notions of *information* and *information processing*, notably in the theory of perception. Both notions acquired scientific respectability when they were given a mathematical definition in a famous paper by Shannon and Weaver (1949). However, although information and information processing must obviously play a vital part in the faculty of consciousness, we cannot reach the nature of that faculty itself in terms of just these concepts - nor the nature of thought, mental images, visual perception, or memory recall.

As another clear illustration of "problems and method passing one another by" one must cite the attempts made earlier this century by the school of radical behaviourism to define consciousness in objective terms, viz. in terms of a subject's observable behaviour or dispositions to behave in certain ways. It was part of a laudable effort, mainly associated with the names of Watson and Skinner, to put psychology on a sound scientific footing. It failed because it was far too restrictive in the type of observables the authors were prepared to consider. In fact, the only systems variables they admitted as legitimate observables were: sensory stimuli, overt responses and learnt associations between the two. No room was left in this scheme for the intervention between stimulus and response of cognitive structures of the kind we shall be considering. The secret of consciousness cannot be unlocked with such simple keys, and it is not surprising that as the result of his own work Skinner eventually came to reject the very notion of consciousness as a spurious one.

The general lack of clear perceptions about the nature of consciousness is also mirrored in a stark diversity of opinions about where consciousness is to be found in the natural world. There are scientists who maintain that if consciousness is taken to include self-consciousness then only humans possess it, and perhaps the apes. Most biologists, though, would credit a greater range of species with some form of consciousness, without, however, claiming that they understand its biological function, i.e. the advantages it bestowed in the evolution of the species. Some go as far down as the protozoa; a few have gone further still, attributing some kind of consciousness even to plants, while some scientists with strong metaphysical inclinations and, perhaps, an admiration for the metaphysics of A.N.Whitehead, have postulated the existence of a 'cosmic consciousness' which endows even atoms and molecules with a conscious dimension. Not uncommonly the notion of a cosmic consciousness has also been embraced as a way of reconciling one's scientific beliefs with one's religious beliefs. Thus according to the Oxford zoologist Hardy (1975) living organisms and their parts are "...carved out of the physical world by the mental element of the universe (to which consciousness and the Divine are related)....".

***

1.4 - ON THE MEANING OF 'CONSCIOUS', 'CONSCIOUSNESS' AND 'INTERNAL REPRESENTATIONS'

Any scientific theory of consciousness must make its terms clear, regardless of whether they denote observable phenomena or merely theoretical entities introduced to explain those phenomena. This calls for technical definitions, but it is a serious mistake to assume that these definitions must *precede* the theory. For, as has been said, in science theory and definitions go hand in hand. The definitions evolve as the theory evolves, for their terms must be terms covered by the theory. Indeed, as Popper and other philosophers of science have repeatedly pointed out, even every scientific *observation* tends to be theory-laden.

The definition of a term determines its meaning. In so doing, it specifies what shall be taken to be necessary and sufficient conditions for any entity to come under that term. Now definitions may be drawn up with different aims in mind and their usefulness or appropriateness will be judged accordingly. In this respect the definitions commonly found in science differ from those found in dictionaries. Dictionary definitions typically try to capture as faithfully as possible the sense or senses in which the respective term is used in everyday life. We may call such definitions *descriptive* definitions. Their aim is to do justice to the language of ordinary discourse. By contrast, the definitions used in science are *prescriptive* definition which declare the *technical* sense in which a given term will be used in the development of the theory. The question of the extent to which such definitions cover the common connotations of the terms concerned, tends to be a secondary consideration. A merely superficial kinship is often enough to have induced the scientist to use the respective word in his theory.

Typical examples are the technical definitions of 'force', 'energy' and 'work' adopted in the science of mechanics. These definitions are precise, unequivocal, mathematically formulated, and they came into being as an integral part of an immensely successful theoretical edifice. None of them cover all the various senses in which the same words tend to be used in ordinary discourse, as when we speak of the 'force' of an argument, of the 'work' done in writing a book, or of the boundless 'energy' of a friend. Nevertheless, their prescribed technical meaning has a sufficient kinship with some of the meanings occurring in ordinary discourse that one readily accepts the use of identical words. However, the science of mechanics would in no way have been impaired if no such kinship had existed. What atomic physicists have dubbed a 'charm' bears no relation whatsoever to the common connotations of that word.

Matters are not quite so straight forward when it comes to deciding on technical definitions for the phenomena of consciousness. Firstly, the definitions must cover the criteria for the occurrence of a conscious experience which have established themselves as useful and reasonably unequivocal in the cognitive sciences, e.g. the person's ability to report that experience. Similarly, they should cover the criteria a clinician would use, for example, to distinguish a state of consciousness from a coma. Failure on either count would only sow confusion in the scientific world itself. Secondly, the definitions must as far as possible cover also the sense or senses in which the noun 'consciousness' or adjective 'conscious' is used in everyday discourse. Failure on this count would not indeed invalidate the theory as part of which the chosen technical definitions occur. But the theory would fail to describe or explain what the world expects it to describe or explain. And the scientist might even be accused of choosing a language apt to deceive about the significance of his work.

Thus in the choice of technical definitions for the phenomena of consciousness we do have to take a close look at ordinary discourse and the way the words 'consciousness' and 'conscious' occur in everyday life.

At this point one could be content just to consult a good dictionary of the English language. But since a great deal of modern philosophy centres on the analysis of concepts, especially concepts relating to mental states, there is more to be gained from this source - provided one bears in mind that for the philosopher engaged in the analysis of concepts the way the words concerned are used in ordinary speech are the ultimate data to which he must bow, whereas, as I have explained, for the scientist they are no more than a secondary consideration. This may now seem an obvious point, but I have found that failure to appreciate this still occasionally tends to alienate a philosophical reader of my writings quite unjustifiably and unfairly.

Scientists are not naturally disposed to look at philosophy for enlightenment in their work, even in brain research. Metaphysics is kept carefully at bay. Theirs is an empirical discipline concerned with the construction of theories which can explain their observations while yet conforming to the tight standards of precision, objectivity and cohesion which good science demands. They tend to show little interest in the broad discussion of logical possibilities that is part of the philosophical tradition: questions as to whether mind is to be conceived as a product of matter, or matter as a product of mind; or whether both are to be conceived as wholly disparate 'substances', as Descartes and other 'dualists' have held; whether mind perhaps compares to matter as computer software compares to computer hardware; whether mind and matter are perhaps just different views of the same thing: like the difference between seeing a cloud from the outside and experiencing it as mist from the inside; or whether mind is perhaps just an epiphenomenon, a concomitant of brain-states that is neither their cause nor their result. The brain researcher just wants to discover what happens in the brain when people think, feel, remember etc. Let these generalized questions wait until he knows the answer to that one, he is apt to say, for the rest will then follow.

However, in modern times philosophers have tended to steer clear of broad metaphysical issues and to devote themselves more intensively to the analysis of concepts and meanings. It is part of the legacy of Logical Positivism and Linguistic Philosophy that Socrates' dart *it all depends on what you mean* has once again struck home. So let us reap the benefit.

I shall draw here on the lucid contribution to the volume CONSCIOUSNESS AND CAUSALITY (Armstrong and Malcolm, 1984), which has come from the pen of the distinguished Wittgensteinian, Norman Malcolm. By contrast with Armstrong's contribution, Malcolm does not seek to formulate some particular theory of the mind, but merely to explore what can be gleaned from the way in which words like 'conscious' and 'consciousness' occur in ordinary discourse. Right at the start he points out a grammatical difference between two uses of the word 'conscious'. And since this difference is important I quote him in full:-

> "In one use this word requires an object: one is said to be conscious of something, or to be conscious *that* so-is-so. One can be said to be conscious *of* a strange odour, of the stifling heat, of a friends ironical smile; and one can be also be said to be conscious *that* there is a strange odour in the house, that it is stifling hot, that one's friend is smiling ironically. The expressions 'conscious of' and 'conscious that' are generally replaceable by 'aware of' and 'aware that'. Being conscious of something or *that* so-is-so, I shall call the 'transitive' use of of the word 'conscious'; and I shall speak of 'transitive consciousness'.
>
> "There is another use of the word 'conscious' in which it does not take an object. If we think that a person who was knocked unconscious has regained consciousness, we can say "He is conscious", without needing to add an 'of' or 'that'. This use, in which someone can be said to be conscious *tout court*, I shall call the 'intransitive' use of the word 'conscious', and shall speak of 'intransitive consciousness'. It may be noted that when 'conscious' is used intransitively it cannot be replaced by 'aware': to be aware is always to be aware *of* or *that*."

Later he also clarifies the notion of 'self- consciousness':

> "In everyday conversation and in literature, a person is said to be 'self-conscious' who *reflects* a lot on his own attitudes, interests, personality, and on his reactions to other people and situations in human life, and also on his responses to music, literature, art and nature."
> "Such self- consciousness if often stimulated and nourished by the observations of others about oneself. One's own person becomes an object of study." However, he adds, the term can also be used in a different sense, which he calls the 'technical philosophical sense'. In this sense it stands for "the normal ability of any persons to report his/her bodily postures, movements, sensations, intentions - such reports being generally true although not based on observation."

In the words he quotes from Amscombe (1981), self-consciousness in this sense is "consciousness that such-and-such holds of oneself".

It follows from what has gone before that this second sense is the more important one in this work, and for the present I shall call it 'self-awareness' - use of 'awareness' being justified since this consciousness now has an object. The notion is also important because some philosophers e.g. Brentano and Locke, have seen in this self-awareness the very essence of consciousness. "Consciousness", John Locke wrote in AN ESSAY CONCERNING HUMAN UNDERSTANDING (1690), "is the perception of what passes in a man's own mind". Similarly, Brentano took it to be one of the general characteristics of all mental phenomena that they are "perceived only in inner consciousness".

Malcolm rejects the thesis that all mental phenomena are perceived in 'inner consciousness' or 'inner perception', on the grounds that it contradicts the normal employment of these expressions. There may be *some* situations in which it can be said that a person perceives (is aware of) his/ her own feelings, e.g. of disgust, contempt or irritation. There are others in which this cannot be said: thoughts and feelings may have qualities of which the subject is unaware, though they might strike someone to whom the thought is uttered: a jealous thought, for example. This observation is correct so far as it goes, but it must not mislead us. It points correctly to the fact the contents of self-awareness may be categorized in various ways, i.e. elicit generalized reactions of different kinds. And when these are expressed verbally (using such labels as 'jealous', 'disgusting', etc.) the subject's own categorizations can be compared with those of an outsider and prove to be rather more limited. But this does not disprove the assertion that the contents of a person's stream of consciousness are invariably *introspectable*, i.e. capable of becoming the object of an inward directed act of attention which will tend to

1.4 - On the meaning of 'consciousness'

be followed by categorizing reactions of one kind or another, including reactions lacking a verbal expression.

In his own contribution to the volume, Armstrong compares this inner sense with a self-scanning or self-monitoring device such as may be found in complex mechanical devices. Malcolm does not dispute that the living organism contains powerful self-scanning devices, but he strongly disputes Armstrong's use of this analogy to draw a distinction between pain and awareness of pain, and to claim that these are 'distinct existences' on the analogy that the oil shortage of a car and the indication of this on the dashboard are distinct existences. As Malcolm sees it, in ordinary life we draw no distinction between 'having a pain' and 'being conscious of a pain' - though, I would add, one might be inclined to use the second rather than the first expression if the pain tends to absorb one's attention. Malcolm supports his criticism with a detailed review of a variety of ways in which we normally talk about our own pains, or ask questions about someone else's pain. With all of this I find myself in agreement. As I see it, we have a sensation of pain, or simply a pain, when the alarm signals issued by internal mechanisms designed to detect harmful or potentially harmful occurrences, penetrate to the level of consciousness. One of our tasks will be to discover what precisely this last step may mean in physiological terms. Armstrong's mistake is to have identified 'pain' with the source of the commotion, reserving 'sensation of pain' for its projection into consciousness - which, as Malcolm points out, conflicts with our common use of these terms.

Malcolm also criticizes, and here again I think rightly, those who claim that the subjective character of experience, the 'qualia' or 'raw feel', is something of which no physical account could ever be given. Once again, in a philosopher's fashion, he looks at the way one might normally talk about this subjective character: the questions one could sensibly ask about it, the answers one could expect, etc. He concludes that the distinction between seeing or hearing on the one hand, and the 'subjective character' of seeing or hearing on the other, is a vacuous one: "Not only cannot I display the 'subjective character' of seeing or hearing to others, I cannot even display it to myself."

As I see it, what people call the 'subjective character' or 'raw feel' of an experience amounts to no more than the distinctiveness of the content of anyone's self-awareness at any point in time, i.e. the distinctiveness of a person's internal representation of the current state of the self. Any such representation is as unique as the self whose current state or condition it represents. Nor would it do its job unless it were able to differentiate categorically between auditory and visual events according to the role they play in the life of that individual. And, as regards the possibility of a physical account of all of this, I know of no principle that precludes a physical account of *types* of occurrence that differ in each of their instantiations.

In connection with the above remarks a few words must be said about the notion of *internal representation* which I have there introduced. This is a concept of which I shall make extensive use, for I shall claim that the phenomena of human consciousness can be fully accounted for on the assumption that the brain has the power to form certain kinds of internal representations, e.g. representations of the nature of the objects perceived in the environment. Here again one must be indebted to the philosopher for exposing some of the dangers inherent in the use of the word 'representation' (cf. Hacker, 1987). The main danger in the present context is the following. To most people *in the arts,* a representation, such as the representation given by a painting or sculpture, is something that presupposes the actual or potential presence of a viewer. For these people, therefore, any talk about the brain forming internal representations of one kind or another may be taken to imply that somewhere in the brain there is a little man or 'humunculus' who inspects these representations.

It is therefore important to realize that scientists and technologists are accustomed to talk about 'representations' in a sense that carries no such implications at all They use this term for example, when they explain how in a flight computer certain external variables, such as altitude, airspeed, rate of climb, etc., are *represented* by certain binary numbers at certain addresses in the computer's RAM. The defining characteristic of these 'representations' is that the binary numbers in question

a) map the external variables concerned (altitude, airspeed etc.), and

b) are used by the computer as part of the data base from which it computes the corrections required in the aircraft's control surfaces, engine speed etc.

I must therefore make it clear from the start that I shall use the expression 'internal representation' in the sense here described (I shall define it more accurately later in this work). Hence any suggestion that my concept of internal representations implies the existence of a humunculus in the brain, would be wholly misplaced.

***

1.5 - FURTHER ASPECTS OF CONSCIOUSNESS

The operational criterion we have now repeatedly met for people being conscious of an event or state of affairs, is that they are able to report that event or state of affairs. But clearly it would be a fallacy to conclude that the faculty of speech is crucial for conscious experience. It would certainly contradict our common perceptions if consciousness were to be denied to any creature lacking the faculty of speech. Hence we must look at the non-verbal powers and operations which the ability to report an experience (e.g. a pinprick) presupposes. As I have already indicated, any self- report presupposes that the subject has formed (in the sense explained) an internal representation of the current state of the self, in this case one which includes the experience (pinprick) in question. This representation covers not only the distinctiveness of the event itself, but also its location

1.5 - Further aspects of consciousness

in the body and the general context in which it occurs. In fact, we know from experience that *whatever is represented in consciousness is represented in the context of the total situation*. Indeed, it is *this* that makes it so meaningful.

The latter consideration leads us back to the *intransitive* sense of consciousness, and so to the criteria a clinician may use to conclude that a patient has recovered consciousness or is gradually recovering consciousness. Full recovery would no doubt be signalled again by the patient's ability to report sensations or feelings. But the clinician would in fact hope for more than this, viz. signs that the patient can again recognize objects, place them in their context, say who he is, grasp where he is etc. Before all of this returns, the clinician may perceive signs of a returning consciousness in the patient opening his eyes, focussing them on objects or scanning the room. These actions are significant in respect of what I have said above, because they are plainly needed for the construction by the patient of an internal representation of his surroundings.

Once again, the faculty of speech in not crucial here. The clinician would be fully satisfied that the patient had regained consciousness if the latter were to get up, get dressed, collect his belongings and walk out.

Thus the most distinctive feature of a fully conscious person is that he shows signs of knowing who, and what, and where he is, how he came to be there, etc. In some sense he has a grasp of the world and his place in the world. The questions "where am I?" and "what happened?" are symptomatic of a brain struggling to recover that grasp after a spell of unconsciousness.

What has the patient himself to say about the state of consciousness that has now returned? How does it appear from the *inside*, so to speak? Is it a particular feeling? Certainly not. We can form no experiential notion of consciousness itself as contingent human state or condition, e.g. as a certain feeling, for the obvious reason that it is never absent when we are awake and thinking. When we are awake it is all-pervasive. We can form an experiential notion of a state of hunger, thirst or tiredness, as contingent conditions, because we consciously experience both hunger and non-hunger, thirst and non-thirst, tiredness and non-tiredness. But we cannot similarly compare experientially a state in which we are conscious with one in which we are not. For we are never conscious of a state of unconsciousness. As Malcolm observes, the statement *I am conscious* never occurs in practice because it has no job to do, for the obvious reason that we can never meet the statement *I am unconscious*.

All one can say from a subjective, experiential, 'inward looking' point of view, is that the world and the self-in-the- world seem to exist only while one is in a state of consciousness, for when consciousness dissolves in the twilight state of drowsiness, the world and the self seem to dissolve with it. What then does one lose at that point? In broadly descriptive terms, the best we can say is that one loses the effective operation of a coherent and (in scope) comprehensive internal representation of the world, of the self and of the relation of the self to the world. Conversely, when one wakes up, one tends to pass through a state of semi- wakefulness in which perceived shapes gradually fall into place until a full representation of the room has reconstituted itself and knowledge returns of where one is. This internal representation is comprehensive in scope, though not in detail. It is a representation of the total situation, global in its extent but deficient in detail. It has many open termini. When I am conscious, I will normally know not only what room I am in, but also in what house, town, country, continent. And I will know that I am on a planet circling the sun, etc. But I don't know what passes outside the door of my room unless I go and look. Major gaps may also occur contingently, as when one wakes up in a hospital after an accident and has no idea where this is or how one came to be there.

This *internal representation of the world and the self-in-the world*, or *global world-model*, as I shall call it for brevity, is a representation of *actualities*. As such it does not cover all that needs to be discussed in connection with consciousness and mental events. Because we plainly also have the power to form internal representations of *possibilities*, of merely possible objects, events or situations. This is the faculty of the *imagination*. A great deal of our thinking is conducted in terms of such *mental images*, as when one tries to figure out how a carpet might fit into a room, plans a trip into the nearest town, envisages the goals one wants to achieve, pictures the possible consequences of this or that act, etc.

Closely allied to this is the faculty of *memory recall*: the internal representation of events which one accepts (acts upon) as representations of past occurrences. One's internal representations of the present, of the here-and-now, are are not representations of a time-slice. Except in severe brain lesions they cover both the recent past and much of the more distant past. We may call this the *historic* dimension of the subject's global world-model. In memory recall, traces left by something that was once an internal representation of the here-and-now, become reorganized in the shape of projections into this historic dimension of the subject's global word- model. However, it is a well-attested fact that a creative element also enters here. It is a *reconstructive* rather than merely *reproductive* process. That is why I have described the faculty of memory recall as closely allied to the faculty of the imagination.

It follows from what has gone before that the faculty of consciousness, indeed the whole realm of mental events, rests on a power of the brain/body system to operate with internal representations of both actual and possible objects, events, or states of affairs generally. In the next section I shall list in an orderly fashion, and briefly discuss, the *three main categories* of internal representations which I shall claim to be sufficient to account for the phenomena of human consciousness and the main categories of mental events. Thus the bulk of my work will focus on the nature of these internal representations, how they are to be defined in functional terms, how these in turn are to be translated into causal terms, how the respective representations come to be formed and corrected, and, finally, how they may come to be realized in the neural networks of the brain.

***

1.6 - THE THREE BASIC CATEGORIES OF INTERNAL REPRESENTATIONS

Although in ordinary speech we predicate states of the consciousness of *persons*, i.e. of the organism as a whole, there are, as I have said, ample reasons for looking at the brain as the responsible organ. I have also said that I believe the phenomena of consciousness can be fully accounted for in terms of the brain's power to form three main categories of internal representation. For brevity these will be called representations of category 1, category 2 and category 3. The sense in which 'internal representation' is here to be understood was briefly indicated in Section 1.4. A precise definition will be given in Chapter 4. Even so, some of the functional relationships we shall be discussing involve the brain/body system as a whole. The definition to be given in Chapter 4 will be a *functional* one. The *structure* of the internal representations will form the main topic of chapters 5-7.

1. The *first* and most fundamental of these three faculties of the brain consists in a power to form a coherent and comprehensive internal representation of the surrounding world, of the body and of the relation of the body to that world. I shall call this the subject's *internal representation of the world and of the self-in-the-world*, or, for short, the subject's *global world-model*. Loosely we may describe it as the 'cognitive presence' of the world in the brain. It amounts to a structure which we must take to intervene between stimulus and response. Representations belonging to this global world-model will be referred to as *representations of category 1*.

Much of this comprehensive and coherent internal representation is based on past experience and on information received in the past, but it is constantly checked against current sensory inputs and enhanced, refined, or corrected accordingly. I shall call this the *reality testing* part of the representational process. In other words, the brain's internal representation of the world needs a feedback which exposes its errors. The notion of feedback here flows naturally from our decision to regard the brain as an *dynamic* system ceaselessly interacting with the environment.

One obvious source of such feedback is provided simply by the success or failure that follows actions based on this internal representation of the world in general and of the immediate environment in particular. However, I shall argue that this is only a minor source of feedback. The brain has a more direct source of feedback here in its power to anticipate coming events (including coming stimuli of different kinds) on the strength of these internal representations of the world, coupled with a power to detect discrepancies between what is anticipated and what actually occurs. Hence we are lead to the conclusion that the brain's major form of feedback here rests on the *expectancies* it forms - where 'expectancy' is one of the terms to which we shall have to give an objective, non-mentalistic definition in due course.

Disconfirmation of expectancies elicit *surprise* reactions (technically called *orienting* reactions), including arousal and shifts in attention designed to facilitate a correction of the internal representations concerned. Indeed, the limitations of a subject's world model can be explored at both a physiological and behavioural level by noting his/her surprise reactions in different circumstances.

It is not difficult to conceive of the brain's power to form internal representations of this kind as having evolved in the evolution of animal life from simple beginnings, such as a basic body knowledge coupled with a rudimentary power to form a cognitive map of the landmarks of the nearby world and of the animal's current spatial relation to them.

That part of the subject's global world-model which relates to the current state of the body, or 'somatic self', will be called the subject's *body-awareness* or *first-order self-awareness*. This comprises representations of the topology of the body, of posture and movement, of major physiological needs, of the location of traumatic internal stimuli, and the like.

I must mention at this point that in systems theory, as in the physical sciences, the term 'state' is used in a sense that differs from the term's common use in the discussion of mental phenomena. Psychologists commonly mean by the *mental state* of a person a condition which persists over a span of time during which it ensues in characteristic behaviour patterns - for example, when a person is said to be in a state of despair or a state of rage. By contrast, in systems theory and the physical sciences the 'state of the system at time t' denotes the values which the variables of the system have at time t, however transient these values may be.

2. The *second* faculty consists of a power to form internal representations, not only of *actualities*, but also of mere *possibilities*, viz., representations of *fictitious* or *absent* objects, events or situations. One of their distinctive features is that these representations are obviously not subject to the reality testing mentioned above. Another well known feature is that they are invariably composed of elements derived from the experience of the actual world. Representations belonging to this class will be referred to as representations of category 2. For reasons already explained, I shall also include in this category representations which are accepted in consciousness as representations of *past* events or situations, i.e. the representations that occur in memory recall.

1.6 - The three basic categories of internal representations

3. The *third* faculty consists of a power of the brain to form representations which register the occurrence of representations of category (1) or (2) as being part of the current state of the organism. I shall refer to them as representations of category 3. This category of representations, for example, would add to the occurrence of a representation A of category 1 or 2 a further representation B(A), viz., a representation of the fact that the occurrence of A is part of the current state of the organism - thus adding A to the brain's global world-model. It may be described as a state or 'set' of the brain in which A enters in an appropriate manner into the determination of responses that need to be correctly related to the current state of the organism as a whole.

Since the representations belonging to this category 3 are representations of relationships in which one of the relata is itself of the nature of a representation, there is a sense in which we can say that they are representations of a *higher* order than the representations of category 1 and 2. I shall shall therefore call them *second-order* representations. The corresponding component of self-awareness will be called *second-order self-awareness*, clearly to be distinguished from the first-order self-awareness discussed above. When attended to as an act of introspection, these representations form the basis of self-reports like *I see a tree* and *I have a pain*, and may thus be identified with what Locke called "the perception of what passes in a man's own mind". I shall say more about their structure in Section 1.8.

This third class of internal representations extends the brain's body-knowledge into a new and non-somatic dimension. In adding this dimension to our concept of self-awareness, the term 'self' in this expression now acquires a meaning which accords with the meaning the term has in ordinary discourse. For, in everyday life we tend to mean by *the self* the subject of our emotions, feelings, thoughts etc.

*

Whether, in an analogous manner, the brain also forms third-order representations (i.e. representations which represent the occurrence of a particular state of second-order self-awareness as being part of the current state of the self), or even representations of a higher order, must be decided on the evidence. To allay possible fears in the reader, I must add that there is nothing in the notion of self-awareness which I shall present that forces this conclusion, or even implies an infinite regress of the kind "I am aware that I am aware that I am aware......". (The notion of self-awareness has been a rich source of philosophical traps. The reader who has definite views on this subject will be well advised to suspend judgement until he has seen how the faculty of self-awareness eventually takes shape in our analysis. Some indication of this will be given in Section 1.8 below.)

I shall treat the brain's internal representation of the world and the self-in-the-world (its 'global world-model') as fundamental to consciousness: in line with remarks already made earlier, I shall postulate that *nothing enters consciousness except by way of the brain finding a place for it in its representation of the world and of the self-in-the-world*. As has been said: whatever is represented in consciousness is represented in the context of the total situation.

According to this postulate, for example, a traumatic (or potentially traumatic) internal event becomes a conscious sensation, in this case a *pain*, only when the brain manages to project its occurrence into its global world-model, viz., by adding this occurrence to its internal representation of the current state of the self. As is well known, the brain can be mistaken in the locality or region to which it assigns the stimulus in this projection. ('Pain' is not always used in a sense that implies a location. One may talk of the pain of grief or the pain inflicted by an insult. But we need not be drawn into the argument whether or not this is to be regarded as a metaphorical use of the term. If we can reach a satisfactory account of localizable pains we shall have no difficulty with non-localizable pains.)

It also follows from the above postulate that the brain's internal representations of absent or fictitious objects can only become *conscious* events if they are supplemented by second-order representations which represent their occurrence as being part of the current state of the self.

*

Most of what has been said above relates to the *faculty* of consciousness, hence to consciousness in the *intransitive* sense. To reach the *transitive* sense, i.e. ' being conscious of' or 'being conscious that' (cf. Section 1.4), I shall draw on the notion of *attending*. Dictionary definitions of this sense of 'conscious' tend to use such phrases as "aware of particular fact or phenomenon" (Webster's Collegiate Dictionary). In all these cases we are talking about something that has come to attract the subject's attention, and attention may be conceived in objective terms as a context-dependent brain-process which enhances selectively the responsiveness of particular brain-structures to particular sets or classes of inputs, thus establishing their dominance in the control of the brain's ongoing activities.

In the brain's internal representation of the world there will be some parts which are attended to at any given moment in time, while the rest is seemingly ignored. This does not, however, mean that the rest is without current influence. Although only a small fraction of one's current model of the world and of the self-in-the-world is attended to at

1.6 - The three basic categories of internal representations

any given point in time, the model remains effective in its entirety. Though the unattended parts remain at the back of the mind, so to speak, they nevertheless enter into the choice of one's activities, albeit only in an indirect or negative manner. For example, only because I have an effective model in my head of the total situation in which I am writing these lines, can I get on with this job without constantly having to glance around to make sure that no one is about to knife me. Again, there may at the time be all sorts of background noises. They don't worry me because they figure in my current world-model as harmless or irrelevant to my present pursuits. I become conscious of them only when for one reason or another they become relevant and so come to attract my attention - perhaps because the noise level becomes too intrusive, or because I hear my name mentioned.

I shall take *introspection* to denote the process of attending to that part of one's global world model which represents the current state or condition of the self.

I shall also discuss the distinction between *voluntary* and *involuntary* actions. I shall take this distinction to relate to the difference between actions which are governed by the subject's global world-model and those that are not (reflexes being obvious examples of the latter). On this interpretation, in turn, I shall base my explanation of how it is that we come to see ourselves both as the authors of our voluntary activities and as creatures endowed with a *free will*.

The technical definition of consciousness

If I am correct in arguing that the phenomena which the common uses of the words 'conscious' and 'consciousness' denote can be accounted for in terms of the three categories of internal representations I have here described, the question arises what technical definition this suggests for the faculty of consciousness within the framework of the theory.

For reasons to be detailed in Chapter 4, I shall recommend defining the faculty of consciousness as a *power of the brain to form internal representations of both category 1 and category 3, but excluding reference to category 2*. Inclusion of category 3 is recommended on account of the fact that the contents of consciousness are always introspectable. Category 2 is excluded because, although the faculty of the imagination is necessary for *thought*, I do not think that it is a necessary condition for what we mean by the *consciousness* in everyday life. On this definition, then, it is a necessary condition for consciousness that the brain should be able to form an internal representation of the world and the self-in-the world which comprises both first- and second-order self-awareness. Thus consciousness on this definition implies self-consciousness. A hypothetical creature which could form internal representation of the surrounding world and the body-in-that world, but lacked second-order self-awareness, might then be described as having *cognition* of the world and the body, but lacking *consciousness*.

However, when we reach the stage in this work at which specific conclusions are reached about the structure and neural basis of these internal representations, we shall also be lead to the conclusion that where a high degree of first-order self-awareness exists, some degree of second-order self-awareness is likely to exist as well. For, both consist of certain categories of states of conditional expectancy which come into being in consequence of identical processes - albeit at different levels of sophistication.

Thus consciousness, as here defined, is not a sufficient condition for *thought*. Thoughts do not necessarily need speech, but they do need the faculty of the imagination, hence internal representations of category 2.

***

1.7 - THE APPLICATION OF SYSTEMS THEORY

To reach a point at which all of the above can be more clearly defined and also translated into physical terms, a great deal of analysis has to be conducted and definite decisions have to be reached. We have to decide what precisely is here to be understood by an 'internal representation', and if the respective definition is given in *functional* terms, the physical meaning of those terms will have to be settled in a precise way and their causal implications exposed. This applies to all terms that will be brought into the discussion. For example, if we have cause to describe some activities as *coordinated*, others as *adaptive, well-matched, appropriate* to a situation, *regulated, integrated*, etc., we may be using terms which appear innocent enough in ordinary discourse, but which yet lack the clarity a rigorous systems analysis needs. Although these and similar terms appear frequently in the biological literature, they are generally left undefined. This will not do in our case. We need definitions which are sufficiently detailed to enable us to infer from each such assertion precisely what it implies by way of spatio-temporal and causal relationships. For, unless we have such knowledge we cannot later infer what conditions the neural networks of the brain must satisfy to produce activities of the kind described by the terms I have cited. And since even 'causal relationship' is a broad term, we shall have to detail precisely what is here meant in the given context. A rigorous scientific analysis demands no less.

All of this is the business of systems theory, for systems theory deals with the abstract characteristics of complex but organized entities. In respect of living systems, this means primarily the analysis of the functional relationships that characterize the nature of their organization.

Now all the terms I have cited - terms like 'adaptation', 'coordination', 'regulation' etc. - relate to one of the most distinctive characteristics of all living systems, viz., the patent *directiveness* or *goal-directedness* of their activities or processes.

It is distinctive of the living organism that almost every activity or process, internal as well as external, appears to be directed towards some particular end or goal, i.e. to be produced in a way that causes it to fit the circumstances in which it takes place in a manner conducive to some biologically significant end-result: the capture of a prey, the construction of a nest, the preservation of the constancy of the internal environment, the body's repair of damaged tissue, the learning of special skills, the production of offspring, and ultimately, of course, the survival of the individual or species.

This directiveness is often called the *teleological* aspect of living systems (from the Greek: 'telos', meaning 'end' or 'goal'). I shall use the same expression, although other writers have preferred the term 'teleonomic' on the grounds that 'teleological explanation' has sometimes been used to describe a rather deficient (and occasionally quite unscientific) way of explaining biological activities in terms of their 'goals' without a clear perception of the nature of a goal-directed activity as such. We can brush this objection aside, for the system-analysis I shall employ is precisely designed to clarify these matters and thus remove likely causes of confusion. *In fact, its main task will be to take a close look at this directiveness, to expose the spatio-temporal and causal relationships in which it resides, and, on the strength of this analysis, to suggest precise physical definitions for all terms (such as those exemplified above) which directly or indirectly relate to this directiveness.* Here lies the key of the formal account modern systems theory can give of the distinctive kind of functional orderliness that characterizes living systems, and the nature of life itself. This part of the work will form the content of Chapters 2 and 3.

The directive activities of living organisms are also sometimes called *purposeful* activities. In this connection students of biology tend to be told that the word purpose can be used in three different senses. It can mean the mental purpose of a human action. This is the psychological sense. A second sense occurs when we talk about the 'purpose' of an organ or anatomical structure, e.g. the 'purpose' of the thyroid gland. Here the word is plainly used as a synonym for *function*. The third sense is the relevant one in the present context and would be described to the student as the 'cybernetic' sense. It occurs when the word is used in an objective sense for the end-result which a directive activity is designed to achieve, e.g. the navigational direction which the autopilot of a ship is set to maintain, or the position to which a limb is moved under the feedback control of the nervous system. However, whereas the biology student is generally expected to be satisfied with such distinctions by example, we need an *explicit* account of these distinctions in the work to come. For example, in most human artefacts, such as robots and homing missiles, and in many biological systems, the directiveness of an activity is produced by way of negative feedback loops. It may, therefore, be tempting to assume that this is necessarily so, and even that teleological concepts could be given a precise definition in terms of feedback mechanisms. Upon analysis, though, both assumptions will prove to be false.

Concepts relating to the directive or 'teleological' character of biological activities or processes span a wide range. Even a seemingly simple concept like *adaptation* belongs to this teleological category. You cannot apply it to non- teleological systems and say, for example, that the planets are 'adapted' to their paths around the sun. Other examples are the concepts of *coordination, regulation,* already cited, and *integration, function, learning, matching, searching, 'trying,* , to mention just a few. Even the nowadays so common notions of *'information'* and *'information- processing'* are generally used only in a context which upon inspection proves to be a teleological one. Indeed, it can be said that the only processes in the universe which require information, are teleologial processes.

All the concepts I have here cited as teleological concepts tend to be left undefined and vague in the literature. For example, people may describe some vital activities as being highly *coordinated* without having a concept of sufficient clarity to let one infer what this implies by way of causal interrelationships - which would be precisely the kind of information one would need to track down the neural mechanisms of this coordination.

In fact, owing to this vagueness the whole teleological aspect of living nature has in the past given rise to numerous sterile controversies, as I shall illustrate in Chapter 3. And past attempts to capture the meaning of various teleological concepts in the form of an acceptable definition have generally tended to lead back to others of the same family. This obviously won't do. We need definitions in non- teleological terms. I have dealt with that problem in various publications (Sommerhoff 1950, 1969, 1974), and have there shown that, viewed objectively and at a sufficient level of abstraction, all directive activities are seen to be based on the same underlying abstract pattern of spatio-temporal and causal relationships. And I have shown that all common teleological concepts can be given a precise and objective definition in terms of that pattern, hence in terms of non-teleological concepts. This clarificatory procedure has also been followed by others, e.g. Nagel (1961). It has been expressed in the language of symbolic logic by Hilgartner and Randolph (1969).

A summary of this procedure will be given in Chapter 3. We shall there also see that this characteristic pattern of spatio-temporal and causal relationships can be expressed in terms of one single and mathematically definable concept: the concept of *directive correlation*. All of this is obviously relevant for anyone concerned to discover the causal relationships that the networks of the brain must satisfy to produce the kind of directive activities or processes we shall encounter in our discussion of the three categories of internal representations I have listed.

Many of the mechanisms responsible for the directiveness of particular vital activities are well understood. As has been said, many of them depend on negative feedback loops, but this concept cannot serve also as a means for defining teleological concepts in non-teleological terms. Other means have to be found. We need a concept which is even more general. And that has been found in this concept of 'directive correlation'.

In terms of this concept of directive correlation, therefore, each of the teleological concepts we require in our analysis, can be given a precise physical definition which will satisfy the strict semantic standards good systems theory demands, while at the same time exposing the underlying spatio-temporal and causal relationships. Important examples of some basic concepts to which we shall apply this in our analysis of consciousness are the concept of *function*, the concept of a response being *appropriately* related to a given situation in a given set of circumstances, and the concept of an organism being in a state of *expectancy* for some future event.

***

1.8 - THE NEURAL CORRELATES OF MENTAL EVENTS

According to the theory I shall develop, the different kinds of mental events that form the stream of human consciousness are all sub-categories of the three basic categories of internal representations I have described in Section 1.6. Conscious sensations and feelings are sub- categories of states of self-awareness; mental images are representations of possible perceptions; desires are internal representations of possible situations linked with expectations of need-satisfaction; memories are representations of possible events or situations which enter into the subject's global world-model as representations of past events, and so on.

It follows that our first step towards tracking down the neural correlates of mental events, must be to arrive at some idea of the physical nature of those three basic categories of internal representations. This, in turn, means that we must first settle in objective terms what precisely is here to be meant by an 'internal representation' (Chapter 4).

My definition of internal representations will be given in *functional* terms, and initially I shall confine myself to the brain's internal representations of external objects Here we shall be guided by two principal considerations:-

   a) An internal representation of external objects must be a state of affairs in the brain which is *used* by the organism as being representative of the objects concerned. That is to say, it must be a state of affairs in the brain which forms part of the data base on which the organism acts in dealing with those objects.

b) It must be a state of affairs which, in some adequate sense and to some adequate degree, corresponds to the features and properties of the object concerned. This is a condition which it must satisfy, not in order to qualify as a representation in the functional sense, but to qualify as a *veridical* and, therefore, *useful* one.

Observation (a) describes the functional aspect of the matter, and I shall make this the basis of my definition of 'internal representation'. In essence, I shall say by way of definition that a state of affairs X in the brain functions as an *internal representation* of an object O if in the organism's dealings with O the brain takes its cues from X rather than directly from the sensory inputs derived from O. (The words I shall use will be different, but this is the gist of the definition.)

However, a functional definition is not enough to go on, since different kinds of 'hardware' can discharge the same function. We also need to translate our functional terms into causal terms, and we need statements about the *structure* of these representations. To arrive at the latter, we shall turn to observation (b) above. And we shall search for hypotheses about the kind of physical states in the brain that could achieve such an adequate correspondence to the objects of the outside world and to their properties or features. Here we shall be guided by two overriding considerations. Firstly, the physical states in question must be of a kind that can accommodate the enormous variety of the objects and object-properties that may have to be represented. Secondly, there must be plausible mechanisms, both for producing the internal representations concerned, and for modifying them in the light of current experience.

The following is an outline of the main hypothesis we shall arrive at, and from which we shall subsequently draw our inferences about the neural correlates of mental events.

Among those psychologists who accept that the brain forms internal representations or 'models' of the world (although they tend to leave the notion undefined), the prevalent view is that these models create expectations of what will be the consequences of any one of the whole spectrum of acts open to the organism, and the organism will initiate those acts that carry the strongest expectation of need satisfaction. This whole body of what-leads-to-what expectancies is taken to have formed in consequence of the subject's past what-leads-to-what experiences. If any of the expectancies fail to be supported by subsequent events, this will then act as an error signal which the brain will use to correct its current model of the world (or that part of the world which is relevant in the given context).

The two questions which this type of account has generally left unanswered concern the physical form in which both these models and these expectancies are realized in the brain.

Apart from insisting on a precise physical definition of 'internal representation' and 'states of expectancy', I shall depart from this view in one crucial way, and in a way which reduces these two questions to a single one. For I shall argue that there is no need to assume that there exist in the brain two distinct entities, viz., both models and associated sets of what-leads-to-what expectancies. My case will be that it is sufficient to assume only the existence of these sets of conditional expectancies, because these sets themselves are capable of acting as the required models. This will be called the *Conditional Expectancy Hypothesis*.

According to this hypothesis, therefore, an organism's internal representations of the world are derived from the organism's past *what-leads-to-what experiences* in its interactions with the world, and they consist of linked sets of corresponding *what-leads-to-what expectancies*. Naturally, we shall examine in detail the evidence that supports this hypothesis. And we shall make a particular point of first defining in precise and physical terms exactly what is to be meant be a state of *expectancy* in this context.

To illustrate in very broad terms: according to this hypothesis the internal model which I have of the room in which I am writing these lines, consists at the physical level of expectancies of what the eyes will see if I cast them to left or to right; what will be perceived if I look behind me; what the hands will feel if I reach out in this direction or that direction; where I have to turn in order to leave the room, etc., etc.

Let me say a little more about this already at the present juncture.

After birth the organism begins to explore the realities of the world it has entered and its capacity to influence them. As it begins to interact with its environment, the most fundamental set of relationships it has to absorb in its brain are the relationships between its movements and the experienced consequences. That is to say, it has to learn the *causal relationships* between the movements open to it and the resulting transformations of its sensory inputs - beginning with such elementary facts as the effect on its visual inputs of a movement of eyes, head and body. The ability to collate such movement information with sensory input-information, is crucial, for example, even for the basic ability to construct a stable scene from the images that flit across the retina as the eyes move in their sockets.

Then comes the tactile exploration of nearby space: learning to extract clues about distance from its visual inputs, learning to touch and grasp, etc. The lessons it learns here are of the general kind: *under conditions A an event B will be followed by an event C* - the most crucial subcategory here being the case when B is a movement open to the organism and C the experienced outcome of that movement. And this continues to be the general character of the lessons we learn in exploring the world, right up to the complex and sophisticated form these lessons take in later life.

The lessons thus learnt result in the formation of enduring brain states which may be described as enduring states of *conditional expectancy*. And the subcategory mentioned may be described as *states of conditional act-outcome expectancy* The states will endure until a failure of the anticipations concerned forces a modification or revision.

Thus the totality of such acquired states of what-leads-to- what expectancy reflects the totality of causal relationships, especially between action and outcome, which the brain has been able to assimilate as the result of experience. In that sense, therefore, this totality of acquired states of conditional expectancy can be said to be *representative* of the realities of the physical world which the organism has encountered.

I have illustrated above the category of causal relationships that is covered by the internal model my brain may form of the room in which I find myself. A different category would be covered by the internal representations which I have defined in Section 1.6 as *body-awareness* or *first-order self-awareness*. This would cover for example, the changes in proprioceptive and kinaesthetic stimuli that would result from a given movement, as well as the changes in the exteroceptively perceived position of body or limbs.

More subtle would be the causal relationships that are covered by second-order self-awareness. This was introduced in Section 1.6. as category 3 of internal representations. It was defined as a category of representations which register the occurrence of a representation of category 1 or 2 as being part of the current state of the self. Let B(A) be a representation which registers the existence of a representation A in the brain as being part of the current state of the self. The category of causal relationships which B(A) would then cover would be relationships which characterize the fact that A is part of the current state of the self. Examples will be given in Section 7.1.

It is along such lines, then that I shall approach the physiology of the different kinds of internal representations on which my treatment of consciousness concentrates.

In Chapter 5 I shall proceed to examine the learning mechanisms involved in the formation of the linked sets of what-leads-to-what expectancies which play such a crucial role in my treatment of the subject, as I have illustrated. I shall argue that these mechanisms can be understood in terms of the basic concepts of classical and instrumental conditioning.

This will be followed in Chapters 6 and 7 with a discussion of the *neural correlates* of the three main categories of internal representations I have mentioned, i.e. the manner in which they come to be realized in the neural networks of the brain. According to the Conditional Expectancy Hypothesis, for representations of *actualities* this now resolves into the question of the neural correlates of states of expectancy of different kinds. A parallel course will be followed in the treatment of internal representations of mere *possibilities*, i.e. the faculty of the imagination (Chapter 7). However, since the products of the imagination are not subject to the error-correcting processes I have described as

1.8 – The neural correlates of mental events

reality testing, the concept of *conditional expectancy* no longer applies and will be replaced by a suitably reduced one, viz., the concept of states of *conditional readiness*. According to the definition I shall adopt, a state of conditional *readiness* differs from a state of conditional *expectancy* in that the non-occurrence of the event prepared for elicits no reactions of the error-correcting type.

The resulting conclusions will then be examined in the light of the available neurophysiological and neuro- anatomical evidence, e.g. observed deficits following known brain lesions, and the published results of micro-electrode penetrations of various regions of the brain. Proceeding from this basis, I shall attempt to cover in outline the general nature of the neural correlates of all the major categories of mental events which I have listed in my introductory paragraphs.

Later in Chapter 6 I shall also consider questions relating to the seat of consciousness and I shall discuss the effects of brain lesions and how these may be interpreted in terms of the brain processes our analysis has highlighted. I shall suggest a set of directives to guide this interpretation (which has always been a notoriously difficult undertaking).

When, in Chapter 8, we turn to the subject of *language* and *verbal representations*, we reach a human competence which I have so far only mentioned in passing, but which we shall also have to analyse in objective terms. This is the brain's power to cast its internal representations, e.g. mental images, into what will be called a *propositional* form. This we shall come to see as a prerequisite for their expression in a *verbal* form. Some of the brain's internal representations may originally already have been acquired in a propositional form, e.g., representations derived from information received in verbal form. Others may originally be present in a quasi-pictorial form, and these will have to be converted into a propositional form before the subject can put them into words. Conversely, information received in a propositional form will often be translated into mental images.

I shall argue that these linguistic competences primarily require a power of the brain to form representations both of class-membership relations and of the logical relationships that can exist between classes - such as the relationships of inclusion, union and intersection. None of this will be found to conflict with the Conditional Expectancy Hypothesis.

In Chapter 9, I shall conclude this part of the work with a general review of the relation between mind and matter, such as it has emerged in the course of our analysis, and I shall look in detail at a number of related questions, such as the unity of the self in split-brain patients and the freedom of the will.

The outline I have here given will have demonstrated to the reader the type of conceptual precision and degree of objectivity I am demanding and the results to which it may lead. This is no sterile pedanticism. To repeat: it is not uncommon today for cognitive psychologists to talk about the brain forming 'internal representations' or 'internal models' of the outside world. But we look in vain for explicit statements of what 'representation' is intended to mean in terms of accurately definable relationships in space and time (including causal relationships). Yet without such clear ideas we cannot draw inferences from such assertions about the neural correlates of these representations; nor, therefore, about the neural correlates of the brain functions involved in our mental life generally, for they are all based on representations of one kind or another.

In the course of all this we shall come to see more clearly in what sense it is true to say that to understand the higher brain functions, one has to think of the brain as a connected whole, as a *system*, and one which itself acts as an integral part of the organism as a whole. This is, what I have described earlier as the *top-down* or *holistic* approach, by contrast with the *bottom-up* approach which starts with the behaviour of individual neurons. As I shall explain in detail later, one specific implication of this holistic aspect of brain function is that it is a mistake to think (as many do) of the path from the visual inputs to the final act of perception as a linear one.

***

1.9 - DEEPER LEVELS OF CONSCIOUSNESS

I have divided this work into three main parts, followed by a fourth which merely consist of a number of appendices that may be helpful to the general reader. Part I (Chapters 2 and 3) deals with the general character of the organization of living organisms, and in particular with the distinctive patterns of causal relationships that mediate the directiveness of their activities. Part II (Chapters 4-9) deals with the nature of consciousness and the neural correlates of mental events in the manner I have outlined above. Part III (Chapters 10 and 11) branches out in quite a different direction. Here I leave the cognitive aspects of consciousness, and look at certain distinctive dimensions of the affective or emotional side of human life. I have singled out two dimensions which are of special scientific interest because they amount to human traits in which man seems to differ radically from the infra-human species. They are relevant in the context of my earlier chapters, because of the special role which I attribute to the faculty of the imagination in the genesis of these traits. They are also of special importance in the general context of the modern world and its movements of thought. I shall reject the 'sociobiological' approach to these phenomena, but nevertheless try to depict them in a general biological perspective, drawing here on some of the work done in my earlier chapters.

The two dimensions of our emotional life which I have singled out are:

1. Those *moral* sensibilities and responses that transcend the phylogenetically determined forms of 'altruism' found in the social species of the animal world, i.e. aspects of morality which plainly transcend what can be explained in terms of genetic cost-effectiveness in a social species. This touches the very core of the ethical beliefs that have shaped the ethos of our cultural heritage and have sustained it against the pressure of tribal and racist ideologies.

2. Man's *aesthetic* sensibilities and responses such as the sense of beauty and the enjoyment of the arts.

In the moral sphere, I have selected here as a matter of particular interest a notable feature in the slow historic development of our civilization, viz., the growing *internalization* and *universalization* of moral values. And I shall take as a relevant example the moral values which the humanist movements in our society share with the Christian, and to some extent also Judaic, tradition.

By the 'internalization' of moral values I mean that, in contrast with the tribal ethos of primitive man, our contemporary ethos is *inward looking*: the ultimate stress is on the *individual* function of morality rather than its more obvious social function. Thus the Christian sees morality as the way towards the salvation of the soul and towards the soul coming into the sight of God. With a similar stress on this individual function, the Humanist sees morality as a way towards self-fulfilment and the full realization of one's potential, both as an individual and as a social being.

By 'universalization' I mean that, again in contrast with any purely tribal ethos, the beliefs that have shaped the ethos of our society regard their moral precepts as valid for all mankind, regardless of race or tribal affinities.

In both the moral and aesthetic sphere we are dealing with human sensibilities and responses which belong to what even the secularist may call the 'spiritual' dimensions of human life. They mark a departure in the evolution of life on this planet which is of special scientific interest for one general reason: unlike the 'altruism' found in the social animal species, these human traits cannot be accounted for in terms of their genetic cost-effectiveness. That they do not fit that bill is obvious from the fact that as Peacocke (1986) has rightly remarked, cost-effectiveness calculations of genetic advantage are not the ultimate criterion by which we judge moral virtue - or, for that matter, cultivate the arts.

These human traits have been able to develop in that seemingly 'Darwin-defying' direction because they are *cultural* phenomena. They are traits which are superimposed on the individual's innate endowment by his or her cultural milieu. And the evolution of this cultural element is not governed by the same kind of selective forces as those that govern the phylogenetic evolution of all animal species including our own. Darwinian theory applies only to the traits of living beings which are transmitted via the genes, whereas cultural goods such as ideas, beliefs and values, are transmitted from one generation to the next by word of mouth, literature, art, example, etc. Neither the religious nor humanist movements of thought that have shaped our culture owe their origin and success to the occurrence and successful propagation of some genetic mutation.

The innate traits of many social species, such as the honey bee and the ant, include certain kinds of 'altruistic' behaviour. Thus the worker bee may sacrifice its life in the defence of the hive so that its genetic cousins may survive. It is a mechanism of kinship survival. Human altruism, which is based on compassion and love of neighbour, and the *internalized* and *universalized* ethos which the Judaic- Christian and Humanist traditions of the Western world have built around it, are plainly not of this 'sociobiological' and necessarily tribe- or race-orientated kind.

Hence any *evolutionist* attempt to understand the roots and rationale of our moral sensibilities and responses is likely to be misplaced, and we must look elsewhere to see these cultural phenomena in a biological perspective. A morality logically derived form the Darwinian theory of evolution would always be a racist one. Similar remarks apply in the aesthetic field. The sense of beauty and role of art in society cannot be explained in terms of genetic cost-effectiveness.

I shall argue that the cultural developments with which we are here concerned may nevertheless be perceived as a *biological* development of a certain kind, viz., as a way in which certain forces have been able to find expression in the growth of our civilization which have their roots in the very nature of life. I have in mind here certain basic reactions which are common to all higher animal species, viz., an aversion to *anxiety*, *conflict*, *frustration* and *cognitive dissonance* - four kinds of disturbance in the smooth functioning of life, each of which, I shall explain, can be defined in objective terms. I shall argue that in the human case these aversions have spawned a host of derivative reactions which are unique in their nature and which have critically influenced the evolution of human culture and the growth of human civilization in consequence of three factors:-

1. The ability of the human imagination to convert these aversions, which we may call negative responses, into positive responses in the form of visions of a world or life-styles in which anxiety, conflict and frustration are minimal - visions of a social order or life-style blessed by security, fulfilment, harmony and peace. Thus the 'negative' motivation to avoid certain conditions, came to be transformed into the 'positive' motivation to achieve certain conditions.

2. The power of the human imagination to exercise itself in the form of empathy - a key factor in man's ability to achieve the conditions of social harmony which are part of his dream of a better world or better life-style. For, empathy enables us to understand, and respond to the feelings and motivation of others. And since empathy also enables us to identify with the suffering of others, it adds emotional depth to our visions of a better world, social order or life-style. In my discussion of this aspect of the matter I shall draw *inter alia* on what my analysis of the representational powers of the imagination will enable me to say in objective terms about the nature of *empathy*.

3. The absence of the restraints of genetic cost- effectiveness in the evolution and propagation of cultural goods. Hence the four aversions I have mentioned and the positive visions they can engender through the power of the imagination have been able to impress themselves on the evolution of our culture to the significant extent to which they have manifestly done so - especially in the form in which they have influenced both the religious and secular systems of thought through which societies seek to rationalize and propagate what they have come to sense as the right and the good. In this connection I shall also discuss the selective forces that would seem to have taken over from the Darwinian criteria of selection in the growth and propagation of the respective value systems and systems of thought.

I shall also stress the importance of empathy in a much wider context, viz., the aesthetic context. For, much of what I have said here about certain *moral* sensibilities and responses applies also to our *aesthetic* sensibilities and responses. The pleasure we derive from contemplating a bunch of flowers bears no relation to the attraction the flowers may exercise on the honey bee. I shall suggest that these particular human sensibilities, too, rest in part on a kind of empathy. I shall argue that our immediate (and still knowledge-free) affective responses to the visible form or composition of an object are *inter alia* determined by two factors. One relates to the nature of information, viz. to the information gathering process and the relative powers of perceived forms or compositions to *communicate* something to the perceiver The other relates to some urge we have at the core of our being to identify with the surrounding world - not only to be able to predict its behaviour, but also to in order to achieve some kind of union and deeper communion with it. Both factors can be seen in a biological perspective.

It is along such lines, then, that I shall try to depict the biological roots and rationale of the moral and aesthetic sensibilities and responses I have mentioned. These sensibilities and responses mark a radical departure in the evolution of life on this planet for two related reasons. Firstly, they have caused the 'laws' of what we understand by a civilized life to diverge radically from the 'laws' that govern the jungle, i.e. from 'laws' that reward the strong while putting the weak to the wall so that their deleterious genes will be removed from the gene pool. Secondly, whereas, in a sense, all animals can be said to be no more than sophisticated survival machines, these particular sensibilities and responses have raised the human potential above that level.

These issues are important also in the general context of modern thought, because the scientists' failure to understand these facets of the human psyche has tended to result in a scientific view of man in which some of the most critical dimensions of human life are missing or explained away - the picture of a naked ape, as Desmond Morris has called it.

***

PART I

LIFE

Chapter 2

ORDER AND LIFE

INTRODUCTION AND SUMMARY

In this and the next chapter I shall deal with some of the basic concepts of systems-theory when applied to living systems and to the general character of their organization. Since this is likely to be familiar ground for only a minority of readers, I shall do so into considerable detail. Most of this turns on the requirements of systems-theory to work only with concepts which can be accurately defined on the observable variables and assumed parameters of the systems under consideration. This covers more than just variables relating to their outward behaviour. It includes any variable whose value can in principle be determined by scientific means, internal variables as well as surface variables.

Certain features of the activities of living systems will here attract our main attention, especially the *directiveness* which I discussed briefly in Section 1.7. When we come to describe certain states and activities in the brain which will prove to be relevant to our enquiry, we shall be characterizing some of these activities as *appropriately* related to the circumstances in which they occur; others will be described as *adaptive*; again others as *coordinated*, or as serving some particular *function*. All of these concepts relate to manifestations of that directiveness. We shall also be talking about *states of readiness* of the brain for some particular events, and about a subcategory of this which will be called *states of expectancy*. And for the purpose of our analysis we shall need a clear perception of what exactly is meant on all of these occasions and what each of the concepts implies by way of spatio-temporal and in particular, causal relationships.

This conceptual clarity is also important if we are to relate our findings to the structure of brain functions in a manner that can be accurately formulated.

Even so, the reader whose main interest lies in the conclusions we shall reach and in the postulates on which they rest, may omit these two chapters at a first reading, referring back to them only when the need arises.

***

2.1 - THE DIRECTIVENESS OF VITAL ACTIVITIES

Human beings are living organisms and living organisms are physical systems in the sense in which this term is now generally understood. But they are physical systems of peculiar kind. For, superimposed on the kind of orderliness that we can also find in non-living systems, e.g. the orderly arrangement of atoms in a molecule or crystal, living systems have another kind of orderliness which we call their organization. It manifests itself in processes which we describe in such terms as *regulation, control, adaptation, coordination, integration,* and *function*. Consciousness is one of the most sophisticated manifestations of this organization, and if we want to understand consciousness in its full biological context, it is imperative to start with a clear idea of the distinctive type of orderliness which characterizes this organization, e.g. a clear idea of the general pattern of causal relationships on which it depends. We need this clear idea not only to be able to give precise physical definitions to the many (in general use still fuzzy) biological concepts which relate to it, but also to get a better perception of the general nature of the organic order in which the life of all human beings is embedded, and of the kind of harmony that it appears to produce within its sphere. For in the final chapters of this book I shall relate some aspects of the human psyche to the nature of this harmony and man's intuitive awareness of it.

At the most general level, this organic orderliness manifests itself in one of the most distinctive characteristics of living organisms, viz., the *directiveness* or *goal-directedness* of their activities or processes. Whatever you see happening in the living world, especially in the higher orders of life, generally seems to be directed towards the production of some biologically significant end-result: the organism crawls, runs, swims, or flies to get from A to B; it gathers food to fuel its energies; it hides to escape detection; it fights to protect its territory; it burrows or builds a nest for shelter; it finds a mate in order to reproduce, and so on. Meanwhile its body performs a multitude of internal regulations to safeguard the constancy of the internal environment and to keep itself going as an effectively operating survival machine. In each case the actions or processes in question are seen to be directed towards the achievement of that end-result in the manifest sense that they are seen to be produced or controlled in such a way that over a variety of circumstances they will match those circumstances in a manner conducive to the end-result in question. If the mouse jumps to the left so will the cat.

This directiveness characterizes not only the behavioural responses and physiological reactions of the living organism, but also the much slower processes of ontogenetic development, maturation and learning. In a slightly more restricted sense, it characterizes also the long-drawn-out processes of evolution. For here, too, we see developments taking place which are able to match changes in the ecological conditions in a manner that is conducive to some biologically significant end-resulte.

This, and no more than this, is what I shall mean by the *teleological* character of vital activities, while by *teleological* terms or concepts I shall mean all terms or concepts that relate to this directiveness or aspects of it.

Many modern artefacts also achieve such directive behaviour, notably artefacts governed by servomechanisms incorporating negative feedback loops. The autopilots of ships and aircraft are typical examples. So are industrial robots and homing missiles. However, we cannot use the concept of feedback loop itself to characterize in objective and sufficiently general terms the nature of directive activities as such. To that end we need a characterization of *what* is achieved by such mechanisms, not an account of *how* it is achieved. And a more abstract analysis is required for that purpose. (Readers who are unfamiliar with the notion of 'feedback loop' will find the feedback system of a simple servomechanism illustrated in Fig. 3.1).

It will be clear from the above remarks that the 'goal- directedness' of organic activities which I have described, is here viewed as an *objective* characteristic of the activities concerned, a characteristic which they share with the flight of a homing missile, for example. Unfortunately, such activities are also often described as 'purposeful' or 'purposive', and these terms have obvious mental connotations. They suggest the intervention of a mind motivated by conscious aims. It is important, therefore, not to confuse our objective sense of goal-directedness with this *subjective* or *mental* sense or purposiveness. The goal-directedness we are talking about does not imply the intervention of a mind at all. Moreover, although rational mental activity may well be regarded as nature's most sophisticated way of producing activities which are also goal-directed in an objective sense, it is not an infallible way. A human action may well be a purposeful one in the above subjective sense without satisfying also the criteria of goal-directedness in the objective sense I have described. One can have a definite goal in mind and yet do all the wrongs things.

***

2.2 - DYNAMIC VERSUS STRUCTURAL ORDER

The story of life can be told at several distinct levels. One of these is the molecular and cellular level. This is the story of the macromolecules of which living cells and tissues are constructed: the story of DNA, RNA, genes, chromosomes, enzymes, and so on. The type of order that comes into focus at this level is mainly a geometrical, three-dimensional order: the orderly arrangement of atoms in molecules and of molecules in paracrystals, cellular membranes, tissue fibres, and other structures. It is a type of order which can be quantified in terms of negative entropy, because the thermo-statistical concept of entropy itself is a measure of the disorder that exists in a physical system at any given point of time.

All the most spectacular advances of biology in this century have been made at this, the molecular level. But this is not the only level at which the story of life can be told.

Inanimate objects are wholly at the mercy of their environment. The extent to which they can resist disruptive forces depends solely on the depth of the potential energy minimum which their physical structure occupies. Any physical force strong enough to lift it out of that minimum and over the surrounding 'energy humps' will prove disruptive.

However, with the arrival of life, a new factor appeared on the scene, capable of ensuring endurance in the face of adverse environmental contingencies: systems appeared which had the capacity over long periods of time (spanning many generations) to undergo *adaptive changes* - that is, changes which were related to the characteristics of the environment in a way that increased the probability of survival. In due course a further improvement occurred: these slow evolutionary changes resulted in the development of mechanisms which could pick up information about the current state of the environment and produce behaviours which could be *instantly* adapted to the changing demands of the environment. Moreover, the resulting activities did not only take account of the environment but also of each other: they became *coordinated* and *integrated*. Responding appropriately to their sensory inputs, these creatures could now move about in search of food, or shelter, or a mate, instantly adjusting their movements to the perceived location of these things and to the varied demands of the terrain. But since all reactions were of the nature of innately fixed and stereotyped responses to the sensory inputs concerned, anything that differed from the norms to which evolution had adapted them was likely to prove disastrous. Thus if the head of a column of Processional Caterpillars is made to join its tail, the creatures will march in a circle until they die of starvation.

In due course the mechanisms of natural selection produced yet another improvement: organisms capable of undergoing adaptive changes during their lifetime - changes which matched the needs the individuals experienced in the ecological niche which the species had come to occupy. That is to say, they could learn from experience.

All these changes were *directive*, hence 'teleological', in the sense I have described: they matched the circumstances in which they took place in a manner conducive to some biologically significant end-result.

In this sense, then, living organisms evolved as harmoniously operating, self-maintaining and self-preserving, as well as self-reproducing, systems of immense sophistication and adaptive powers. A higher kind of order had come into being - the dynamic order of a system in which all parts work together towards a common goal. The higher up we move in the scale of life the more impressive this kind of order becomes: impressive both in complexity and performance.

The story of this *organic* order is no longer a story of potential energy minima and entropy changes. It is the story of countless interlocked, mutually coordinated, integrated and ingeniously orchestrated controls, regulations, checks, balances, error correcting mechanisms, and adaptive adjustments which, each in its own way, help the organism to maintain its structures and functions, and support its powers to meet the vicissitudes of its life with a vast battery of appropriately matched reactions and skilled behaviour patterns. And since, as I shall show, this type of order rests on distinctive patterns of causal relationships, we are dealing here with relationships in both space and time, i.e. with a four-dimensional rather than a three-dimensional type of order.

This dynamic order is also the feature to which the living organism owes its distinctive kind of unity. As Claude Bernard wrote more than a century ago: "It is the subordination of the parts to the whole which makes of the complex creature a connected system, a whole, an individual". This is true enough. But it has to be realized, of course, that the subordination we are here talking about is again a teleological relationship. It is the subordination of the proximate goals of the individual activities of the parts of the organism to the requirements of the whole, and thus to the ultimate goal of their combined activities. The clarity of our perceptions in this whole field, therefore, must depend on the clarity of our perceptions of teleological relationships as such.

At the physiological level this subordination of the activities of the parts to the demands of the whole manifests itself *inter alia* in the intricate mechanisms through which the body achieves *homeostasis*, i.e. a constancy of the internal environment. Familiar examples are the constancy of the body temperature and of the blood concentrations. In the 1930s many of these aspects of organic function were for the first time brought to the attention of a wider public by Cannon's famous book THE WISDOM OF THE BODY (Cannon 1939).

By that time even more impressive feats of regulation and control had also come to light in the field of embryology and morphogenesis, mainly in amphibians. People were frankly perplexed by such discoveries as the extent to which it is possible to reshuffle or even ablate cells in the early stages of an amphibian embryo, and still finish with a perfectly normal and viable adult specimen. Even up to the 64-cell stage the embryo can be divided along any plane into two halves and, yet, two normal adults will develop. Conversely, two early balls of cells can be fused together and a single normal adult will nevertheless result. Again, the amphibian limb bud may be divided into two or three parts and each part will grow into a complete limb; and - perhaps even more surprisingly - the developing eye bud can be transplanted to locations below the dorsal skin and the skin above the transplant will become transparent and develop into a cornea.

Such adaptive processes are not confined to the embryonic stage. In all higher animals there continues to exist a measure of adaptive plasticity well after birth and throughout most of the normal lifespan, except that this capacity diminishes with age so that eventually what will knock a young person down will knock an old person out. The most impressive development of this ontogenetic adaptability is, of course, our power to learn from experience.

Yet, although the mechanisms of all this are often well understood, the whole question of the *formal structure* of the teleological characteristics of vital activities has remained an area of which the average biologist has only a hazy perception. If asked to translate statements containing teleological terms into statements containing only non-teleological terms, he would be hard put for an answer. Indeed, the question itself may strike him as rather too dangerously abstract. Unfortunately one cannot avoid abstract thought if one becomes involved with questions as general and fundamental as those that occupy us in this volume: and we shall find that to cope with these we need above all a set of precise and objective concepts that can be defined in non-teleological terms and yet capture the essence of the teleological relationships we are interested in.

Past attempts of academic philosophers to analyse teleological concepts, have often resulted merely in a reduction of the concepts concerned to others of the same family. Indeed, this has throughout been a major stumbling block. In his recent TELEOLGY, for example, Woodfield (1976) reduces them to the basic teleological concept of (organic) *function*. The philosophy of mind known as *Functionalism* also accepts *function* as a primitive concept. This is unacceptable to good systems-theory. It needs technical definitions of teleological concepts which are couched in strictly non- teleological terms, e.g. in terms of clearly stated patterns of causal relationships. The concept of *function* is no exception.

***

2.3 - THE RANGE AND IMPORTANCE OF TELEOLOGICAL CONCEPTS

In ordinary discourse we may well be satisfied that we know what is meant when the movements of the Cheetah are said to be instantly *matched* or *adapted* to the changing demands of the terrain, when they are described as highly *coordinated* or *integrated'*, or to have *organic unity;* and when its limbs are said to move in perfect *unity* or perfect *harmony*. Again, we think we know what is meant when some physiological processes are said to be highly *regulated* or *controlled*, or when people speak of the *function* of an organ. But do we really have a clear idea of what exactly is being asserted in each case?

2.3 - The range and importance of teleological concepts

All these concepts are teleological concepts because they all relate to one or other manifestation or aspect of directive activity. And even among biologists they have remained fuzzy concepts. Thus although most biologists may be satisfied that they know what is meant by describing some animal movements as 'coordinated', few are likely to agree on the precise criteria that distinguish coordinated from uncoordinated movements, or what inferences can be drawn about the causal relationships involved when a set of movements are said to be coordinated; or what difference in causal relationships would make us say in one case that *A is adapted to B* and in another that *B is adapted to A*. At a more general level, few biologists are likely to agree on the objective criteria that distinguish a biological activity which is goal-directed in the sense I have illustrated, from one that is not. Again, few would be able to state of any given teleological concept what it implies in non-teleloligcal terms.

This fuzziness of teleological concepts does not in general impede the work of biologists routinely, but it is unacceptable in the accurate analysis of brain functions we are aiming at - nor, indeed, in a serious discussion of the more abstract characteristics of the organic order of living systems as a whole.

The general public, of course, tends to take the teleology of the living world very much for granted. In everyday life we invariably tend to classify the activities of animals, not according to the specific movements they execute, but according to the *goals* we attribute to those movements - as when we speak of 'flight from danger', 'attack', 'pursuit', 'escape', 'search', 'defence', 'nest- building', 'seeking a mate', 'caring for the offspring', etc. Again, our emotional reactions towards the animal world are mainly governed by the goals we attribute to their activities. We fear the power of a bull only because of the aggressive goals of which we hold it to be capable. And when your dog dies, you do not mourn the irreversible cessation of its metabolic processes: you mourn the cessation of the particular kind of purposeful responses that made the creature the good companion it was. Indeed, our emotional attitudes towards our animal friends, be they dogs, cats or horses, depends in a large measure on the extent to which we can read purposes into their behaviours - and preferably purposes with which we can identify - as when we say of a dog that "she craves attention", "wants to be taken for a walk", "has a guilty conscience", etc.

This directiveness is also the one characteristic of observed life which invites us to think of living things as somehow endowed with a soul (the original 'anima' of 'animate matter'). Regardless of whether this response is scientifically sound, the fact remains that it forms the basis of any kind of empathy we are able to feel for the living world and the basis of our sense of union with that world - a topic to which I shall return in Chapters 10 and 11.

On all these counts, one can say that it is the directiveness of vital activities which gives to the distinction between living and non-living systems the outstanding significance it has in our daily affairs. In this sense it is to us a far more important quality than any of the scientific distinctions between living and non-living objects that we are taught at school - such as growth, reproduction, respiration, water and food intake, disposal of waste products, responsiveness to stimuli and the like. Whatever the schoolbooks may say, if you were to meet some unknown object on a foreign planet, your first instinct would be to prod it to see if its response would be a blindly inert one or a somehow directive one. And if the latter, you would place it on the scale of life, or class it as friendly or hostile, according to the direction and degree of directiveness of its responses. You would not examine its metabolism.

***

2.4 - PAST MISCONCEPTIONS

Conceptual confusions have bedeviled discussions about the meaning and significance of teleological phenomena ever since Aristotle first drew attention to the "absence of haphazard and conduciveness of everything to an end" in living nature. They have left their mark not only on the history of biology but also on the history of philosophy and theology. Largely responsible for these confusions has been the fact that this aspect of living matter often seemed hard to reconcile with the blind determinism of the laws of physics and chemistry. Thus the 'Vitalists' of the 18th, 19th and beginning of the 20th century argued that it required the existence of forces in living matter which differed in kind from any of the forces found in non-living matter - forces which in some sense seemed to be forward looking. Bergson called it the 'elan vital' of living organisms, Driesch their 'entelechy'.

Other thinkers have gone further and argued that the teleology of living nature demonstrates the presence of Mind or Supreme Intelligence in the universe. This vision, too, has a venerable history. No one has expressed it better than Virgil after reflecting on the life of the honey bee and on the manifest purposefulness and harmony of the teeming activities in the hive:-

> "These acts and powers observing, some declare
> That bees have portion in the mind of God
> And life from heaven derive, that God pervades
> All lands, the oceans, plains, the abyss of heaven
> And that from Him flocks, cattle, princely men,
> All breeds of creatures wild receive at birth
> Each his frail, vital breath; that whence they came
> All turn again dissolving." (4th Book of the Georgics)

To-day there are still many who who see in such images and inferences the foundations of a Natural Theology. I would not wish to dispute that such imagery has a valid place in religion as part of the image-driven way in which religion helps man to relate himself to the nature of life and the conditions of human existence. But it is a fallacy to hold that we are dealing here with phenomena from whose objective description logical inferences can be drawn about the existence of some transcendent mental or spiritual entity. That, in the objective sense I have described, there exist a multitude of teleological systems in living nature, cannot be denied. But anyone who would wish to make this the cornerstone of a Natural Theology would have to show in the same objective terms that in addition to these individual teleological systems there exists one single comprehensive teleological system that embraces them all - and, indeed the whole universe.

In the early years of this century, and mainly in response to the astonishing discoveries that had been made in the field of embryology, another school of thought found some favour. This attributed the 'organic wholeness' of an organism to an 'overall plan' which had some kind of transcendental power to restrain the development and activities of the organism and its parts.

However, already in those days, leading biologists set themselves against these 'organistic' schools of thought and warned about the lack of explanatory value of ill-defined terms like 'organic whole', 'organic part' and 'organizing relation'. More than half a century ago, Woodger (1929) wrote:

"It is easy enough to see 'intuitively' what is meant by these terms........; the difficulty is to make these notions precise in order to enable us to see how we can use them for scientific purposes. Intuition is the indispensable cutting edge of intellectual enquiry, but the ground won is not consolidated until it has passed from the stage of intuitive apprehension to that of logical analysis."

Yet for many years to come his remained a voice in the wilderness. Until after the Second World War, when cyberneticists and robot engineers began to apply mathematical analyses to the variety of directive activities they were seeking to achieve in their artefacts, the whole realm of concepts relating to organic order and organization remained a foggy one indeed.

In sharp opposition to the Vitalists and Organicists stood a radically mechanistic school of biological thought which dismissed the whole concept of goal-directed activity in living nature as a figment of the imagination. According to this school, the notion of directiveness in vital activities is a mere by-product of a human tendency to project man's own conscious purposiveness and rationality into the environment - in other words, a mere anthropomorphism or primitive animism. Indeed, there were times when even such innocent statements as that an animal was *trying* to escape from its cage might earn an examination candidate a negative mark. This outright rejection of all teleological concepts was at its most influential in the first half of this century. Since then is has had to retreat step by step as logical and mathematical analysis came to establish beyond doubt that the goal-directedness of vital activities can be characterized in objective terms. At the same time the development of feedback-controlled servo-mechanisms and robotics furnished the world with machine examples of activities which were patently goal-directed in an objective sense - and yet clearly ran their course without the intervention of 'mind'. Few things can be more patently goal- directed than the flight of a homing missile. Even so, the old arguments have not yet been wholly laid to rest. Time and again tracts still appear which claim that this directiveness in living nature points to a conscious design. Some even claim that this insight is the hallmark of "a new biology" (Peacocke, 1986; Augros and Stanciu, 1987).

***

Chapter 3

THE DEFINITION OF TELEOLOGICAL CONCEPTS IN NON-TELEOLOGICAL TERMS

SUMMARY

In this chapter we turn to the question of the common pattern of causal relationships in which the directiveness of vital activities resides. We need this knowledge if we are to discover what causal relationships the networks of the brain must satisfy to produce the directive activities in which our investigation will come to take an interest. I shall also use it for defining in strictly non-teleological terms the teleological concepts we shall need in this work. In Section 3.1 I shall describe some of the general characteristics of directive activities, and in Section 3.2 I shall examine the common pattern of their underlying causal relationships. This common pattern will be called a *directive correlation*. A formal definition in strictly physical terms of this type of correlation will be given at the end of that section. It will also then be illustrated with a mathematical model.

After some further comments and elucidations (Section 3.3) I shall then use this concept in Section 3.4 to give prescriptive definitions in causal terms for the teleological concepts which we shall need in our analysis of the nature of consciousness. Definitions along the same lines will be given in Appendix A to a further set of teleological concepts because of their general interest.

***

3.1 - SOME GENERAL CHARACTERISTICS OF DIRECTIVE ACTIVITIES

Consider the action of a hawk diving onto its prey. In what formal pattern of relationships lies the directiveness of this action? Clearly, it does not just lie in the fact that the bird's flight-path happens to have a direction that in the given circumstances causes it to hit the prey. Rather it lies in the fact that the bird's movements are governed in such a manner that not only does it hit the target under the given circumstances, but would also have hit the target if this had been displaced to an extent that would have required a modified flight path. Thus if the prey had been situated one or more yards further North, South, East, or West, the hawk's flight-path would in each case have undergone the appropriate variations. That is to say, *there exists a set of possible variations in the location of the target, each of which would have required a modified action on the part of the bird, and for each of which the required modification would have been produced by the bird.* Naturally there will be limits to the variations permitted. The wider these limits, the greater might be said to be the *degree* of the directiveness concerned.

The same applies to the details of the execution of the flight-path: every disturbance in the air that threatens to throw the hawk off the required course is promptly met - or even forestalled - by finely adjusted compensations in the movements of wings or tail. That is to say: the variations in the circumstances for which the bird is able to compensate include variations in conditions affecting the execution of the required action.

In consequence of these qualities of the bird's movements, a hit on the target is achieved with some measure of certainty despite the many contingent factors that might have conspired to prevent it, and despite all the aspects of the situation that would have made this outcome a highly improbable event in the first place. After all, what could be more improbable than that some object dropping out of the sky should hit just the one small object on the ground that can satisfy its metabolic needs? In other words, it is precisely this quality of the bird's movements that raises the occurrence of a hit on the target above the level of chance-coincidence, and makes it more than a fluke. On the other hand the outcome is not guaranteed: a *mismatch* of the bird's movement in relation to the position of the target is a physical possibility - through clumsiness, distraction, or failing eye sight, for example. That is to say, the action variables of the bird are not *intrinsically* coupled to the location of the target by some nomological relationship: their appropriateness in relation to the target is *contingent* on certain special causal relationships' being satisfied. In the present case these are established by the *negative feedback loops* that govern the control of the action: broadly speaking, the bird is able to perceive errors in its flightpath; these perceptions set up error signals in the brain, which in turn influence the control mechanisms of the flight. Fig. 3.1 gives a flow-diagram of a basic feedback system of the type one might find it in the servomechanisms of autopilots, robots etc. The feedback is called *negative* because it is used to counteract deviations of the actual outputs from the commanded outputs.

To illustrate the contrast between the bird's flight and the ballistic flight of inanimate objects, the hawk's attack on its prey may be contrasted with the movements of (say) a boulder bouncing down the hill. Neither in the case of the boulder, nor in that

3.1 - *Some general characteristics of directive activities*

of the hawk, is the end-result a matter of blind chance. But the boulder arrives at the bottom of the hill in consequence of the invariant force of gravity, whereas the bird's actions are produced by forces which its internal mechanisms cause to vary in a manner conducive to a specific end-result by virtue of the special causal relationships I have mentioned, and these themselves are contingent on the current motivational state of the bird. (The difference between goal- seeking and equilibrium-seeking will be discussed in Section 3.3.)

Generalizing all this, we can describe the common objective feature of typically goal-directed actions as follows: *they are all produced in a way which causes them to be conducive to the occurrence of the event described as their 'goal' **not only** in the given circumstances **but also** over a significant range of **alternative** circumstances, each demanding a **different** action.*

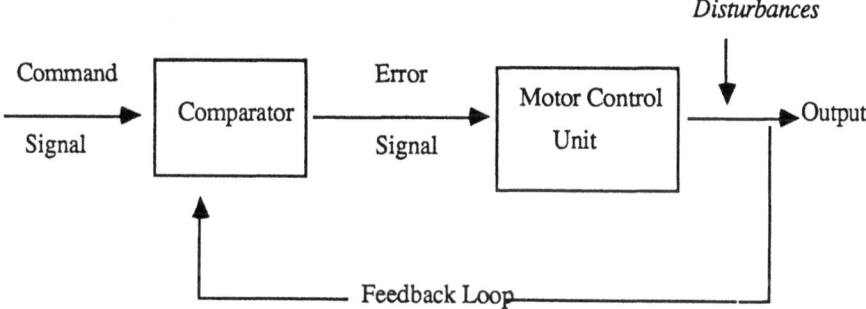

Figure 3.1. *Servomechanism with simple feedback control.*

In this work that kind of correlation between action and circumstance will be called a *directive correlation*: we shall say of such actions that they are *directively correlated* to the circumstances in which they take place. Clearly, we have here a very general expression for the kind of correlation between action and environment which bestows upon the higher organisms their unique brand of independence from the caprices of the environment, viz., a power to meet these caprices with appropriately matched responses. And clearly, too, we have in this correlation between action and circumstance a system-property whose formal structure can be investigated, regardless of how little we know about the specific mechanisms that bring it about in any particular case. But before we can define this concept formally we must track down the common pattern of causal relationships on which this correlation rests (Section 3.2). And more needs to be said even before we can reach that stage.

That it is the above feature which commonly makes one classify certain activities as goal-directed may be illustrated by the kind of test that one would intuitively tend to apply if one wished to test some robot activity for its goal-directedness. For example, if you were to find a 'black box' on a table which periodically fired a dart at a target on the wall, hitting the bullseye every time, and you wanted to know whether this was because the box had the ability to produce a goal-directed activity - or whether it was because the box had been deliberately aligned in this way by a person or persons unknown, you would do the obvious: you would either move the target or move the box, and then watch the results. That is to say, you would change the requirements for a successful shot and test whether the box would produce the necessary modification of its action.

Again, consider a homing missile which is targeted on an object which is both a source of heat and a source of noise. Suppose we do not know whether it homes on the one source or on the other. How could the objective difference between these alternatives be described without reference to the missile's internal hardware or software? Surely in these terms: if the missile is programmed to home on the source of noise, then any variation in the location of that source (while leaving the location of the heat source unchanged) would have produced a corresponding modification of the flight path, whereas a variation of the location of the heat source (while leaving the location of the noise source unchanged) would not have done so. And the converse would be the case if the missile had been targeted on the heat source.

The examples discussed above related to goal-directed *behaviour*. But it is not difficult to see that the characterization of directiveness I have given applies to a great variety of biological processes, the slow and very slow as well as the fast. For example, it applies to the slow ontogenetic adaptations that occur as the result of learning from experience; also to the processes of morphogenesis, regeneration and repair. For in all these processes we are dealing with changes which are produced in a manner that causes them to match the circumstances in which they occur in ways conducive to some specific end-result.

In the same objective sense we can speak of 'directive' processes in the phylogenetic evolution of a species. The mechanisms of natural selection, too, bring about changes - in this case changes in the genetically transmitted characteristics of the species - which correlate with the character of the animal's ecological niche in a manner serving some particular end-result or 'goal' - in this case the optimization of the chance to survive in competition with other species. And here again, the results produced by these directive processes have a high degree of *a priori* improbability. Indeed, it can be said that the sense of wonder we feel at the 'marvels' of living nature which to-day are so vividly brought home to everybody by the brilliant documentaries shown on television, is essentially a surprise reaction evoked by the *seeming improbability*, and hence *unexpectedness*, of the phenomena we witness. And in each case these phenomena have come about, despite their inherent improbability because they have been the end-product of a directive activity of one kind or another - in this case, generally, of an evolutionary kind.

When, in Section 3.2, we reach a more formal definition of directive correlation, we shall also see that those cases will automatically come to be excluded in which the end-result under review follows simply because the system is defined in terms which rule out a mismatch between action and environment (for example, a toy hawk sliding down a taught wire attached to the target). This is as it should be, because goal-directed activities are always conceived as activities in which failure is a logical possibility: a mismatch between action and circumstance is open to the system. Thus the outcomes of goal-directed activities contrast on the one hand with what is unavoidable and, on the other, with what is accidental. This contrast is both significant and illuminating. Although I must leave the details till later, it is a contrast worth noting right from the start.

Other features that may strike one as common to goal-directed activities will, on closer inspection, be found to be either trivial in comparison, or to be merely derivative from features already covered by the definition of directive correlation. Some writers have taken it to be characteristic of goal- directed activities that when the goal is reached the action ceases, whereas if the goal is not reached, the action persists. This is a fallacy: the notion of the action ending already presupposes a notion of goal-directedness. For, what ceases is merely an activity *having that event as a goal*. Activity as such does not cease, it merely switches to the next goal in the overall behaviour pattern of which it forms part. Having caught the prey, the hawk does not become inert: it will either begin to devour it or it will carry it to some safer place. In fact, most common behaviour patterns comprise a hierarchy of goals and sub-goals: they consist of steps, each having a proximate goal whose attainment serves as a starting point for the next step, all steps jointly serving the unifying goal of the behaviour pattern concerned - which itself may, in turn, amount to no more than a step towards a yet more distant goal. Activity may only stop when final consummation is reached, exhaustion sets in, or sleep demands its turn. As Tinbergen (1951) has shown, the behaviour patterns of wild life typically consist of a continuous flow of activity in which the completion of each stage serves as a trigger for the next.

***

3.2 - THE COMMON PATTERN OF CAUSAL RELATIONSHIPS

In the last section I have described some general features of directive activities and I have claimed that their directiveness resides in a distinctive pattern of spatio- temporal and causal relationships. In the present section I shall isolate those relationships. For a more detailed discussion the reader may be referred to one of my earlier publications, e.g.

Sommerhoff (1969) or to the treatment given in the language of symbolic logic by Hilgartner and Randolph (1969).

What we have so far established about the nature of goal-directed activities may be put as follows: they are all produced in a way which results in the occurrence of the event described as their 'goal' (or in a positive contribution to that occurrence), not only in the given set of circumstances, but also over a significant range of alternative circumstances, each demanding a different action if that event (or that positive contribution) is to occur. We must now seek to articulate this and see what causal relationships it presupposes.

The above description is couched in terms of *hypothetical* variations of the circumstances in which the actitvities take place. In the following pages such hypothetical variations will be called *disturbances* of the system.

What is to be meant by a *positive contribution* to the occurrence of the goal-event, can be left undefined, for it can without difficulty be specified in each individual case to suit the context. For example, in some contexts, an increase in the probability of the goal-event occurring may conveniently be counted as a positive contribution; in others it may be the removal of an obstacle to that occurrence, or the completion of a necessary condition for its occurrence. Again, in the case of a pursuit action, both the correction of a deviation from the required direction, and a decrease in the distance between pursuer and the object of the pursuit, may each be taken to count as a positive contribution.

In many concrete instances an action is not so much directed towards the occurrence of some concrete event, as to the production of some particular state of affairs or condition: concealment for example. We might then speak of a *goal-condition* rather than *goal-event*. But there is no need to treat these cases as separate categories.

To proceed, consider first an isolated goal-directed act, for example a rifleman shooting at a target. If you prefer a mechanical example, the patterns of relationships I shall discuss below apply equally in the case of an automatic anti-aircraft gun shooting at a target located by its radar.

Let the rifleman pull the trigger at time t. The direction of his shot will then have been determined by his view of the target a split second before he pulls the trigger. Let this be called the time $t-dt$. As we shall see, we cannot ignore this small time lapse between the sight of the target that determined the action and the actual execution of the action. It plays a critical role in the pattern of relationships we shall analyse. Further, let E denote the relevant environmental variable (here the position of the target) and let A denote the action (pulling the trigger while the rifle is pointing in a certain direction). Let G denote the goal-event (hit on the target). Finally, since the target will only be hit if the action A stands in an appropriate relationship to the target position E, let this required relationship for the time t be written as $R(A_t, E_t)$. This expression, therefore, is taken to denote the necessary condition for a successful outcome of the action.

3.2 – The common pattern of causal relationships

The causal relationships between the variables I have mentioned may then be symbolized as in Fig. 3.2. By the existence of a *causal relationship* between two variables, is here meant that the value which one of these variables has at a given point of time is (at least partially) determined by an *antecedent* value of the other variable.

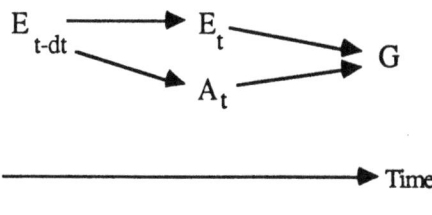

Figure 3.2.

A crucial observation is the following: the rifleman's aim at time t, and the location of the target at time t are *independent* variables in the technical sense in which that term is understood in the physical sciences. That is to say, A_t is not *intrinsically* determined by E_t. A mismatch of the aim is open to the system (which would not be the case, for example, if the target were fixed in an appropriate position to a beam attached to the rifle). Since this notion of independence is important, I must add a few words about it before we move on.

Two physical variables are said to be *independent* in this technical sense, if the value which either has at any particular point in time does not determine the value which the other variable has at that *same* point of time. This is also sometimes expressed by saying that the two variables are *logically independent* or *nomologically unrelated*. It implies, for example, that *arbitrary combinations of values of the two variables concerned may be conceived as possible initial states of the system*. This does not, of course, preclude the possibility of the value of either variable influencing *subsequent* values of the other.

All of this clearly applies in our present example: the rifleman can be conceived to start with any arbitrary relationship between the direction of the target and the direction in which the rifle points. To illustrate: in Newtonian physics, the position and velocity of a freely moving particle are independent variables, because the value of neither variable at any given point in time is influenced by the value which the other has *at the same point in time*. Hence arbitrary combinations of these values may be conceived as possible initial states of the system. By contrast, the force acting on a particle and the particle's acceleration are not independent variables, since in Newtonian mechanics the acceleration is always proportional to the applied force. Hence the Newtonian theory does not permit us to conceive of arbitrary combinations of values of these two variables as a possible initial state of the system. However, although the position and the velocity of Newtonian particle are independent variables, once the initial position and velocity of the particle have been determined, and the system is allowed to run its course, the velocity of the particle at any subsequent point of time can be calculated from its position. If the force imparts a constant accelaration a, and the motion runs parallel to the x-axis, then the position and velocity of the particle at time t are given by the familiar equations

$$v_t = v_0 + at$$

$$x_t = x_0 + v_0 t + at^2/2,$$

which enable v_t to be calculated from x_t if the other values are known. The example makes the important point that *independent variables can become correlated, but only if they have shared causal determinants:* in this case v_0. (Shared causal determinants are also important in the case of other kinds of correlation, e.g. statistical correlations. Thus it is common knowledge that when two variables are found to be statistically correlated, it does not follow that either is the cause of the other: the correlation may be due to a shared causal determinant. For example, the incidence of lung cancer and coronary thrombosis are independent variables in the technical sense explained above. But they are known to be statistically correlated, and they come to be so because they have a shared causal determinant in subject's smoking habits.)

It was worth giving the actual equations above, because they illustrate in what form causal relationships tend to enter into the equations of the physical sciences: viz., by way of formulae which express the value of a variable at time *t* as a function of the initial (t_0) value(s) of the same and/or other variables of the system. They also illustrate the defining characteristic of what is known as a *deterministic* physical system. For a physical system is said to be deterministic if the values of the variables defining the current state of the system are all single-valued functions (see Appendix B) of the initial state of the system and the time variable - as clearly illustrated in these formulae. (Sometimes, namely in so-called 'stochastic' systems, this condition is only satisfied if the current state of the system is defined, not in terms of its physical variables directly, but in terms of the current probability distribution of the possible values of these variables. But that is irrelevant in the present case.)

To return to our rifleman: we can now see that the following statements jointly describe the objective features characteristic of the directive correlation a skilled rifleman can establish between his action and the location of the target:-

i) A_t and E_t are independent variables.

ii) A_t must stand in some particular relation to E_t, if it is to make a positive contribution to the occurrence of the goal- event. Let this relationship be called a *goal-related requirement*, and let it here be designated by the expression $R(A_t, E_t)$ (In my earlier publications these goal-related requirements were called 'focal conditions' of the directive correlation.)

iii) A_t here comes to satisfy that relationship because of the way in which the rifleman causes his aim to be determined by E_{t-dt}, where dt denotes the time lag involved in perceiving the position of the target and in processing that information.

iv) There exists a significant range of possible disturbances of the system (here variations in E_{t-dt} each of which would have entailed a different E_t and called for a different A_t if $R(A_t, E_t)$. is to remain satisfied; and for each of which the rifleman would have produced the required A_t. Note that E_{t-dt} here plays the role of the shared causal determinant which any correlation between independent variables requires, as explained.

Although for simplicity we have here considered a single action, the above considerations apply equally to a continuous activity, e.g. the controlled flight of a homing missile. The only difference is that now the relationship $R(A_t, E_t)$. becomes a function of the time variable. It is worth noting in this connection that the causal link from E_{t-dt} to A_t forms the main *feedback loop* on which the missile's control system will be relying.

Now the important point to be made is that *the general pattern of relationships given in (i) - (iv), and symbolized in Fig 3.2, can be accepted as characteristic of goal-directed activities in general and as the mark of what is called 'directive correlation' in this work.*

However, to arrive at a definition of directive correlation which can be applied to the whole wide range of directive activities, we must introduce the following generalizations:-

1. We must allow that the goal-related requirement - $R(A_t, E_t)$.in the above example - may embrace more than two variables. In other words, we must allow for directive correlations between more than two variables.

2. We must omit from our definition the distinction between *action* variables and *environmental* variables, since, for example, a set of simultaneous movements may be directively correlated *inter se* as well as to the environment - this in fact, being distinctive of what we tend to describe as *coordinated* activities. It is also typically

an example of a case in which the directive correlation may embrace more than two variables.

3. We must allow for cases in which the shared causal determinant does not consist of the antecedent value of one of the directively correlated variables (as happened to be the case in the examples we have looked at).

With these generalizations in mind, a formal definition of directive correlation may then be given in terms of the following generalized pattern of causal relationships:-

DEFINITION:

*The current values of a set of two or more independent physical variables U,V,... are **directively correlated** with respect to the subsequent occurrence of an event or state of affairs G, if and only if*

a). the joint occurrence of these values contributes positively to the subsequent occurrence of G because, and only because, they satisfy some particular relationship R(U,V,...);

b). there exists a discrete and non-empty set D of possible prior disturbances of the system, each of which would have changed each of the current values of U,V,...; and

c). the change produced in any one of these variables would have been detrimental to R(U,V,...) were it not for the change(s) produced in the other(s).

For the case of three directively correlated variables, Fig. 3.3 symbolizes the pattern of a set of causal relationships that would satisfy our definition. J_0 denotes the value at time t_0 of a shared causal determinant, whose hypothetical variations constitute the disturbances mentioned in the definition.

It is obvious that our definition does not demand that the variables concerned are simple variables of a numerical kind. i.e. scalar variables. They could be vector variables (as in fact they were in my examples). More generally, they could be variables defined in set-theoretical terms (cf. Appendix B), e.g. variables denoting members or subsets of a set of discrete elements (e.g. the discrete alternatives open to a rat at the choice-points of a maze). The object of the definition is not to extend our powers of computation and

3.2 – The common pattern of causal relationships

prediction, but to clarify conceptually the nature of the teleological character of vital activities and processes and to give us a precise set of concepts to work with.

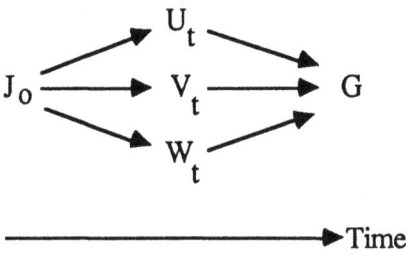

Figure 3.3.

SUPPLEMENTARY DEFINITIONS:

Assuming that some measure can be defined on the full range of possible disturbances of the system for which the above conditions are satisfied, this measure will be called the *degree* of the directive correlation. The event or state of affairs G will be called the *goal-event* or *goal-condition* of the directive correlation, while the relationship $R(U,V,...)$ will be called a *goal-related requirement*.

A MATHEMATICAL ILLUSTRATION

For those readers who are familiar with differential calculus it will be instructive to illustrate the concept of directive correlation by setting up a simple mathematical example of a system which satisfies the conditions of a directive correlation as defined above. To this end assume that U, V, and W in Fig. 3.3 are three independent numerical variables in a deterministic physical system and that the value at time t of each of these variables can be expressed as a differentiable function of J_0, i.e. of the value which a certain variable J has in the initial (t_0) state of the system. Next assume that

$$F(U_t, V_t, W_t) = 0$$

is a goal-related requirement in respect of some 'goal-event' G. Finally, assume that the variables mentioned satisfy the following equation over a significant range of variation of J_0 and that none of the terms on the left hand said vanish identically:

$$\frac{dF}{dJ_o} = \frac{\partial F}{\partial U_t}\frac{\partial U_t}{\partial J_o} + \frac{\partial F}{\partial V_t}\frac{\partial V_t}{\partial J_o} + \frac{\partial F}{\partial W_t}\frac{\partial W_t}{\partial J_o} = 0.$$

The equation implies (a) that each of the variables U_t, V_t, W_t will undergo changes in consequence of disturbances in J_0; and (b) that whatever variation one of these three variables undergoes in consequence of such a disturbance, its effect on the occurrence of G is cancelled out by the variations resulting in the other two variables. In accordance with our definition, therefore, the variables U_t, V_t, W_t are *directively correlated* in respect of the 'goal event' G, with J_0 as the shared causal determinant.

This simple mathematical model is instructive, because it demonstrates particularly clearly that directive correlation, hence goal-directedness as interpreted above, is a *physical systems-property* in the truest sense of the phrase; hence that our perception of directiveness in living nature is not a misguided projection into the environment of our own human rationality, not an anthropomorphism, as some biologists have assumed in the past.

However, the value of this mathematical model is purely heuristic. The variables we shall have to deal with are not of the simple numerical, continuous and differentiable kind assumed in the model.

***

3.3 - FURTHER COMMENTS

To illustrate the wide applicability of the notion of directive correlation, let me show how it covers also much slower processes than the behavioural activities we have considered so far. As a suitable example consider the case of a chameleon changing its colour to match that of the environment:-

i) Clearly we have here a process which one would describe as being directed towards the achievement of a special end- result or 'goal', namely to render the animal inconspicuous and thus to optimize its chance to escape the attention of potential predators. The goal-related requirement here is that the colour of the animal should match that of the environment.

ii) The two variables mentioned in this requirement, namely the colour of the environment and that of the animal, are independent variables in the sense defined above. That is to say, the colour which the animal has at any given point in time is not determined by the colour which the environment has at that same point in time; a mismatch is physically possible; and the system can be conceived to start from any arbitrary combination of colours. Given time, though, the required matching of these colours will come about; the required relationship will come to be satisfied.

iii) There exists a set of possible disturbances of the system which satisfy conditions (b) and (c) of our definition, viz., lasting variations in the colour of the environment conceived to have occurred at a point in time preceding the present by an interval sufficient to permit the required colour changes of the animal to come about.

The degree of this directive correlation depends on the time interval taken into consideration. The smaller the interval conceived, the smaller will be the possible variations in the colour of the environment to which the animal could have made the required adjustments.

I have taken this particular example of a directive process of the ontogenetic kind, because it makes it obvious that exactly the same formal conditions are satisfied also in directive processes of a phylogenetic kind, i.e. in adaptive changes of the kind that occur in the genetic evolution of a species. For, in terms of these formal relationships, it is a small step from the protective coloration of a chameleon, which is the result of on-the-spot adjustments, to the protective coloration of a species of caterpillars which is acquired genetically and not capable of on-the-spot adjustments. It is obvious that, despite this difference, the directive processes here involved can be characterized by the same formal pattern of causal relationships as in the previous case. Only the relevant time-span would be a very different one. The hypothetical changes in the colour of the environment would have to be placed many generations back in time if the required changes are to come about in the colour of the species.

*

These examples, together with that of the rifleman, illustrate the three **main categories of directive correlations** which we meet in the living world, namely:-

a) Those produced by the slow, long-drawn-out adaptive processes of evolution. We may call these the *phylogenetic* correlations. Their main characteristic is that, owing to the nature of the evolutionary process, they can exist in significant degrees only in time-slices which span a great number of generations of the biological species concerned.

b) The adaptive changes that occur in the development of the individual before and after birth, and especially those that occur in the processes of maturation and learning. These may be called the *ontogenetic* correlations. We can find these directive correlations in significant degrees within time- slices which may have to span no more than minutes or hours in some cases, but months or years in others.

c) The rapid adaptive changes and instant adjustments that occur at what may be called the *executive* level of vital activities - by which I mean the fast regulatory and control functions that govern the dynamics of the higher organisms, especially their skilled motor activities. The case of the rifleman and of the hawk diving for its prey, were typical examples. Here we have directive correlations which can already be found within time-slices spanning only seconds or fractions of a second.

*

Why do we commonly tend to contrast directive activities with the 'blind' activities of non-directive systems? Why use the expression 'blind'? Because a system capable of producing goal-directed activities can only produce the specific type of appropriate response which this goal-directedness demands if it has access to an adequate flow of *information*. In the example of the rifleman (cf. Fig. 3.2) this flow of information was provided by the causal link between E_{t-dt} and A_t. This was the information carrying channel. And it must be pointed out in this connection that what passes along this channel must be capable of having as great a *variety* as the variety of values of E over which the postulated directive correlation is satisfied.

Only *directive* activities or processes require the causal link to which the notion of information here relates. Hence the use of the expression 'blind' in the case of non-directive systems merely serves to acknowledge the absence of information-carrying channels of the kind we have here considered. In fact, the ultimate significance of what we call 'information' lies in the role it plays in directive correlations, as illustrated in the above example. That is why this notion is only applicable in a teleological context, and why I have said that we do not have a clear notion of information in any particular case unless we also have a clear notion of the relevant teleological context. In the now classical treatment of information by Shannon and Weaver (cf. Appendix E), this teleological context is generally understood and obvious. But this is by no means always the case when one sees the notion applied in the literature, for example, in Gibson's theory of perception (cf. Section 5.7).

*

3.3 - Further comments

It may help to illustrate the analysis by pointing out that the following would be examples of directive activity in the objective sense given by the definition of directive correlation: a chick pecking at a grain - but *not* the imitation of that action by the simple clockwork birds you can find in toyshops (since their actions would not match a displacement of the grain); a ship or aircraft guided by an autopilot - but *not* a truck running on rails; a bird building a nest - but *not* water hollowing out a cavern; men digging away a mountain - but *not* nature creating one; and the evolutionary adaptation of a plant or animal species to its environment - but *not* the blindly deterministic evolution of the solar system. Excluded, too, would be all instances of animal activity in which a biologically significant change results merely accidentally or fortuitously. Excluded, for example, would be the case of a caterpillar acquiring a protective coloration by falling into a green paintpot.

*

The formulae I have given also enable us to draw a distinction between *goal-seeking* (in the objective sense I have defined) and *equilibrium seeking*. This is a relevant point, because it was his failure to distinguish between these two cases that flawed von Bertalanffy's (at the time influential) analysis of goal-directedness in terms of his concept of 'equifinality' (von Bertalanffy, 1933). 'Equilibrium-seeking' describes the behaviour of physical systems held to be in a state of stable equilibrium. Now by definition a state of *stable equilibrium* is one in which small displacements of the system from the given state result in forces tending to restore the original state. But, by contrast with the case of a directive correlation, these redressing forces result *nomologically* and not by way of a causal link providing a requisite amount of information flow: any displacement of the system occurring at a given point of time determines the restoring force *for that same point in time* . That is to say, the correlated variables do not satisfy the condition of independence which we stipulated for a directive correlation; (see the definition of 'independence' given in Section 3.2).

*

Consider the servo-controlled flight of a homing missile programmed to hit a target T. This hit may be described as the goal of the flight in the objective sense of the term as established in our definition of directive correlation. Now assume that such a hit invariably makes a loud noise of a special kind. Then it is obvious, that in respect of the corrections the missile executes in response to the different kinds of contingencies it meets on its flight, the case, from an objective point of view, is indistinguishable from the case of a missile the goal of whose flight is to produce that particular kind of noise in that particular location. In fact, it follows from our definition that if an action is directively correlated in respect of a goal-event G, it will also be directively correlated in respect of any *consequence* of G. And this may seem an odd implication of our definition. But it is correct. The data which enable us to decide that an action is directed towards a goal-event G_1, do not suffice to decide whether G_1 is pursued as an end in itself, or merely because it entails the occurrence of a further event G_2. To formulate that difference in objective terms, we would have to consider a different category of disturbances, namely disturbances in consequence of which G_1 would no longer entail the occurrence of G_2.

*

Throughout this chapter we have exclusively interpreted 'goal-directed' in an *objective* sense. To avoid confusion, therefore, it is wise to return for a moment to the fact that, in the human case there is also a *subjective* or *mentalistic* sense in which we can speak of 'goal-directed' or 'purposive' actions. I mean the sense in which the terms merely serve to convey that the actions are determined by conscious psychological purposes, i.e mental images of some desired end-result.

I have mentioned that this ambiguity has been a frequent source of confusion in the past. The confusion springs from the fact that the perception of animal activities which are goal- directed in the objective sense defined above, by their very nature invite analogies with our own consciously purposed activities. It was for this reason that in the first half of this century a number of biologists who were aware of the limitations of the analogy, but also lacked a clear idea of the objective characteristics of animal activities that invite it, felt driven to proscribe outright the use of teleological terms in the description of animal behaviour, condemning it as a mere anthropomorphism and as a species of animism. But biology has never actually been able to get along without teleological concepts, for, even basic terms like 'function', 'control', 'regulation', and 'adaptation' are at heart teleological terms, although they look more innocent because they invite no parallels with mental activities.

As I have also pointed out, the fact that a particular human action is purposive in the subjective sense does not entail that it is also goal-directed in the objective sense defined above. Reasoned behaviour is no doubt the most sophisticated way in which the human animal manages to produce behaviour which is directive in that objective sense. But human reason is fallible and does not therefore invariably produce actions which are appropriately related to the circumstances in which they take place. To avoid confusions of the kind I have discussed it is expedient to reserve one set of terms for consciously purposed activity and another for activities which are directive in the objective sense we have defined. In the remainder of this work, therefore, I shall reserve 'purposive' and 'purposeful' for the first, and 'directive', 'goal-directed' or 'directively correlated' for the second.

***

3.4 - APPLICATIONS: THE TRANSLATION OF TELEOLOGICAL INTO NON-TELEOLOGICAL TERMS

The most important general application of the concept of *directive correlation* lies in the means if offers to substitute precise and strictly objective teleological concepts for existing fuzzy ones, and to do so by way of introducing technical definitions for those concepts which are couched in precise and strictly non-teleological terms, and which yet manage to capture the core of the meaning they would seem to have in normal discourse. As I have explained, this is essential when one deals with such intricate conceptual problems as the ones that face us in our current enquiry. In the present section I shall confine myself to the handful of teleological concepts we shall need in our analysis of the phenomenon of consciousness. A further set of teleological concepts, of interest in the broader context of this work, will be given the same treatment in Appendix A.

1. The concept of an action being *directed* towards the occurrence of an event or condition G.

DEFINITION:

> An activity is **directed** towards the occurrence of an event or condition G if it is *directively correlated* to the circumstances in which it takes place, with G as the relevant goal-event or goal-condition.

2. The concept of *appropriateness*.

In everyday life we say that someone has taken an *appropriate* step or *appropriate* measure if he has taken a step or measure which we deem to be an effective one in the pursuit of some particular goal or desirable outcome. An appropriate step towards cashing a cheque is to go to a bank. Clearly, the context in which we here use 'appropriate' is always a teleological one: there is always an explicit or implicit reference to some goal-event - in this case the cashing of a cheque.

DEFINITION:

*An action is **appropriate** in a set of given circumstances if it satisfies a goal-related requirement (as defined in Section 3.2) in the context of a goal-directed activity.*

3. The concept of *adaptation*.

As used in everyday life, and generally in biology too, the concept of adaptation is also a teleological concept. Once again there is always some implicit or explicit reference to an end-result or goal. Things are adapted for some end: you 'adapt' a car for use in the desert by making it impervious to sand-dust; plants are 'adapted' for life in dry climates by having developed thick leathery leaves which reduce water loss; when I pick up my pencil the movement of my arm is 'adapted' to the location of the pencil on the table; and so on. In some specialist applications 'adaptation' has lost its teleological connotation - for example, when physiologists speak of 'adaptation' in a nerve cell. Such special uses of the word will be excluded by our definition.

In general 'adaptation' denotes a binary relationship: when we think of something as being 'adapted' to something else, we are thinking of a relationship between two variables (or sets of variables) only. An important feature of this relationship is that it is *asymmetrical*: 'A is adapted to B' does not mean the same as 'B is adapted to A'. This difference is easily taken care of in the definition of adaptation whihc I shall adopt:-

DEFINITION:

*The value of a variable A_t is **adapted** to the value of a variable B_t if there exists a directive correlation between A_t and B_t which has an antecedent value of B as the shared causal determinant required by the correlation.*
(Conversely, the value of B_t would be said to be adapted to the value of A_t if there exists a directive correlation between them which has an antecedent value of A as the shared causal determinant.)

This definition clearly captures the asymmetry to which I have already drawn attention earlier on. Fig.3.4 represents it in diagrammatic form.

To illustrate: when we say that the colour of the curtains was adapted to the colour of the wallpaper, we imply that if the colour of the wallpaper had been different our choice of colour for the curtains would have been modified accordingly. Conversely, when we say that the colour of the wallpaper was adapted to that of the curtains we imply that we had the curtains first and that the wallpaper was chosen to matched to these.

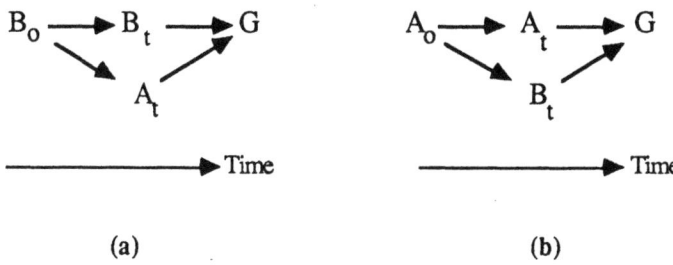

Figure 3.4 - (a) A_t adapted to B_t (b) B_t adapted to A_t.

4. The concept of *function*.

The concept of biological function is another very general teleological concept, and it is of particular interest because even in the heyday of mechanistic thought in biology scientists never managed to get along without it. One just could not manage to discuss a living organism without discussing the functions of its components parts and organs, just as one might want to enquire after the function of some part of a machine. Typical examples of such functional statements are:-

> *The function of the heart is to circulate the blood.*
> *The function of the bird's posturing is to attract a mate.*
> *The function of the police is to enforce the law.*
> *The function of the spark plugs is to ignite the fuel mixture.*

That we are here dealing with a teleological concept can hardly be doubted. You cannot talk about the function of a star in the galaxy or of the mountains on earth. As Sorabji (1972) has pointed out, it is an ineradicable part of the notion of biological function that to fulfil a function is to serve some biological good, to do something of biological value. However, Sorabji was mistaken, as was Woodfield (1976) in his turn, in concluding that the notion of function could not therefore be translated into non-teleological terms. They were mistaken because the concept of directive correlation enables us to do just that.

Part of the reason why the notion of function has caused much confusion in the past lies in the fact its teleological implications are fairly subtle and indirect. To make them explicit let us turn again to a machine example. We are less likely here to be confused by preconceived biological ideas. What, then, do we mean when we ask about the function of the spark plugs in a petrol engine? Clearly we want to be told what contribution they make to the operations of the whole machine. But it is not any *accidental* or *contingent* contribution that we are interested in. It is the contribution they were created to make, the contribtion *they are there* to make. In other words, we want to be told the *contribution to the whole to which they owe their presence in the engine, and (by implication) their design.*

These conclusions are readily translated into biological terms: when we ask about the function of the heart we expect to be told those contributions of the organ to the performance of the whole body *to which it owes its presence* as a feature of the body -- given the forces of selection that have governed the evolution of the body and its parts. And there is nothing in this formulation that cannot be given a precise and explicit expression in non-teleological terms with the aid of the concept of directive correlation. For, the nature of the teleological element should now be obvious: the presence of the relevant organ or mechanism as a feature of the organism (and by implication its of operation) is assumed to be the result of an *adaptive* process (in the sense defined above) which had the efficient performance of the body, as its goal-condition.

DEFINITION:

> *The function of a structural or dynamic component of a system is that contribution to the performance of the whole to which that component owes its presence in consequence of adaptive processes which had that performance as their goal.*

Clearly, this definition covers also the notion of 'function' when applied to the components of a machine.

5. States of *Readiness*.

I must finish with a couple of concepts which have a major part to play in our analysis of sensory perception and of the brain's internal representations of the world (Chapter 5). They are (1) the concept of a system being in a state of *readiness* for an event, and (2) that of a system being in a state of *expectancy* for an event. I have placed them in that order because I shall define a state of *expectancy* as a special case of a state of *readiness*.

DEFINITION:

A physical system is in a state of readiness for an event X, if it is in a state which will facilitate the making of an appropriate response to X, should X occur.

I must add that I take the term *readiness* here to be synonymous with *preparedness*, and generally prefer the former only because it is the shorter. The meaning of 'appropriate' was defined above.

6. States of *Expectancy*.

The notion of states of expectancy will play a crucial part in our explanation of how the brain uses its sensory inputs to build internal models of the outside world. We need an objective definition here especially because in everyday use 'expectancy' and 'expectation' tend to denote a *mental* state and are not in this subjective sense acceptable terms for the purposes of the present analysis. To avoid confusion I shall reserve the term *expectancy* for the objective state of affairs I shall define below, while reserving 'expectation' for the mental state.

DEFINITION:

An organism is in a state of expectancy for an event, if

i) it is in a state of readiness for that event, and

ii) this state will undergo adaptive modifications if the event in question fails to occur (or fails to occur at some specific point of time, or within some specific interval of time).

This is a strictly objective definition since the notions of 'readiness' and 'adaptive modifications' have been given strictly objective definitions in the preceding text.

The following important difference between the concept of *readiness* and that of *expectancy* should be noted: a system can simultaneously be in a state of *readiness* for two mutually exclusive events: when I go on holiday I can pack clothes both for rainy days and sunny days. But a system cannot simultaneously be in a state of *expectancy* for two mutually exclusive events. I cannot both expect Uncle John to arrive, and not to arrive, on the 4 o'clock train.

*

This concludes my analysis and clarification of the handful of teleogical concepts I shall need in the remainder of this work. By giving us prescriptive definitions for these concepts in non-teleological terms, it has made them safe to use in the work to come. At the same time it has exposed the general pattern of causal relationships which are implied in activities or processes having a teleological character. It has also given us certain general insights into the nature of organic order. These will again prove valuable when in part III of this work we look at certain cultural aspects of human nature and try to see these in a general biological perspective.

***

PART II

BRAIN AND CONSCIOUSNESS

Chapter 4

CONSCIOUSNESS AND SELF-CONSCIOUSNESS

SUMMARY

After a brief account of the dilemma that has faced psychology from the start, and which our treatment of consciousness aims to resolve, I shall return to the meaning of the words 'conscious' and 'consciousness' (Section 4.2). I shall argue that consciousness rests on certain representational powers of the brain, i.e. states of the brain which can function as a kind of surrogate of the objects, events or situations (actual or merely possible) to which the organism has to respond - not unlike the sense in which certain binary numbers in a computer controlling the flight of an aircraft may be described as *representations* of the aircraft's current altitude, airspeed, rate of climb etc. (cf. Section 1.4). I shall maintain that the faculty of human consciousness and the main categories of mental events can be fully accounted for in terms of three main categories of internal representations formed by the brain. These will be discussed in Section 4.5, after we have examined and defined the concept of internal representation in Sections 4.3 and 4.4 respectively. The question of a technical definition for consciousness will then be examined in Section 4.6.

The problem of the physical structure of the internal representations concerned will be left to Chapter 5.Specific hypotheses will then be introduced and examined in the light of the evidence. These will form the basis of our final conclusions about the neural basis of consciousness in general, and of various categories of mental events in particular (Chapters 6 and 7).

***

4.1 - THE PSYCHOLOGIST'S DILEMMA

The nature of mental events remained the province of philosophy until, a hundred odd years ago, psychology began to emerge as a separate discipline. From the start, however, the new discipline failed to cope with the critical fact that our inner life is not observable by the methods of science and what we discover through introspection cannot qualify as hard evidence in a strictly scientific sense. This harsh reality created a dilemma which to this day has left psychologists deeply divided between those who give priority to the methodology of the natural sciences and accept the resulting limitations, and those who regard it as the duty of psychology to speculate about all aspects of the psyche, even if it means using fuzzy concepts, mere metaphors, or mental concepts defined merely in terms of other mental concepts.

In the early days of psychology, the study of consciousness was regarded as the central object. The whole 'stream of consciousness', as James was to call it, was under investigation. Notably the Würzburg school concentrated on attempts to explore thought processes and volitions under experimental conditions. But the heavy reliance of this school on introspection soon provoked a reaction and a growing demand for stricter objectivity. Another obstacle was that then (as even now), psychologists who talked about consciousness generally failed to clarify what exactly they were talking about. Thus voices were soon raised which claimed that only observable behaviour was a fit subject for scientific investigations and admissible as scientific evidence. At the same time the work of Pavlov on conditioning in dogs demonstrated how the association between observable stimuli and observable responses could be studied systematically. Eventually, in the 1920s, Watson and his behaviourist school came to see in consciousness and 'mind' no more than a state of the organism apt to bring about certain types of behaviour. However, since in these behavioural terms they failed to reach the essentials of what is commonly meant by consciousness, it was a short step to conclude that consciousness was not a fit subject for strictly scientific investigations. To this day that has remained the attitude of some of the more radical disciples of this school. Indeed, "Consciousness", Skinner came to declare, "is a non-existent entity invented for the sole purpose of producing spurious explanations."

According to this extreme view, all statements about mental events are to be construed as statements about behaviour or dispositions to behave in some certain way. Talk about the sensations of hunger or thirst, for example, is to be construed as talk about the disposition to eat or drink. And the main task of psychology is to explore the laws that govern the associations of observable responses with observable stimuli and to study how these associations develop under the impact of reward or punishment.

Thus the brain came to be seen by this school as no more than a conditionable mediator between stimulus and response, and this was taken to apply to all creatures alike. "What goes on inside your head", Skinner has said "is no more important than what goes on inside a pigeon's.",..."We think we determine our actions but our environment does",... "There is no such thing as free will" (Skinner, 1978).

These methodological precepts were asserted to be only ones that could yield a truly scientific psychology fit to rank as an extension of the natural sciences, thus enabling these sciences at last to encompass the totality of human life.

The behaviourists' aim was sound enough. Its implementation was not. As has been said, far too narrow a view was taken of the kind of variables that are admissible in the theories of the natural sciences. These variables do not have to be individually and directly observable in the literal sense. The only condition they have to satisfy is that from their definitions, from the place they occupy in the scientist's formulae and hypotheses, and from their assumed values, it must be possible to draw unequivocal inferences (statistical or otherwise) about what are, in principle at least, observable events. Most of the variables that occur in atomic physics, for example, are of this kind. Hence to qualify as a science in the strict sense, psychology need not eschew the introduction of hypothetical variables or constructs - such as the 'internal representations' I shall be discussing in this chapter

Although Skinner remained an exponent of the radical form of behaviourism I have outlined above, the 1930s and '40s showed a growing tendency to relax these rigid precepts, especially the prohibition of hypothetical variables. Hull (1943), for example, constructed an elaborate theoretical system on the basis of such hypothetical variables as *drive* and *drive reduction*. And in the 1950s the liberalization of the original behaviourist doctrine continued apace. Even introspective self- reports came to be admitted as legitimate evidence by regarding them as special forms of behaviour. The awkward problem of the uncertain meaning of the words used in these self-reports tended to be brushed aside.

Alongside these developments new attempts had been made in the 1930s to reinstate consciousness as fit subject-matter for scientific studies. These concentrated mainly on finding objective criteria for consciousness. Some writers equated the faculty of consciousness merely with the ability to differentiate. Others suggested the possession of language and the ability to communicate as a criterion, since they saw mental processes essentially as symbolic processes and only language, they claimed, makes symbolic representations possible in the brain - a view which seems oblivious of the fact that much of our thinking is conducted in terms of internal representations of a quasi-pictorial kind, i.e. in terms of mental images. If these attributes are accepted as definitions of consciousness, and sometimes they seem to be treated as such, the doctrine would deny consciousness to the neonate, the deaf-mute and any animal devoid of language.

Nevertheless, the ability to report an event has firmly established itself in experimental psychology as a sound indicator of a subject being conscious of the event.

My comment, therefore, would be that it is acceptable as *sufficient evidence* for consciousness, but is not to be identified with the state of consciousness itself.

More recent attempts to describe consciousness in non-mental terms have tended to produce little more than generalizations which could hardly claim to reach the heart of the matter and often lacked clarity as well. Typical examples: "a device that determines what action system shall be dominant" (Shallice 1972); "a high level operating system with a recursive model of itself" and " purely a class of parallel algorithms" (Johnson- Laird, 1983); and "a kind of metaphorical space in which we can represent ourselves" (Jaynes, 1979). Other writers have tried to deal with the problem by equating consciousness with awareness. Thus Battista (1978) describes it as "a system by which people become aware". But if 'awareness' is given no independent and objective definition, this equation does not help since the common dictionary definition of 'awareness' is "being conscious of something". Occasionally one sees 'awareness' defined as reactivity, but that is obviously much to broad a definition, since on this view any reactive organism has consciousness.

Descriptions of consciousness in terms of internal representations have also been advanced, but their authors have generally failed to clarify the meaning of 'representation' when used in this context. Hence the dangers inherent in the notion of 'representation' which I mentioned in Section 1.4. were not always clearly perceived and guarded against. I mean in particular the danger of regarding an internal representation as some kind of projection onto some internal screen (which would then have to be scanned by some kind of internal eye). It leads to the erroneous idea that the concept of a 'representation' always implies the actual or potential presence of an observer of that representation - in this case therefore the existence of a little humunculus in the brain who did the observing. I will cite one case in particular, because it allows me to make a couple of additional points.

In his THE EVOLUTION OF HUMAN CONSCIOUSNESS Crook (1980) cites with approval Ryle's demolition of the idea that the mind is something separate from the body, a 'ghost in the machine' (Ryle 1949). Mind, Ryle said, is no more a thing apart from the body than is the University a thing apart from the libraries, laboratories, lecture halls and colleges of which it is composed. Crook realizes that consciousness and cognition are largely a matter of the formation in the brain of internal representations of the outside world. But then he falls into the trap of the 'humunculus fallacy' by assuming this internal representation to be of the nature of a projection onto some internal screen. I quote:

> "In higher organisms one prime function of this system (the cognitive system) is the maintenance of consciousness as a kind of screen upon which both novel information and the background presentational continuum can be projected." (p. 24)

"This interior analogue of the exterior world is actively 'seen';" (p. 29)

"Consciousness of the world comprises a form of re- presentation of the current perceptual input on a mental screen, so that a constant awareness of the monitoring process arises." (p.35)

The analogy with a screen is a fatal one, because it leads to an infinite regress: it implies that there must be some inner eye which then scans this screen, and presumably projects its perceptions onto an even more deeply recessed screen, which in turn is scanned.... etc.

Nor can I go along with the writer when he turns to the subject of self-awareness. For he virtually equates self- awareness with self-evaluation or self-esteem, thus ignoring one's consciousness of the current state of the self, which is a *cognitive* function and not an *affective* one. To quote:

"A characteristic of objective self-awareness is that it operates primarily in a comparative mode. Self-evaluation is based on the existence of a set of standards by which each person regulates his self-esteem and sense of social well- being. " (p.313)

In consequence, the evolution of human consciousness becomes for Crook little more than the story of the development of social relationships as adaptive mechanisms in the evolution of the human species. A different view, but also a social one, is that taken by Nicholas Humphrey (1980), who sees the main biological value of self-consciousness in the insights it affords to animals with complex social lives into the motivation of other members of the tribe.

***

4.2 - THE MEANING OF CONSCIOUSNESS

In Sections 1.4 and 1.5 I have reviewed some of the main senses in which the words 'conscious' and 'consciousness' occur in normal discourse. I drew attention to Malcolm's distinction between the *transitive* and *intransitive* sense of 'conscious'. In the former, the word has an object, as in *I was conscious of a strange noise*. In the latter it has not: *he was fully conscious during the operation*. We also noted that 'consciousness' in the transitive sense may be equated with awareness. Pulling all this together, we can list the following three notions of consciousness as the most commonly occurring ones.

1. *Consciousness understood in the intransitive, objective, and clinical sense in which being conscious is contrasted with being in a coma or asleep.*

In this sense of 'consciousness' the criterion for an individual having fully regained a state of consciousness after coma or sleep, would seem to be that he shows evidence of once again knowing (or seeking to know) who he is, where he is, how he came to be there, what state he is in, etc. As has been stated, the ability to report such things may be regarded as sufficient evidence for consciousness to have returned. But it is not to be identified with it. Everything may have returned except the faculty of speech, for example. And, as has been said, the clinician would probably be fully satisfied that the patient had recovered consciousness if the latter were to get up, get dressed, collect his belongings and walk out. I shall discuss later what this activity must show to be distinguishable from sleepwalking.

This clinical sense of consciousness tends to be rendered in medical dictionaries in such phrases as " having present knowledge or perception of oneself, one's acts and surroundings" (Steadman's Medical Dictionary, 24th edition). It is plausible, therefore, to equate consciousness in this intransitive, objective and clinical sense with a state of the individual in which his brain disposes over a coherent and comprehensive internal representation of the world and of the self-in-the-world. The fact that "Where am I?" and "What happened?" are the most characteristic questions asked by anyone gradually returning from a temporary loss of consciousness, is positively revealing in this respect. Again, I have said that a return to consciousness after sleep is marked for us by things gradually falling into place. This 'falling into place', of course, is precisely the gradual reconstitution as a functional entity of a comprehensive and coherent internal representation of the world and the self-in-the-world.

2. *Consciousness understood in the transitive sense in which it occurs in such phrases as*
 I was conscious of a strange noise or I was conscious that I was being followed.

In all these cases 'conscious' has an object, and this is invariably something which has attracted the person's attention. This sense of 'conscious' is rendered in Steadman's Medical Dictionary as "something occurring within the perceptive attention of the individual".

This sense of 'conscious' therefore relates to the fact that not all parts of the brain's internal model of the world enter into the determination of behaviour on an equal basis. The internal model is subject to *attentional* reactions in the form of *selective* operations in

4.2. - The meaning of consciousness

the brain which cause some parts of the model to dominate in the control of the brain's ongoing activities - presumably in consequence of a selectively enhanced responsiveness of particular sets of neural units to particular classes of inputs. The outputs which these inputs thus come to dominate may be more or less comprehensive. They may dominate no more than the direction in which we cast our eyes.

These selective operations are context-dependent reactions under motivational control. In general they reflect what the brain has learnt about the significance of particular stimuli or objects in particular contexts. However, the non-facilitated inputs, i.e. the unattended background of the current situation, will not in general remain without influence, if only in an indirect or negative sense. To repeat an earlier illustration: while I am sitting at my desk writing these lines, my eyes do not wander about restlessly like those of a bird sitting on its perch, compulsively scanning the surrounding world for signs of danger. Because I know that my immediate surroundings are of a kind that harbour no threat to my person.

I have implied that the capacity of being conscious of something presupposes a capacity of the brain to form an internal representation of the world and the self-in-the-world. For, I have stated this to be the field from which the attentional processes mentioned above select. That this is correct is attested by the fact that whatever we are conscious of we are always conscious of *as part of the total situation*. If we are conscious of a strange smell, we are conscious of it as a state of affairs in the room (or other area) in which we find ourselves. At the very least we are conscious of it as an experience which is part of the current state of the self.

3. *Consciousness understood in the subjective or phenomenal sense in which the term denotes the subjective character of experience, i.e. one's perception of the distinctiveness of the totality of what is happening to oneself at a given point in time - including the distinctiveness of one's feelings and thoughts, and what is sometimes called the "raw feel' of one's sensations.*

For many people, including the philosophers Bretano and Locke, this is the primary sense of consciousness. For them consciousness means primarily what one experiences subjectively in a state of wakefulness as revealed in introspection.

This sense of consciousness may therefore be regarded as a special case of the transitive sense of consciousness described in the previous paragraphs, viz., when the attentional processes in the brain are directed at that part of the brain's internal representation of the world and the self-in-the-world which relates to the current state of the self.

It covers what in Section 1.4 was described as the *philosophical* sense of *self-consciousness*. It is also described by the Oxford English Dictionary as the *philosophical* sense, and there rendered as "*the perception of one's sensations, feelings, thoughts.*"

Something of all the three connotations I have listed is encapsulated in the definition given by Collins English Dictionary (1979 edition) for one sense of 'consciousness': "*denoting or relating to a part of the human mind that is aware of a person's self, environment and mental activity, and that to a certain extent determines its choice of actions.*"

*

The three connotations of 'conscious' or 'consciousness' I have discussed above, will be the main ones I shall seek to cover in my analysis. Other connotations can generally be related to these three (e.g. a 'class-conscious' society).

Presently I shall postulate that the various facts to which these three connotations relate can be fully covered by the assumption that the brain has the power to form *three main categories of internal representations*. These will not individually correspond to the three connotations I have discussed, but they will cover them jointly. A preliminary outline of these three categories of internal representation was given in Section 1.6. However, before we look at these in greater detail, I must first clarify and define what precisely is to be meant by an *internal representation* in the present context. This will be the main object of the next two sections.

***

4.3 - INTERNAL REPRESENTATIONS OF THE OUTSIDE WORLD

In the present section I shall focus on just one category of internal representation as the best way to introduce the subject. This is the brain's internal representation of the outside world and of the objects, events or situations that are part of that world. This discussion will then be followed in Section 4.4 by a functional definition of what precisely is to be meant by an internal representation in these and related contexts.

On the existing evidence, it seems that already many infra- human species appear to have the power to form some kind of internal representation or 'cognitive map' of their environment, i.e. some kind of cognitive structure which intervenes between stimulus and response. Behaviourists, of course, have been predisposed to deny this, but the evidence seems incontrovertible. As Sokolov (1963) has shown for dogs and Premack & Woodruff (1978) for apes, these animals show kinds of behaviour which are uninterpretable except on the assumption that they form an internal representation of their environment. Earlier, Tolman (1948) had shown that when a rat learns to run a maze, there is good evidence to suggest that the animal forms some kind of internal map of that maze. One can infer this, for example, from such phenomena as *latent learning:* if the rat is given a chance to acquaint itself with the maze *passively*, e.g. by being repeatedly pulled through it on a trolley, or by repeatedly running around a given maze without food present, it will subsequently be found to master the maze more readily in active and rewarded trials. Moreover, when learning is complete, the animal proves to be able to reach the goal by means of actions which it had not in that context tried before, e.g. by dragging itself along if it is injured, by swimming to the goal when the maze is flooded, or by making an appropriate detour when some passage is blocked. These last observations show that, in learning to run the maze, the animal does not simply learn to associate particular responses with particular external stimuli. Contrary to behaviourist orthodoxy, it does not simply learn stimulus-response (S-R) associations.

One of the most telling demonstrations with rats was devised by Olton (1979). He constructed a radial maze consisting of a central area from which eight runways extended radially like the spokes around the hub of a wheel. Equal amounts of food were placed at the outer end of each runway, but invisible from the centre. Rats introduced to the central area in single trials on successive occasions (no food being replaced once taken) then showed after familiarization that they could remember which arms were empty as the result of previous trials: only rarely did they enter an arm in vain. Olton followed this up with experiments which showed that the choices of the experienced rat could not be due to any external cues and were explainable only on the assumption that the animal had formed some kind of internal map of the maze.

The ability to form internalized maps of their hunting grounds or territories has been demonstrated for a variety of birds, and even much lower species seem to be able to form some rudimentary representation of spatial relationships, at least in their vicinity. A well-known example is the Digger Wasp. The female of the species lays its eggs in a shallow burrow dug into the ground, and before flying away in search of provisions, she will execute some exploratory flights around the nest, apparently noting the landmarks. For if the wasp is subsequently taken some distance away from the nest in a dark box, she has no trouble returning to it when released. Again, she can fly away from the nest but find her way back *along the ground* when she drags back the paralysed caterpillar that is to serve as food for her offspring.

That some birds can also form internal representations of categories of objects, and not merely of a local topography, has been demonstrated by Herrnstein *et al* (1976). They showed that pigeons exposed to a variety of pictures in a Skinner box could learn to respond for food only if the picture shown contained a 'tree' despite the different forms in which these trees came.

In the human case it is evident that on a much broader, more penetrating and sophisticated scale (and possibly in quite different ways), our brain forms comprehensive internal representations, not just of the topography of the environment, but indeed of the total situation in which we find ourselves at any given point in time and of the totality of the world in which we are moving - including a representation of the relation of the physical self to that world. Moreover, this includes representations of the properties of the familiar objects of that world, of relations between them, of their likely behaviour, and much else. For, our internal representations of the situation in which we find ourselves, and of the world beyond, comprise all that we commonly call our knowledge of the world. Thus the visual stimuli that reach our eyes when we look at an object call up in the brain a representation of the object which embodies what we know about that type of object in general, and perhaps also about this individual object in particular This representation will be mainly in the form of expectancies of how the appearance of the object will transform as we approach it, walk around it or retreat from it; how the object will respond if manipulated in various ways; what we shall feel if we touch it; etc. We may describe these as *what-leads-to-what expectancies* based on past *what-leads-to-what experiences*. Later I shall suggest that the brain's internal model of the world in fact consists of just such sets of expectancies.

It follows that the brain's internal model of the world is bound to be conjectural: it is the brain's assumption, as it were, of the nature of the reality that currently confronts the subject. This conjectural character is highlighted when the sensory inputs are ambiguous: a dark shadow in the dusk may be 'seen' either as the figure of a man or as that of a tree. The sensory inputs support either interpretation. Ambiguous figures like Wittgenstein's 'Duck-Rabbit' (now you see it as a rabbit, now as a duck) or the wire cube known as the Necker Cube (the face you now see as the front of the cube you presently see as the back) are not experienced as ambiguous entities: they are seen either as the one thing or the other. This fact indicates that in these cases the brain has made its conjectural decisions on grounds lying below the level of consciousness. They contrast with the many occasions when conjectural decisions are made at a conscious and rational level. Indeed, it could be said that the whole body of our scientific knowledge is but a complex and coherent set of conscious conjectures (called 'hypotheses') about the nature of the world, and one which we accept on rational grounds because our observations confirm the predictions that flow from them, and we know of no observations that would falsify them.

The advantage of this whole mode of operation by the brain is obvious. In so far as the brain's internal representations are based on past experience, they are able to cover properties and characteristics of objects, events, or situations which go beyond what is immediately presented to the senses. They are the brain's sophisticated way of bringing past experiences to bear upon the interpretation of present inputs, thus enabling the organism to jump from the *appearance* of objects to the *reality* that had become manifest in previous encounters with the same or outwardly similar objects. The method also enables the brain to produce a continuous behavioural output despite temporary interruptions of the sensory inputs (e.g. during eye blinks or passing distractions).

The limits of a person's internal model of the environment can be explored by strictly objective methods. For, the perception of any event that does not fit the model will tend to elicit *surprise reactions*, technically known as *orienting reactions* (cf. Section 5.4), and these can be monitored at different levels in terms of both behavioural and physiological responses, e.g shifts of attention and mild states of arousal. The detection of such discrepancies between what is expected and what actually occurs serves the brain as an *error signal* calling for a revision or 'updating' of its internal model. The absence of such surprise reactions when a particular event occurs, does not of course prove that the event in question is included in that person's internal model. Its occurrence may simply not have been registered. On the other hand, a surprise reaction if some particular event fails to occur, would prove that the event in question was expected, hence part of the brain's internal model.

The brain's internal model of the world is exceedingly comprehensive in scope, though inevitably patchy in matters of detail. I know not only what room I am in, but also in what house, in what town, in what country, in what continent of what kind of planet and in what kind of universe. At any particular time I may not attend to any of this. But it is all taken into account in my actions, if only implicitly and in a negative sense - as I have illustrated above. It is also easy to ignore in experimental psychology. That is to say, it is easy to forget than when a subject is asked to press a button, the whole of the situational context enters into the processes that trigger and shape this simple action.

One consideration must be uppermost in our mind when we come to speculate how the brain's global world model may come to be physically realized in the brain, viz. the enormous variety of what the brain is able to represent internally, and the enormous dimensional requirements this makes. The brain's internal representation of the world has far more dimensions that just spatial and temporal ones. For, it includes representations not only of the objects as distinguishable entities, but also of their properties, mutual relationships and their likely reactions if interfered with. Thus the state of affairs in the brain which functions as an internal representation of the world must be of a kind that permits of an adequate 'logical multiplicity', to use Wittgenstein's expression.

The brain's internal model of the world is of the nature of an *intervening variable*, not merely in the trivial sense that it intervenes between the subject's sensory inputs and motor outputs, for a lot of happenings in the brain do that, but in the specific sense that it has to be taken into account as a separate variable in the explanation of human behaviour. It entails, for example, that one has to envisage two distinct categories of learning changes in the brain (each based on its own kind of feedback loops), whereas the simple S-R theories of learning propounded by the behaviourists envisage only a single category. The two categories are (a) learning changes that result in a modification of the internal model, and (b) learning changes that result in a modified way of responding to the model. If a child consistently disappoints my expectations, I may either revise my view of what kind of child it is, or revise my strategy of dealing with a child of the kind I assumed it to be.

Indeed, cognitive psychology only came into its own when psychologists abandoned simplistic S-R theories and came to realize that (voluntary) human behaviour cannot be understood except on the assumption that the brain uses its sensory inputs not as direct determinants of behaviour, but as information for updating its model of the world, the latter operating as the actual data base from which it computes the behavioural responses required in any given situation. One of the functions of *attending* to an environment object is to perfect that data base at a point where it matters in the given circumstances.

However, cognitive psychology has generally failed to follow up these insights by a systematic attempt to understand the nature of this internal model in precise and objective terms, thus to lay the foundation of a proper science of consciousness and of its neural correlates. For example, in his widely read MENTAL MODELS (1983), Johnson-Laird goes no further than describing the internal representation of external objects as 'structural analogues' of those objects. And it is not untypical of other work in this field to leave matters at such an unsettled level.

***

4.4 - THE FUNCTIONAL DEFINITION OF 'INTERNAL REPRESENTATION'

Any scientific theory about the nature of the brain's internal representations of the outside world must satisfy two conditions: it must be formulated in precise and objective terms, and it must fit the facts. The first of these conditions must be satisfied first, because without such a formulation the second condition cannot seriously be put to the test. Our next task, therefore, must be to find a precise and objective definition of what is to be meant by an 'internal representation'. We can approach this problem either from a *functional* or a *structural* point of view. That is to say, we can define an internal representation as a state of the brain, either in terms of the role this state plays, or in terms of its structure and the kind of analogue of the outside world that this structure achieves. For reasons to be given presently, I shall follow the former course.

Historically, writers have leant towards the structural or information-processing approach. In this country, for example, the first detailed suggestion that the behaviour of the higher orders of life can only be explained on the assumption that the brain forms

4.4 - The functional definition of 'internal representation'

internal models of the world, dates from Kenneth Craik's THE NATURE OF EXPLANATION, published in 1943. The following two passages illustrate how he conceived those models and their usefulness:-

> "By a model we thus mean any physical or chemical system which has a similar relation-structure to that of the processes it imitates. By 'relation-structure' I do not mean some obscure non-physical entity which attends the model, but the fact that it is a physical working model which works in the same way as the processes it parallels, in the aspects under consideration at any moment..."

> "If the organism carries a 'small scale model' of external reality and of its own possible actions within its head, it is able to try out various alternatives, conclude which is the best of them, react to future situations before they arise, utilize the knowledge of past events in dealing with the present and future, and in every way to react in a much fuller, safer, and more competent manner to the emergencies which face it."

From these slight beginnings the idea of internal models gradually spread. But the notion itself did not gain in clarity. The above description of the internal model as "a physical working model which works in the same way as the processes it parallels" does not bear being taken literally. Others have thought of the brain's internal representations as in some sense a *symbolic* one, but this too is an unfortunate concept, because it is of the essence of a symbol that it stands for something else without mapping the structure of that something else - which is obviously not the case here. Some workers in AI have tried to remedy this by devising a system of relational symbols that could serve as a descriptive framework for representing knowledge about the outside world, including a set of rules which would permit the required symbolic representations of the outside world to be derived from the raw data of sensory inputs. Minsky's 'frames for representing knowledge' is a good example (Minsky, 1975). But there is nothing to be found here that could serve as a model of what physically happens in the brain. Nor have those who have operated with the notion of symbols furnished us with a set of objective criteria a state of the brain would have to satisfy to qualify as a symbol of something else.

Because of the paramount importance of our visual senses, most of these efforts have confined themselves to the *visual* perception of the world. Probably the most detailed and best known modern attempt in this direction is that presented by the late David Marr in his VISION (1982). Marr accepted the general idea of internal models in these words:

"Modern representational theories conceive of the mind as having access to systems of internal representations; mental states are characterized by asserting what the internal representations currently specify, and mental processes by how such internal representations are obtained and how they interact."

The author then defines internal representations as:-

"...a formal system for making explicit certain entities or types of information, together with a specification of how the system does this."

To illustrate what he means by a formal system he cites Arabic, Roman and binary numeral systems as examples of formal systems representing numbers. Thus once again the notion of internal representation is left totally unclear, for the analogy here offered is quite unacceptably remote.

Moreover, Marr regards the formation of these internal representations as the product of a *serial* process, i.e. of a *string* of dedicated neural modules, each devoted to some specific information extracting process. Neither the evidence nor the logic of the case suggest that the brain works in that passive and serial information- processing way. Visual perception is part of a process of dynamic interaction between brain and environment in which the brain participates as an integral whole, and as the result of which the brain's internal model of the outside world gains in currently relevant detail and precision. For example, just to create the internal representation of a stable scene from the flitting images that jump across the retina as one moves eyes, head, or body, the brain has to integrate and evaluate the totality of the information it has available about those eye-, head-, and body-movements. For only thus can it decide, for instance, whether a given shift of an image across the retina is due to a movement of the subject or a movement in (or of) the scene. All this amounts to a *parallel* processing routine, not a *serial* one.

I shall return to the mechanisms of perception in a later chapter (Section 5.5). In the present context it is a digression, for our present concern must be with the question of how we are to define the concept of internal representation in terms that satisfy our demands of objectivity and precision. To this end it will suffice for the moment to confine ourselves to internal representations of single external objects.

If this notion is to make any sense in the context of brain mechanisms, it must stand for a state of affairs in the brain which is *used* by the organism as in some sense representative of the object concerned, and which also, in some sense, *adequately corresponds* to the features of that object - since otherwise it would be used in vain. We may call these two requirements the *functional* and *structural* requirement respectively, since the first requirement relates to the role which the representation plays in the

4.4 - The functional definition of 'internal representation'

determination of the organism's activities, while the second relates to the extent to which the representation maps the features of the object represented and to the manner of its mapping.

Now clearly, only the *functional* requirement can be used for a *definition* of internal representation if we are to allow for the possibility that the representation in question may be erroneous - which we must since we know that our senses can deceive us. Again, only a *functional* definition can cover the case of internal representations of *fictitious* objects, such as those that occur in the exercise of the imagination. In order then to arrive at a functional definition of 'internal representation', let me put the following question:

QUESTION: Consider the general case in which some biologically significant outcome depends on the organism responding in some appropriate way to some particular object O in its environment (e.g. a hungry monkey reaching out for a banana). Then, what conditions must a physical state of affairs X in the brain satisfy to justify the assertion that X is functioning as an internal representation of O?

ANSWER: If X is used as as a representation. And it is used as such if, in the production of appropriate motor responses, the brain takes its cues from X (uses X as its 'data-base', if you like), instead of taking them directly from the sensory inputs received from O. That is to say, if in the determination of the brain's motor responses the state of affairs X acts as surrogate for O in a sense not unlike the sense in which a map of a town may be used as a surrogate for that town when one plans how to get from the hotel to the museum.

This answer leads us to the following generalized definition:-

DEFINITION:

*A physical state of affairs X in the brain will be called an **internal representation** of an object O throughout any time interval in which, whenever a response has to be produced that is appropriately related to O, the brain is set to behave as if it would suffice to produce a response that is appropriately related to X.*

The term 'object' is here to be understood in so general a sense that the definition will cover representations not only of *actual* objects, events or situations and their properties or relations, but also representations of *merely possible* objects, events or situations, such as fictional objects or events, or some desired situation or state of affairs. Because in a given situation one can respond either appopriately or inappropriately to mere possibilities as well as actualities. A physical definition of the meaning of 'appropriate' in biological contexts was given in Section 3.4.

One point needs hardly to be mentioned here: obviously this definition does not imply that the required relation between the brain's outputs and brain-state X need in any way *resemble* the required relation between those outputs and O. For example, if O denotes a nearby physical object and the required response is an arm movement in the direction of that object, the brain would not achieve much by producing an arm movement in the direction of X! It is also obvious that the definition disposes of the misleading claim, mentioned in Chapter 1, that any idea of an internal representation implies the existence in the brain of an observer of that representation.

*

So much for the *functional* character of the brain's internal representations. The question of the *structural* requirements they must satisfy will be taken up in Chapter 5. I shall also then advance certain hypotheses about their physical realization in the brain and about the mechanisms involved. These hypotheses will at first be formulated only in very general physical terms and examined at this level in the light of the general empirical evidence as well as a variety of purely logical considerations. The question of their realization in *neural* terms will be taken up in Chapter 6.

First, however, I must discuss the three main categories of internal representations which I take to be accountable for the phenomena of consciousness and the nature of mental events *generally*.

***

4.5 - THREE MAIN CATEGORIES OF REPRESENTATIONS

I have claimed that the faculty of human consciousness in all its main aspects can be adequately accounted for if we accept that the brain has the power to form three main categories of internal representations. They were the following:-

CATEGORY 1: Representations amounting to a *comprehensive and coherent internal representation of the world and the self-in-the-world.*

So far we have considered only the brain's internal representation of the outside world. But everything that has been said in this connection can be readily extended to

4.5 - Three main categories of internal representations

cover also the brain's internal representation of the body as an object in that world. And it can be extended to cover internal representations of the body's topology and of the current state of the body, such as its posture, movements, particular physiological needs etc. This internal representation of somatic variables is commonly known as *body-knowledge* or the *body-schema*. The latter notion has become well-established in brain-research since it was found that variables representing posture and movement are extensively mapped in certain cortical areas, notably the parietal lobes. The posture- and movement- information that reaches these areas comes from interoceptors, such as proprioceptive and kinesthetic receptors in muscles, tendons and joints. To form a coherent whole, this information has to be integrated with information derived from exteroceptors, e.g. from the visual perception of the body and its limbs. For reasons which will become apparent presently, I shall call this capacity for body-knowledge the faculty of *first-order self-awareness*. Here, of course, the 'self' denotes only the *somatic* self, i.e. the body, its posture, movements, major physiological needs etc. The concept will be expanded presently.

If we add the brain's internal representations of the outside world to this first-order self-awareness, we arrive at our first broad category of internal representations, viz. one which may be described as a *comprehensive and coherent representation of the world and the self-in-the-world,* or, for short, as the brain's *global world-model.* I have explained in Chapter 1 that the term 'model' is here used purely for brevity as a synonym for 'internal representation' - the latter now to be understood in the sense defined in the last section. The predicate 'global' has been added solely to underscore the comprehensiveness of its scope.

I shall regard this as the fundamental basis of consciousness, for I shall postulate that *nothing enters consciousness except by way of the brain finding a place for it in its global world model.* This postulate is suggested *inter alia* by the simple observation that whatever we are conscious of we are conscious of as a part or detail of the total situation.

However, this global world model does not relate solely to the *current* state of the world and the body-in-the -world. It also has an *historic* dimension in the form of representations relating to the past: to the *history* of the objects, events or situations which it encompasses. Without this dimension we could form no representations of either change or continuity. Thus one generally knows not only where an object is, but also where it was a moment ago, and perhaps where it was yesterday or the day before. One knows not only where one is, but also how one came to be there - and a lot more besides: one's very sense of identity hinges on one's internal representations of one's past life. I shall look at this historic dimension more closely after I have dealt with the faculty of the imagination and have turned to the faculty of memory recall (Chapter 7). These faculties hang together, as the next item will acknowledge.

CATEGORY 2: Representations of *merely possible* objects, events or situations and their properties or relations.

To this *second* category belong not only the representations of fictitious objects, events or situations formed in the *imagination*, but also some of the images arising in the mind in the re-living of past experiences and events witnessed in the past, i.e. the images occurring in *episodic memory recall*. For this faculty is closely linked with that of the imagination, a fact of which psychologists have become increasingly aware ever since Bartlett (1932) demonstrated that memory recall is a *reconstructive* rather than purely *reproductive* process. That memory may deceive us is commonplace knowledge and the bane of all police investigations. But it was only through the seminal work of Bartlett that the *creative* element in memory recall began to be taken seriously and the link between memory and imagination came to be fully appreciated by the theorists.

Of course, the role of the imagination extends far beyond this. In the imagination we review the possible consequence of our actions and create the objects of our desires, plan the steps that can lead to their gratification, picture the dangers that may confront us, and much else besides. A great deal of our thinking is conducted in terms of such mental images. We try to picture in the mind how a certain carpet might fit a particular room, of how best to get from A to B. We tend to use mental images also in our logical reasoning; indeed, according to the empirical studies of Johnson-Laird (1983), to a greater extent than is often held to be the case. The distinctive feature of the images occurring in memory recal lies in the way the brain responds to them, viz. as representations of the past.

In broad terms, therefore, this second category may be described as representations of *merely possible* rather than *actual* objects, events or situations. The reader will recall in this connection that our functional definition of 'internal representation' covers both these cases.In objective terms this difference will be reflected in the fact that these representations are not subjected to reality testing. That is to say, the states of the brain which are involved in imagining an object, event or situation, are not subject to corrective feedback under control of the sensory inputs. They are not governed by the same kind of corrective feedback loops as those that are involved in the perception of actual objects, events or situations.

CATEGORY 3: Representations *which represent the occurrence of a representation of category 1 or 2 as being part of the current state of the self.*

By this category of representations I mean the following. When one sees an object, one also knows *that* one is seeing the object, in other words, that the perception of the object is part of the current state of the self. One may not be conscious of that fact, for one may not attend to it. But it is still part of the global world model - precisely because one can become conscious of it merely by one of those selectively facilitating processes in the brain that constitute a shift of attention. This particular shift of attention would be an instance of what we call *introspection*.

Let R_1 denote the internal representation formed in the perception of some particular external object, say a tree. Now in producing a self-report of the kind *I see a tree* the brain is clearly producing a response which is based on an awareness of the fact that R_1 is part of the current state of the self. This presupposes that the brain has formed an internal representation of the fact that R_1 is part of the current state of the self. Let R_2 denote this internal representation. We can conceive of R_2 as a state of affairs in the brain which here intervenes between R_1 and the respective verbal outputs of the brain.

Since R_2 is the internal representation of a relationship in which one of the relata is itself an internal representation (viz. R_1), I shall call this class of internal representations *second-order representations,* and the corresponding faculty of the brain will be called *second-order self-awareness* or *self- consciousness*. It supplements the *first-order self-awareness* which consists of the body knowledge described above. And it raises the internal representation of the self from the representation of an entity whose current state is defined in terms of purely *somatic* variables, to that of an entity whose current state is defined in terms of both somatic variables and variables of a kind that we would commonly describe as mental variables. The relationships that cause these representations to be representations of a *unity*, here called 'the self', will be discussed in Chapter 7.

We are clearly here covering that aspect of consciousness which is commonly described as *subjective experience*, i.e. the field of what is *introspectable*, the field of what can become the object of an 'inward directed' act of attention, the field of what is involved in the "perception of what is in a man's own mind" (cf. Section 4.2 above). It is the field of what we call our 'inner life'.

It follows from our postulate that nothing enters consciousness except by way of the brain finding a place for it in its global world model, and that internal representations of fictitious or absent objects, events, or situations enter consciousness only if they are covered by this faculty of second-order self-awareness. Since we cannot assume that this latter condition is invariably satisfied, it follows that the present theory allows for the possibility of behaviour being influenced by internal representations of this category of which the subject is not conscious - an assumption commonly made by psychoanalysts.

Finally, let me recall a fact already mentioned in my first Chapter: there is nothing in the present theory that forces the conclusion that these *second-order* representations must in turn be subject to *third-order* representations which express the fact that they occur as part of the current state of the self - and which would in turn be subject to *fourth-order* representations carrying this one stage further, and so on. Thus there is nothing in our hypotheses that would saddle us with an infinite regress of kind *I am aware that I am aware that I am aware*.......... The problem of avoiding such an infinite regress has haunted many a theoretician of self-awareness in the past. The brain does not automatically develop in any direction that is logically possible. It develops in the directions that have proved to be of practical benefit. Any tentative steps beyond this would lack the necessary reinforcement. Where the cut-of point lies in practice must be decided on the evidence.

I might also add that this second-order self-awareness may at a first glance seem to be an exceedingly complex and sophisticated function, but when we come to look at it more closely in Chapter 7 I shall suggest that its realization in the brain is readily envisaged, since, on the hypotheses that will presently be introduced, it consists merely of a special class of states of conditional expectancy. Hence, in any creature in which first-order self-awareness is well developed we can also expect to find some measure of second-order self- awareness, i.e. self-consciousness.

***

4.6 - THE TECHNICAL DEFINITION OF CONSCIOUSNESS

If scientists are to agree on what they are talking about when they mention consciousness, and if this is to be relevant to what we mean by that word in everyday life, they need a technical definition of consciousness which manages both to satisfy the standards of precision and objectivity good science demands and to cover as much as possible of the common connotation of the word.

Since I have asserted that the phenomena of consciousness (and the general nature of mental events) can be adequately covered in terms of the three kinds of internal representations discussed in the last section, a definition in terms of one or more of these categories of representations would seem to be an obvious choice.

The three categories were:

1. A comprehensive and coherent internal representation of the world and the self-in-the-world. This includes the body knowledge which I have called *first-order self-awareness*.

2. Representations of merely possible objects, events or situations and their properties or relations. These are the elements of the imagination.

4.6 - The technical definition of consciousness

3. **Representations which represent the occurrence of a representation of the first or second category as being part of the current state of the self.** These constitute what I have called *second-order self-awareness* or *self-consciousness*.

   I also added a postulate which is of obvious relevance in the present context, viz., that *for a subject to become conscious of any object or event, it is a necessary condition that the brain shall have been able to incorporate that object or event in its internal representation of the world and the self-in-the-world.* In other words, *nothing enters consciousness except by way of the brain finding a place for it in its global world-model.* To repeat an earlier illustration: according to this postulate a traumatic event in the body becomes a *pain*, i.e. a conscious sensation, only by way of the brain representing its occurrence as part of the current state of the self in its global world-model.

   Having thus stated a *necessary* condition for consciousness, the question arises whether our planned technical definition of consciousness should accept this also as a *sufficient* condition.

   I suggest not. Most people would regard reflexive self-awareness and the perception of what passes in one's own mind as an essential part of what we commonly mean by consciousness. This also applies to philosophers who agree with Locke's view of consciousness, and to the psychologists who treat the ability to report an event as a working criterion for consciousness of that event. On these counts, therefore, a power of the brain to form internal representations of the third category should be regarded as an essential part of the faculty of consciousness. On the other hand, it does not seem to me compelling to treat the second category, i.e. the faculty of the imagination, as an equally essential ingredient. Hence I shall suggest the following definition:

DEFINITION:

By the faculty of consciousness is to be understood a power of the organism to form internal representations of both category 1 and category 3 (as defined above).

I take it here to be - it being understood that this is coupled with an ability to submit these representations to selective processes of the kind described above as *selective attention*.

Creatures capable of forming internal representations of category 1, but *not* of category 3, would then be said to have cognition of the world but *not* consciousness.

Since this definition excludes the faculty of the imagination from the necessary conditions for consciousness, it also excludes the faculty of thought, as well as the re-living of past experiences (episodic memory recall), and the faculty of language.

It should also be noted that the definition given above makes no distinction between the attended and unattended part of the brain's internal representations. This differs from the concept of consciousness entertained by some psychologists, e.g. Carl Gustav Jung (1928), who use the term *subconscious* for the unattended parts. The subconscious on this understanding covers everything that becomes conscious (as they understand the term) by mere shifts of attention. Jung was protesting against those writers who relegated to the *unconscious* everything of which at any given time the subject is not aware because he is not attending to it.

By contrast, I would prefer to use *subconscious* for those processes in the brain of which we are not conscious but which govern the entry of events into consciousness (e.g. the hidden processes that make us suddenly see the answer to a question). And I would describe the unattended parts of the brain's global world model simply as *background-consciousness* or *context- consciousness* (e.g. the internal representation of the room in which I am writing these lines which assures me that I need not glance around for possible threats to my person while I am getting on with the job).

I would claim that the account I have given of the internal representations here described as the foundation of human consciousness and mental events generally, not only satisfies the standards of objectivity and conceptual precision I have demanded, but also covers adequately the common connotations of the words 'conscious' and 'consciousness', the transitive as well as the intransitive, and the objective (clinical) as well as the subjective (cf. Section 4.2). And it accords with the most common test used in psychology for an event being a conscious one, viz., the ability of the subject to report that event.

However, when it comes to applying this notion of consciousness to the animal world, there is a price to pay for this inclusion of self-consciousness as an essential ingredient of the faculty of consciousness. Although the omission of representations of category 2 makes the definition more readily applicable to the animal world, it is difficult to design tests for any degree of self-consciousness (by contrast with tests for representations of category 1). The tests sometimes suggested for 'self-consciousness', viz. an animal's ability to recognize itself in a mirror, are obviously misplaced in respect of the sense of self-consciousness I have defined above. An alternative test, viz. evidence that the animal can understand the motivations of other members of the tribe (Premack and Woodruff, 1978) is rather nearer the mark.

It is sometimes asserted that no objective definition of consciousness and self-consciousness can capture subjective experience in the sense of capturing, for example, *what it actually feels like* to be in pain. But I would claim that the account I have given does just that. According to my account the contents of self-consciousness, as defined above, are as distinctive as the self to whose current state they relate. What we call the *quality* of a feeling, is no more than this distinctiveness.

I must add that on the same grounds my account of self-consciousness also rules out the idea of any person having a sensation *identical* to that of another. I have my feelings when a tooth hurts, you have *yours*, and the only link between them is that we have learnt to attach the same name to the respective categories of feelings on the strength of certain communicable criteria, e.g. the location (tooth) to which the brain has assigned the stimulus in its internal representation of the current state of the self and the aversive reactions which that stimulus arouses.

It is not, of course, to be assumed that in a state of wakefulness, all one's activities are governed by the brain's global world-model. Only some are so determined, and the category of activities which are so determined I shall identify with what we commonly describe as *voluntary* activities. In addition, the brain's motor outputs include a wide range of involuntary and semi-voluntary reactions, such as eye-blinks and other reflexes, also the semi-automatic control functions that operate in the execution of acquired skills like typing or riding a bicycle.

When we gradually awaken from sleep, it seems that first- and second-order self-awareness recovers before the rest of the global world-model so that we can pass through a phase in which some of the sundry internal representations that appear to be formed during the active (REM) phases of sleep come to be registered by the brain as states of the self even before the brain's global world-model is fully reconstituted functionally. This is how I see the recollectable parts of the brain's dream states. Whether these representations must be regarded as spurious, or whether they play a constructive part in the internal adjustments that undoubtedly occur during sleep, still appears to remain an open question.

By contrast, in *sleepwalking* it seems that a partial model of the world has reconstituted itself and taken effect in the continued absence of self-consciousness and, therefore, in the continued absence of consciousness as defined above.

***

Chapter 5

THE PHYSICAL STRUCTURE OF THE INTERNAL REPRESENTATIONS

SUMMARY

Having defined internal representations in terms of their *function*, we now reach the next step in our effort to track down the neural correlates of consciousness: the question of their *physical structure*. Once again I shall begin by confining myself to internal representations of *external* objects, events, or situations, and I shall look at the question of how the brain can use its sensory inputs to build up such representations and how it tests their validity.

As one guideline here I shall take the consideration that the physical structure of these internal representations must be of a kind that can accommodate the enormous variety of objects and their properties or relations that may have to be represented, i.e the question of the potential *dimensionality* of these structures.

A second guideline will be the observation that these internal representations must be correctable by way of error exposing *feedback* from current sensory inputs. I shall reject the hypothesis that the only feedback the brain has available here comes from the success or failure of the behavioural responses that flow from its internal model of the world. I shall argue that the brain has a more powerful and effective form of feedback available in its general power to anticipate coming events on the strength of this model, i.e. by way of the *expectancies* that flow from, or reside in the model. Deficiencies in the model show up through the occurrence of failed or 'violated' expectancies. Appropriate steps to remedy these discrepancies and update the internal model are then initiated in the brain through the *surprise reactions* which such events elicit – the so-called *orienting reactions* (Section 5.4). Typically these include shifts of attention to the unexpected events.

A third guideline will be the need for hypotheses about these structures and their formation in the brain which are plausible in terms of known learning mechanisms.

The main categories of expectancies that are relevant in the above context will be of the nature of *what-leads-to-what expectancies,* most commonly *act-outcome expectancies,* i.e. expectancies about the consequences of the different kinds of actions (complex as well as primitive) that are open to the organism. One can readily conceive how such what-leads-to-what *expectancies* come to be formed in consequence of past what-leads-to-what *experiences.*

States of expectancy are here to be understood in the physical sense in which they were defined in Chapter 3. Thus it is not implied that they are states of which the subject must be conscious.

Those psychologists who accept that the brain forms some kind of internal model of the outside world, generally recognize also the importance of failed expectancies as error detectors. I shall go a step further and claim that, to explain the facts, there is no need to assume the existence in the brain of *both* a model of the world *and* associated states of expectancy: *the relevant states of expectancy themselves are capable of functioning as the required model.* For example, the weight of an object can be adequately represented in the brain by expectancies relating to the effort required to lift it. And I shall postulate that this is in fact the way they function.

Thus we shall arrive at the conclusion that the brain's internal representations of the features, properties and mutual relations of external objects consist in the main of linked sets of *what-leads-to-what-expectancies* which are based on past *what-leads-to-what experiences.* These *states of conditional expectancy* are elicited by the optical images of the objects concerned, and form a 'set' or 'operational mode' of the brain to which the required behavioural responses then become conditioned. This conclusion will be called the *Conditional Expectancy Hypothesis.*

We shall accept this hypothesis on several grounds. It is the most economic; it satisfies the requirement of an adequate dimensionality and of the ready correctability of the brain's internal model of the world; it suggests plausible mechanisms for the formation and updating of this model; it is formulated in terms which we have been able to define in a physical language; and it seems to fit the empirical evidence.

In the present chapter I shall introduce the Conditional Expectancy Hypothesis (Section 5.2), explain its implications and cite both the theoretical reasons and empirical evidence that support it (Sections 5.3 and 5.4). The learning mechanisms involved will be discussed in Section 5.5, while the remaining sections will deal with the specific case of visual perception and shape recognition. The question of what inferences we can draw from all this about the neural correlates of the brain's internal representations of the world, and of what lines of research should be followed to uncover them, will be left to Chapter 6. Clearly, this has now become a question of looking for the neural correlates of states of conditional expectancy and of the brain's way of discovering (and dealing with) discrepancies between what it anticipates in any given situation and what actually comes to pass.

5.1 - GENERAL STRUCTURAL CONSIDERATIONS

Having defined internal representations in terms of their function, we must turn to the question of their physical structure. The following are the three main considerations I shall take here as a guide:-

1. *The requirement of adequate dimensionality.*

An important fact about this structure can at once be derived on purely logical grounds from the consideration that the physical states of the brain which function as internal representations of the outside world must have a dimensionality capable of accommodating the immense variety of objects, events or situations and their features and mutual relations that need to be represented. They must have what Wittgenstein has called an "adequate logical multiplicity". Contrary to what some writers seem to suggest, therefore, these internal representations can hardly be conceived to be of the nature of a flat projection onto some 'screen' at the back of the brain - some two-dimensional cortical sheet, for example.

2. *The requirement of testability and correctability.*

The optical image of a car elicits in the brain the representation of a solid object having all the properties we have learnt to associate with objects of that appearance. Thus the brain's internal representations of external objects go beyond what is immediately given to the senses. It follows that they are of the nature of an *interpretation* of the sensory inputs and, therefore, of a *speculative* kind. In this respect they resemble the hypotheses formulated in science, and, like these, they stand in constant need of being *tested* for their validity and corrected if the results demand. One obvious test for the validity of these interpretations is whether the behavioural responses which they elicit will prove to be successful. However, the brain would be poorly served if the positive or negative outcome of an activity would be the only feedback available to it. For, that outcome may only manifest itself at some distant point of time. To be able to update its internal model of the world continuously, the brain needs a faster feedback on a running basis.

It derives such feedback from its general power to anticipate coming sensory inputs on the strength of its model of the world, coupled with an ability to detect any discrepancy between the anticipated sensory inputs and the actually occurring ones - discrepancies to which it will then respond in an appropriate manner, e.g. by attending more closely to the sources of those inputs. Most of the expectancies here concerned will be of the nature of *what-leads-to-what expectancies* based on the subject's *what-leads-to-what experiences* relating to the consequences of the movements and actions open to the subject - what Irving (1971) has called *act-outcome* expectancies. Simple examples are the expectancies that *if* the eyes are moved to the right the visual scene will shift to the left; *if* the head is moved sideways nearby objects will shift relative to the background; *if* one pushes against this object it will resist; *if* one pushes hard enough it will roll over, etc. All these elementary expectancies are based on past experiences. At a more sophisticated level, of course, what-leads-to-what expectancies may also be indirectly acquired by absorbing information communicated to us by others through the spoken or written word.

If any of the expectancies are disconfirmed, this will indicate to the brain that its sensory inputs have been misinterpreted and that a revision of that part of its world-model is called for. Typically, the brain's first reaction to an unexpected event will be a very general one. At the behavioural level it will manifest itself as a *surprise* reaction, or (in a stronger sense) *startle* reaction, generally accompanied with *shifts of attention* to the source of the discrepancy. This is part of a syndrome already mentioned as *orienting reactions*. It may have observable physiological components as well as behavioural ones, e.g. changes in the galvanic skin resistance (but see Section 5.4). How such discrepancies come to be detected in neural terms, will be discussed in Chapter 6. To anticipate the leading hypothesis: the orienting reactions are lying in wait all the time. In familiar situations they are suppressed in consequence of the *habituation* that occurred during the familiarization. But unexpected experiences release them because no such habituation has taken place.

The sets of what-leads-to-what expectancies I have discussed may be called sets of *conditional* expectancies. For, a state of conditional expectancy may in objective terms be defined as a state of the organism in which, in a given context, there exists a set A of one or more mutually exclusive contingencies ($a_1, a_2, \ldots a_n$) and a set B of events ($b_1, b_2, \ldots b_n$), such that

$$\begin{array}{llllll} \text{the occurrence of } a_1 & \text{causes an expectancy} & b_1 \\ \text{the} \quad " \quad " & a_2 \quad " \quad " \quad " & b_2 \\ \ldots\ldots\ldots\ldots\ldots\ldots\ldots\ldots\ldots\ldots\ldots\ldots\ldots\ldots \\ \text{the} \quad " \quad " & a_n \quad " \quad " \quad " & b_n \end{array}$$

The items appearing on the left hand side may be called the *conditionals*, while those on the right hand side may be called the *consequents*.

It should be noted in this connection that, although a state of conditional expectancy as here defined is merely a state which will result in a particular expectancy under particular circumstances, it is nevertheless an *actual* state of the organism which will in some way differ physically from a state in which that conditional expectancy does not exist. It may be described as an *operational mode* of the brain, and is akin to what in psychology is called a *set* of the brain. It should also be noted that, whereas the *actual* expectancies elicited under the required conditions will typically be context- dependent and transient, a state of conditional expectancy itself may last a lifetime. Throughout my life I may have the conditional expectancy that if I put my finger into a flame I shall hurt myself.

Let me further remark that the 'if....then' expectancies include the 'when....then' expectancies, since the latter merely denote the case when the conditional of the expectancy is assumed to occur sooner or later. Nor need we treat unconditional expectancies as a separate category, since these can be construed as conditional expectancies in which the conditional is a certainty, e.g. because it has already been realized.

3. The requirement of plausible mechanisms.

Ideally any hypothesis about the physical realization of the brain's internal representations of external objects, events, or situations will be of a kind that enables us to explain in plausible terms how these representations come to be formed and how at any point in time the appropriate ones come to be elicited or 'called up' by the current sensory inputs. We want a plausible account, for example, of how the brain's internal representation of the generic character of a car comes to be formed, how it is stored in the brain and how it comes to be brought into play by the optical image of a car.

Expectancies based on personal knowledge are frequently referred to by psychologists as the *subjective probability* of an event occurring. I must therefore stress that the objective definition of *expectancy* given in Chapter 3 offers us a much more definite concept, and that 'expectancy' throughout this work will be understood in that objective sense. The concept is still entirely person-related and must not, therefore, be confused with another concept. viz., the objective probability of an event occurring. The reader should also recall that according to our definition of a state of expectancy, such a state need not be one of which the subject is conscious.

***

5.2 - THE CONDITIONAL EXPECTANCY HYPOTHESIS

The requirements I have listed above flow on merely logical grounds from our central assumption that the brain forms internal models of the outside world, from the role we have attributed to these models, and from the assumption that the testing and up-dating of these models is one of the routine feedback processes occurring in the brain.

Our next step must be to look for the most economic hypothesis that can meet these requirements, putting aside all extraneous considerations, such as the desire for a hypothesis that would readily lend itself to computer simulations.

Now the simplest hypothesis about the nature of the physical counterparts of the brain's internal representations of external objects, flows logically from the following considerations:

As we become acquainted with objects and discover their properties in the course of handling them, manipulating them, etc., the resulting experiences give rise to what-leads-to-what expectancies of the kind I have illustrated above, and it is obvious that whatever we discover in this way about an object can be characterized by a corresponding set of expectancies. In other words, all the properties and special features of external objects which are, in principle at least, observable (directly or indirectly) can be adequately characterized by an appropriate set of conditional expectancies. Most of these expectancies will be of the act-outcome type. These may range from simple expectancies, like the shifts of retinal images to be expected from moving the eyes, to complex ones such as the overall consequences to be expected from manipulating an external object in one way or another. However, they may also relate simply to the events to be expected next as some familiar sequence of events runs its course in a given set of circumstances.

It is clear that the universe of conditional expectancies which the brain can form in the manner I have described admits of as great a variety as the universe of objects, event or situations, and their features, properties or relations of which the brain needs to form an internal representation.

Hence the very economic hypothesis lies close at hand that these sets of conditional expectancies can themselves act as the brain's internal model of the world - especially as this would also give us a plausible account of how this model comes to be formed and corrected when found to be deficient. The details of this account will be examined in Section 5.5, and I shall argue there that it requires no stronger assumptions than the orthodox assumptions of classical and instrumental conditioning.

Thus we are lead to the following simple but, as we shall see, powerful hypothesis, to be called the *Conditional Expectancy Hypothesis:*--

HYPOTHESIS:

The brain's internal representations of the world consist of linked sets of conditional expectancies of the what-leads-to-what kind, which are in the main based on past what-leads-to-what experiences of the world.

5.2 - The conditional expectancy hypothesis

It is to be understood that these linked sets of conditional expectancies may include expectancies relating to the subjective consequences of an action, e.g. expectancies of need satisfaction.

According to the above hypothesis, then, to have a model of a room in one's head is really just to have a very comprehensive set of conditional expectancies in the head about, for example, what visual impressions will result if the eyes are cast to the left or to the right, or if one were to turn the head; what the hands would feel if one were to reach out in this direction or in that; what effort would be required to move the objects concerned; what obstacles one would meet if one were to walk straight ahead, etc., etc. And all these expectancies would really be of the nature of an interpretation by the brain of current sensory inputs on the basis of past experiences, experiences incurred since infancy in the perception and handling of similar objects, supplemented or corrected by the experiences one is having currently.

Sometimes expectancies may guide behaviour in more than one way. Thus when I drive along a country lane I may be guided either by what Hintzman *et al.* (1981) have called a 'route map' or by what they have called a 'survey map'. In the first case the controlling factor is a sequence of expectancies about what lies beyond the next corner or beyond the next hill. These expectancies come to be elicited solely by the perception of my present location. In the second case the expectancies are derived from a maplike representation of the area which I hold in my head and on which I mentally chart my progress. The nature of such mental maps will be discussed when we turn to the faculty of the imagination (Chapter 7). These two procedures need not be mutually exclusive. They may both be operative simultaneously in an interlocking way.

It is important to stress again at this point that 'state of expectancy' is here to be understood in the strictly *objective* and physical sense in which it was defined in Section 3.4. It is a state of *readiness* which is realized in the form of internal anticipatory reactions of a certain kind (which I shall examine in detail later), and which is subject to modification if the expected event fails to occur. It is plausible to assume that the anticipatory reactions themselves will typically take the form of a preparatory potentiation in the neural networks of the brain of either a particular category of inputs (e.g. when they take the form of internal shifts of attention) or of a particular category of outputs (e.g. when the organism actively prepares to meet some expected contingency). Failing evidence to the contrary, this potentiation has to be conceived in *relative* terms: the enhanced responsiveness may be due either to the sensitization of particular structures or the desensitization (inhibition) of rival structures.

According to this account, therefore, the physical nature of the brain's internal representations of the world, has the general form an *operational mode* or *'set'* of the brain which may be conceived as consisting of an enhanced neural responsiveness at a variety of particular brain locations.

No doubt the most compelling *a priori* argument for the Conditional Expectancy Hypothesis is the argument from dimensionality. As has been said, the variables available to the brain for the construction of effective internal representations (as defined) of the outside world, must be able to cover all the manifold attributes and interrelations of the objects or events of which we can become conscious. They must possess as great a potential *variety* (in the technical sense used in information theory) as have the objects, events, and their attributes or relations, which we know the brain to be able to incorporate in its internal model of the world. Moreover, these variables must not only be able to represent the current values of the external variables they are representing, but also their *intrinsic physical character* or *type*: their 'spatiality' if they are spatial variables, their mechanical character if they are mechanical, their optical character if they are optical, etc. For example, those of the brain's internal representations that represent the spatial variables of the outer world, must also mirror their geometrical properties. This is a tall order. But it is a tall order which our hypothesis can comfortably meet, since in practice this intrinsic character is invariably revealed in terms of our what-leads-to-what experiences, either those of everyday life or, in more intricate cases, those brought to light in the scientists' experiments. The geometry of the surrounding world, for example, is embedded in sets of conditional expectancies relating to the way its appearances transform as we move about in it. Thus the internal representation of this geometry requires no structures in the brain mapping Cartesian coordinates, nor tacit knowledge of the principles of Euclidean geometry.

The above considerations have been wholly ignored by those theorists who hold that the brain's internal representation of the world consists merely of hierarchies of 'analyzers' which cause some specific neuron to fire when some specific property is present in the environment, like the signal lamps in a pilot's cockpit when the engines overheat. The point here is that such signal lamps merely convey that something has occurred, but themselves convey no information about the character of what has occurred. The pilot has to infer this from his knowledge of the system. These theorists, therefore, by-pass the main problem.

The reader should also note at this point that in our daily life statements expressing conditional expectancies are the most common way of characterizing the properties of any object, or the intrinsic nature of any physical variable. Asked to explain the meaning of *inertia*, for instance, one will tend to explain this in terms of the behaviour to be expected from the object if a given force is applied to it.

Another weighty argument lies in the fact that, given the nature of the brain's internal model of the world as outlined above, we can find in the mechanisms controlling the orienting reactions a ready answer to the question of how the brain detects, and comes to correct, discrepancies between its model of the world and the realities it encounters (cf. Section 6.3).

5.2 - The conditional expectancy hypothesis

The very general importance of what-leads-to-what expectancies in human life is obvious quite apart from the present context. But also in the animal world they play a fundamental role. This is indicated by a fact well known to animal psychologists, viz. that animals associate two things only if the relationship between them is *predictive*. For example, if an animal is presented with a signal that is only *sometimes* followed by food, it will not associate the signal with food. It may even learn that the signal predicts the *absence* of food. Again, if in a series of trials a rat is given a food reward in a hopper after a particular tone has been sounded, and if, after this simple lesson has been learnt, the animal is then repeatedly given food in the absence of a tone, but food that has now been injected with an agent producing vomiting, then a return to the original setup will show a diminished inclination of the animal to run to the hopper after the tone has been sounded. This shows that the first set of trials did not simply produce a simple set of learnt S-R associations, but a set of *expectancies* which were modified by the experiences the rat had to endure subsequently.

Further supporting arguments for the Conditional Expectancy Hypothesis will be given in Section 5.4. Let me already add at this point, though, that I know of no other hypothesis about the brain's internal model of the world which is formulated in a physical language (as is ours), and meets the demands I have listed (a) in respect of the required logical multiplicity, (b) in respect of the faithfulness with which the brain's internal representations of external variables must mirror the intrinsic character as well as variety of those variables, (c) in respect of the required testability and correctability of the representations concerned, and (d) in respect of the mechanisms that bring past experience to bear on the representations called up by current sensory inputs. Nor have I come across any evidence that would falsify the hypothesis.

*

The Conditional Expectancy Hypothesis has virtually been waiting in the wings since Tolman (1948) suggested that rats in learning a maze build up a kind of cognitive map of the maze composed of expectancies concerning the consequences of turning this way or that way at the choice points of the maze. However, Tolman's notion of a state of expectancy was never defined with sufficient clarity to enable inferences to be drawn about the physiology involved and to allay the suspicion that 'expectancy' here was merely a purloined mental concept. It was even suggested at the time that Tolman was merely projecting into his maze-learning rats what intuitively he felt would happen if a human being were placed in a similar situation - in other words, that the whole theory was an anthropomorphism. Nor, according to his critics, did he give adequate explanations of the mechanisms through which expectations came to be reinforced or weakened during the learning phase. Similar objections have been raised to later revivals of the theory, such as the perception theory of Hochberg (1968,1981). Minds also sometimes failed to meet owing to the mistaken belief, held by some workers, that the internal representation of external space requires only exteroceptive sensory inputs, whereas in truth (and according to Tolman) it is based both on these *and* on internal information about movements intended or executed. And, of course, behaviourists never liked the theory because it postulated the existence of independent variables intervening between stimulus and response.

In our case, the main objection against Tolman, viz. that his notion of expectancy is a purloined *mental* notion, has been removed by the objective definition of expectancy I have given in Chapter 3. The other objections against his theory which I have mentioned will be effectively countered by the account of the mechanisms involved which I shall suggest in Section 5.5.

One obstacle I have encountered in arguing the Conditional Expectancy Hypothesis, is the reluctance of some scientists to accept hypotheses which do not lend themselves readily to a symbolic representation in terms of flow charts as a first step towards computer simulations. Nor does the hypothesis instantly generate new programs along established lines of research. For it challenges the very premises of some of that research and, above all, it changes the questions that need further investigation if a final solution of the problem of the brain's internal model is to be found. For example, it shifts our interest to the form in which states of expectancy come to be realized in the brain, to the importance of detecting and investigating anticipatory responses in the brain (cf. Section 6.2), and to the systematic exploration of a subject's internal model of the world, e.g. by way of an exploration of a subject's orienting reactions, either at the behavioural or the physiological level.

***

5.3 - CATEGORIES AND ASCENDING ORDERS OF EXPECTANCIES

The foundation of the brain's internal model of the world is formed by our *exploration* of the world from infancy onwards. Exploration is an active process, hence the what-leads-to-what expectations formed here relate to the experienced consequences of *moving* eyes, head, body, or limbs and of *acting* on the environment in sundry ways. These are act-outcome expectancies. The activities of *testing* and *experimenting* belong to the same category of activities specifically designed to enhance or correct our model of the world. But, of course, the model can also grow in consequence of activities executed in the pursuit of any other goal. And *passive observation*, in the sense of merely watching and listening also adds to our model of the world by revealing regularities in that world: thus we come to expect a thump when we see a heavy object dropping to the floor. The

5.3 - Categories and ascending orders of expectancies

conditionals of the conditional expectancies acquired in this particular way, therefore, relate to perceptions, not actions. Of course, active and passive acquisitions of knowledge may be combined in the same process, e.g. in the process of *search* and in *orienting* reactions, such as turning the head in the direction of a perceived noise. However, in the rest of this chapter I shall concentrate mainly on the act-outcome expectancies since these seem to me to be the most fundamental.

Personal experience will invariably have some effects below the level of consciousness and this will give rise to states of expectancy which also remain below the level of consciousness. By contrast, expectancies based on conceptual knowledge and rational inference will be in the conscious realm. Some expectancies of this kind will be based on information received by word of mouth or the printed word: my expectations of the effect of a medicine may be based on what the doctor told me. It follows from the nature of these particular contributions that the internal representations which they produce in the receiver of the information will initially have a linguistic form. Whether they will be retained in that form, or be converted into visual imagery, is likely to depend both on the nature of the information and on the individual disposition of the receiver.

How the structure of representations cast in a linguistic form relates to that of representations cast into a quasi- visual form, will be discussed in Chapter 8. The extent to which the linguistic form in fact prevails in our internal representation of the world and our 'cognitive maps' is a much debated topic in contemporary psychology. It is not an easy matter to investigate, since this may differ greatly from individual to individual, possibly depending on the relative contributions made by the left and right hemispheres to the subject's model of the world. Engineers, artists, physical scientists and mathematicians tend to think more in visual and diagrammatic terms, philosophers more in propositional and verbal terms. And it is an open question to what extent this individual bias precedes or follows a subject's choice of career.

What about the internal representation of the occurrence of such singular and partly subjective events as a loud bang? This may not appear to be easily translated into conditional expectancies. But it presents no major problem once we realize that its internal representation may be divided into two components: (a) an internal representation of the fact that this stimulus is part of the current state of the self, and (b) the change which my interpretation of this event will have effected in my internal representation of the world around me (e.g. inferences about the location and source of the noise.) The first of these will be dealt with when we come to examine the nature of self-awareness (Section 7.1), the second needs no further comment. Both amount to modifications of the brain's global world model and of the expectancies of which it consists.

The Conditional Expectancy Hypothesis covers the brain's internal representation of the abstract properties of objects as happily as it covers the representation of concrete details. For in terms of conditional expectancies, the difference is merely a question of how general or specific are the conditionals and consequents of the expectancies concerned. In fact, the expectancy hypothesis gives the formation of the brain's internal representations of the world a definite bias towards representing abstract rather than concrete characteristics. Because the development of responses as the result of learning tends to proceed from general responses to progressively more specific and discriminating ones, rather than *vice versa*. In the common order of things, any reaction the brain may form to a novel stimulus (other than the standard orienting reactions) tends to generalize over a broad spectrum of similar stimuli until a subsequent demand for finer discriminations produces the required particularizations.

Internal representations of the general character of a class of objects or events, amount to what may be described as *abstract prototypes* or *templates* of the classes of objects or events concerned. Such 'templates' play an important part in any attempt by the brain to make sense of what it perceives. Thus, when the image of a car strikes the retina, the nucleus of the elicited representation is likely to be formed by an abstract prototype of the kind of vehicle here concerned – a generalized response formed on the basis of past experience and acquired knowledge of similar objects. It will consist of conditional expectancies representing the general properties of that class of object. The specifics relating to *this* car rather than any other, would be added by way of conditional expectancies elicited in addition by the corresponding specifics of the perceived scene (e.g. expectancies representing the speed and direction of its movement), plus rather more straightforward categorizing reactions like those relating to the car's colour.

The importance of such abstract representations has been highlighted by Marr (1982) for the case of object recognition. In his studies of the visual system Marr came to the conclusion that in the recognition of different kinds of objects the brain avails itself of highly abstract representations of the object's characteristics - representations which Marr called 'frames' or 'symbols'. For example, in recognizing a moving object as a dog, the brain may work from a level of abstraction which is virtually equivalent to a stick-drawing of the animal.

It should further be noted that our account of internal representations in terms of conditional expectancies applies also to the representation of *relations* between objects as much as it does to their *properties*. In logic, properties and relations may both be formally viewed as *classes*, a relation being viewed as a class of ordered pairs (triples, n-tuplets) of members (see Appendix A). Thus the relation of to the left of may formally be viewed as denoting a class of ordered pairs of objects, viz. pairs of objects so situated in space relative to the perceiver that a leftward movement of the eyes (head, body) is required to shift fixation from the second object cited in each pair to the first. Hence such a class of ordered pairs, too, can be represented in the form of a set of conditional expectancies of an act-outcome type.

*

5.3 - Categories and ascending orders of expectancies

I have described how the attributes of external objects, events, or situations represented in the brains' model of the world may range from the concrete to the very abstract. Since they may also differ in complexity, there are a number of criteria we could use to arrange the conditional expectancies concerned in some kind of *ascending order of complexity*.

One such order will prove to be of special interest when we come to consider the neural correlates of conditional expectancies. This relates to expectancies concerning the visual consequences of self-induced movements of eyes, head or body. For in some of these cases we have neurological evidence about the brain-locations at which the relevant anticipatory reactions occur (cf. Section 6.2).

At the lowest level, the movements in question consist of simple eye-movements (saccades, tracking- or pursuit movements) and the conditional expectancies formed at this level concern the resulting shifts of the retinal images. These have a major part to play in shape recognition. Visual shape recognition rests on elementary brain mechanisms capable of detecting such visual primitives as colours, changes in contrast, bars or lines, terminations, tilted lines, angled lines etc. (Marr, 1982; Treisman and Gormican, 1988). But the spatial character of these elements and their significance in the organism's transactions with the outer world, e.g. the trackability of lines and the importance of contours in the particularization of shapes, comes to be known to the brain only in the course of its *active* engagement with such visual elements. And it comes to be known here in the form of what movements of the eyes lead to what optical results - or, conversely, what optical results (e.g. centration) require what movements. In short, this knowledge comes to be registered in the brain in the form of a variety of what-leads-to-what expectancies based on what-leads-to-what experiences. Thus we can conceive that the characteristics of a particular figure or (flat-form) shape make themselves known to the brain in the first instance (i.e. at a first encounter, perhaps in infancy) in the form of the eye-movements required in following its contours and/or the eye-movements required to jump from one salient point of the shape to another. As the subject becomes familiar with a particular shape, the latter's characteristics will come to be stored in the form of a particular set of what-leads-to-what expectancies based on the what-leads-to-what experiences met in the active exploration of the shape by the eyes. In due course this set of conditional expectancies becomes conditioned to the optical image of the shape, so that they are now automatically elicited by the optical image of that shape, and the shape now comes to be instantly recognized for its distinctive features, i.e. without the need of active exploration. Serial processing has been supplanted by parallel processing, as it were.

At the next higher level, movements of head relative to the body would also be taken into account, and movements of the body at the level above. Jointly the conditional expectancies formed at these levels would suffice to present the brain with the representation of a *stable* scene despite the fact that all visual information comes merely from images that flit across the retina as our eyes dart from side to side, the head moves relative to the body, and the body relative to the environment. The world does not spin around us as the eyes rotate in their sockets. The importance of act-outcome expectancies in this process is attested by such elementary observations as that when the eye-ball is moved passively (e.g. by pressure of a thumb) the scene *does* dance about. In fact, the need for the brain to construct a stable scene from the images that flit across the retina is an incontrovertible proof of the importance for the brain of collating sensory information with own-movement-information. Without this it could not discover, for example, whether the movement of an image on the retina was due to a movement of the eyes or movement of the object.

At still higher levels in this ascending order of complexity we have expectancies which relate to the manner in which the perceived scene transforms as we advance or retreat, move to the right or the left, etc. These play an important part in the perception of solid objects, of their depth and mutual relations in three-dimensional space.

Above these levels we reach the expectancies relating to the consequence of even more complex movements and actions, e.g. expectancies concerning the consequences of manipulating objects in one way or another. At this level the human brain is able to form representations of non-spatial properties, such as weight, brittleness etc. - also of the uses to which objects can be put, and many other attributes that prove significant in our transactions with the surrounding world.

Finally, at the highest levels we reach the conscious expectancies derived from conceptualized knowledge and conscious inferences from that knowledge.

In general these higher levels of expectancy will supplement rather that displace the lower level expectancies, for they relate to different categories of experience. For example, I may reason that I must have left my glasses on the kitchen table. If I then go and look, I shall not be surprised to find them on that table. But, prior to my spotting the glasses, while the eyes are still scanning the plain surface of the table they will be arrested (if only flittingly) by any object on that surface by the unexpectedness of the stimuli that strike the retina as they reach the object's position on the table, until attention is finally captured by the object I have been looking for.

The example also illustrates that attention is governed by a multitude of factors, the occurrence of an unexpected event being only one of them. Ultimately it fastens on an object owing to the *significance* which the object has for the subject in the context of the current situation and the subject's current state of motivation. And according to the present theory that significance resides in the set of conditional expectancies which represent for the brain the features and properties of the object, and the expectancies of need- satisfaction which may in turn be elicited by one or other of the recognized features or properties - the latter expectancies tending to be the decisive ones.

What I have said here about the *significance* of a perceived object, does, I believe, also answer a question which has sometimes embarrassed psychologists. Namely, the question of the *meaningfulness* of perceived objects, and how stimuli which are merely

5.3 - Categories and ascending orders of expectancies

patterns of light, or sound or pressure can have meaning. It certainly answers the question of how the perceived shape of, say, an apple acquires meaning. What about the 'meaning' of a written symbol like a letter or a word. Here I believe the meaning (significance) lies in the expectancies the perception of the letter or word elicits about the job which that letter or word was intended to perform by whatever agency placed it there. Indeed, the expectancy that the letter or word was there to do a job, and not merely by accident, seems to me a precondition of the perceiver accepting it as a carrier of information. In short, from the standpoint of the Conditional Expectancy Theory, the question of *meaning* does not seem to pose any special theoretical problems.

*

Basic responses are made to *categories* of stimuli. In the words of Lenneberg (1967):

" All vertebrates are equipped to superimpose categories of functional equivalence upon stimulus configurations, to classify objects in such a way that a single type of response is given to any one member of a particular stimulus category.... ....Frogs may jump to a great variety of flies and also to a specific range of dummy- stimuli, provided the stimuli preserve specifiable characteristics of the 'real thing......Most primates and probably many species in other mammalian orders [also] have the capacity to relate various categories to one another and thus to respond to *relations* between things rather than to the things themselves; an example is to respond to the largest of any collection of things........In summary, *most* animals organize the sensory world by processes of categorization, and from this derive *differentiation* or *discrimination* and an interrelating of categories, and the perception of and tolerance for *transformations*."

What I have added by way of the Conditional Expectancy Hypothesis is this: in so far as the categorizing reactions elicited by the stimulus configuration are part of the brain's internal model of the world, their primary reaction component consists of a set of conditional expectancies which is elicited by the stimulus configuration and reflects the features, properties and significance of the stimulus-objects concerned. And behavioural responses which are based on the brain's model of the world (the 'voluntary' responses according to our definitions) will be responses that take their cues from this complete set of internal reactions.

***

5.4 - EMPIRICAL EVIDENCE AND ADDITIONAL ARGUMENTS

1.The interpretation of sensory inputs

The end-product of the act of perceiving some external object is an internal representation of that object which will be part of the brain's global world model. This internal representation may be called an *interpretation* by the brain of its sensory inputs, and I have stressed the *speculative* nature of this interpretation. As examples illustrating this speculative nature I have cited Wittgenstein's familiar Duck/Rabbit (now you see it as a duck, now as a rabbit) and the equally familiar Necker Cube (now you see it one way, now you see it the reversed way). Gregory (*op.cit.*) has shown that in the absence of supplementary cues the latter reversal even occurs with a wire cube which is held in the hand and rotated. When seen the reverse way the rotations are seen as elastic deformations and the subject may even feel as if his wrists were broken! In fact, the literature of experimental psychology is replete with examples of a similar kind.

Again, on a dark evening your entire way of seeing the scene before you may be balanced on a knife edge when, for example, the brain has to decide whether the vaguely outlined shade in your path is a tree or a human figure.

The examples also illustrate that, although we are conscious of the end-product of an act of perception as here conceived, we may not be conscious of the internal processes controlling the interpretations accepted by the brain: the brain does not project them into its global world model as part of the current state of the self. Generally, you just see these objects either the one way or the other, and no conscious process or rational thought enter into the determination of the alternative that happens to defeat its rival at any given moment of time.

In their historic experiments with inverting prisms worn as spectacles, Kohler and Erichmann (Kohler 1964) first demonstrated the powerful internal cohesion of the brain's representations of the external world, and much work in this field has been done since. Subjects fitted with such prisms would gradually adapt, and in the course of a few weeks come to see the world again the way they did originally. But one of the surprising results of these experiments was the apparent block- like nature of this perceptual adjustment. With up-down reversal, for example, subjects might at some stage report that they saw snow steadily falling downwards past trees that were still seen upside down. With left-right reversals they might at some stage see a car on the correct side of the road and facing the right way, but see its number plate still in mirror writing.

According to the Conditional Expectancy Theory this block- like character of the brain's internal representations resides in the linked set of conditional expectancies which

5.4 - Empirical evidence and additional arguments

the visual stimuli activate. Typically these linkages are between what- leads-to-what expectancies formed in consequence of the linkages encountered in past what-leads-to-what experiences - as has been explained earlier on.

Another aspect of the visual system that is relevant here is its ability to *separate figure from ground*. To take a simple case, what brain-processes enable my eyes to pick out a circle from a motley collection of figural elements, and respond to it as a unit? On the expectancy theory, the primary cause lies in the order that characterizes the points forming the circle. As I shall discuss in greater detail in Section 11.1, this orderliness manifests itself to the brain in the fact that what the eyes meet in tracking any part of the figure's contour reduces the unexpectedness of what they meet in tracking the next part of the contour. (In informational terms, it reduces the information conveyed by that next part of the contour. The figure has *redundancy* - cf. Section 11.1). Hence the characteristics of the circle come to be represented in the brain by a linked set of conditional expectancies relating to what the eyes will meet when they pass from any part of the figure to any other part. Through the conditioning I have described this well-formed set of conditional expectancies comes to be elicited *en bloc* by the ensemble of figural elements that form the circle and thus amounts to a categorizing reaction elicited by that ensemble as a whole. If this categorizing reaction then takes control over the subject's attentional reactions (and perhaps behavioural reactions as well) the figure will in effect have become separated from the background in the sense in which this term is understood in the present context. And what has here been said about a simple figure like a circle, applies equally to more complex figures. Always, I suggest, the critical factor is the degree and type of *order* in the elements of the figure; any such order can be represented internally in terms of linked sets of conditional expectancies characteristic of that order. I suggest then that it is the activation of these linked sets, and the control they come to assume in consequence, that constitutes the separation of the figure from the ground. Since these linked sets depend on past experience, it also follows that *familiarity* with the type of figure concerned will be a decisive factor.

The general importance of expectancies in the interpretation of sensory inputs, and their dependence on exploration in unfamiliar shapes, is also demonstrated in the recognition of shapes which are exposed only very briefly, as in the tachistoscope. It accords with the present theory that shape recognition in these conditions is generally defective except when the patterns are familiar ones. And in some cases the influence of expectancies can be observed directly. For example, an ambiguous figure which might be either the letter 'B' or the figure '8' will be seen as a 'B' in a trial series in which letters are shown, but as an '8' in a series devoted to numbers.

A revealing phenomenon also worth mentioning in this context is that of visual *completion*. If a subject is shown a figure, say the figure 'H' on a uniform background, and the experiment is arranged so that part of the retinal image of the H falls on the blind spot or on a visual scotoma, i.e. a part of the visual field which happens to be blind, the subject will not see a gap in that figure, but will literally see the whole figure that he has become accustomed to. These 'default assignments', as Minsky has called them, are hard to explain except on the assumptions that the processes of visual perception largely consist of the 'calling up' of familiar models, and that these models consist in expectancies: the blind spot fails to interfere with the perceptions because it is unable to contribute anything that would contradict the expectancies aroused by the stimuli impinging on the rest of the system.

All this is part of the processes through which the brain forms its internal representation of the outer world. As has been mentioned, it is now widely accepted that many animal species form some kind of internal spatial map of their environment on the basis of experience. The global nature of that internal model has been demonstrated for rats. Proceeding from experiments first performed by Maier in 1929 (see Gallistel, 1980), many investigations have shown that rats perform maze running and other obstacle-bound tasks more efficiently if they are set up in a room with which the rat has become familiarized: the room itself is used as an auxiliary frame of reference (and sometimes even as the main reference frame).

In the human case, the employment of reference frames of one kind or another appears to occur even in the discrimination and recognition of single shapes - at least according to psychologists who have looked at the question of why familiar objects look so different when they are rotated into unfamiliar angles. It seems that when we perceive a shape we assign to it a top, a bottom, a left edge and a right one, using our own axis as a reference direction, and this is all part of what we recognize when we meet the shape again.

As I have already explained, our theory of perception suggests a simple explanation: we are dealing here with cases in which the brain derives its internal representations of the spatial relationships that characterize the shape from the movements which the eyes have to execute to jump from one salient point to another, and the brain has registered these movements mainly in relation to the head's axis.

2. Direct manifestations of states of expectancy in the brain

Examples of animals learning to expect events have abounded ever since the systematic study of instrumental conditioning began early this century. The textbook example is familiar enough: if a rat is repeatedly given an electric shock from which it can escape only by jumping over a fence, and on each occasion this shock is preceded by a warning signal, it will soon come to jump as soon as the signal occurs. Not all demonstrations have to be so drastic. Krushinsky (1965) has shown that, if pigeons are first familiarized with a food stimulus which moves with constant velocity along a straight line, and if a screen is then interposed behind which the stimulus temporarily disappears, the birds can extrapolate the point at which the stimulus will reappear.

In the human case, states of expectancy in the brain, or the converse, viz. reactions to the unexpectedness of an event, can be observed at different levels and with different methods. The following examples illustrate this diversity.

Electrophysiological data have for some time demonstrated the reality of states of expectancy in the brain in the form of preparatory reactions for events to come or tasks to be performed. In his studies of EEG recordings on the frontal cortex Walter (1964) noticed slow potential waves associated with expectancy. In experiments in which a light flash was regularly followed by a click to which the subject had to respond by pressing a button, his recordings showed that the evoked potential following the flash comes to be succeeded by a secondary surface-negative wave which will last until the click occurs. If the latter is consistently omitted this wave will disappear, but it reappears if the click is again introduced, and then lasts until the click occurs. The flash has evidently become a warning signal which causes the subject to prepare for the response. Again, if the subject has been accustomed to a double click and only a single click is given, some cortical areas show an evoked potential response when the second click is expected. From similar observations E.R. John (1967) concluded that in these and other cases the brain appears to construct a facsimile of an expected (internal) event at the expected time of occurrence.

Getting ready for a required movement also comes under the heading of *expectancies*. In 1964 Kornhuber and Deecke reported slow cortical potentials prior to the initiation of voluntary movements, which they called *Bereitschaftspotentiale*. Prior to this, Jasper and Penfield (1949) had observed a desynchronization and blocking of rythms in the sensorimotor areas before movement. Since these early reports, a growing number of studies have demonstrated the importance in the higher brain functions of states of readiness for receiving a stimulus or making a response, and it has become common now to speak here of readiness sets (Evart et al., 1984).

These sets have been studied in a variety of contexts. Thus Gevins *et al* (1987) have shown that spatially distributed readiness sets may be essential precursors of accurate visuomotor performance. These particular sets appeared to have both invariant left frontal components and hand-specific central (sensori-motor) and parietal (body knowledge) components. Speaking generally, these preparatory processes for voluntary movements appear to be widely and bilaterally distributed in the cortex, and in a number of observed cases they began about one second before the movement (for references, see Pfurtscheller and Berghold, 1989).

Cerebral blood flow studies have suggested the occurrence of preparatory reactions also at the metabolic level. Thus Szekely and Montgomery (1984) have demonstrated an increased metabolic activity in parts of the superior prefrontal, midfrontal and anterior parietal cortices which could be shown to be sensory specific foci of attention.

Then again, they can be observed in electro-physiological studies on muscles. For example, Lacquaniti and Maioli (1989) have observed a variety of anticipatory EMG responses in their studies of the activity of catching a dropping ball. Some of these began already the moment the ball was released. Later components, representing the major build-up of muscle activity, occurred nearer the moment of the ball's impact. Their results have suggested to them that an internal model of the ball's properties is initially used to set the amplitude of muscle anticipatory contraction, while kinesthetic and cutaneous information obtained during the act of catching is subsequently utilized to calibrate this and other motor parameters.

Widely distributed preparatory reactions can also show up in single neuron activities when these are observed with the aid of implanted microelectrodes. I shall list some of these in Section 6.2. They demonstrate the widely distributed presence in the brain of neural units which come to be sensitized to the reception of particular perceptual inputs in advance of their occurrence.

Evidence about the impact made on the brain by *unexpected* events is also relevant in the present context, and some of that evidence has come from the study of a variety of event- related EEG potentials, notably the 'P3 wave', also called the P300 wave. This is a positive-going wave in the surface potentials which external stimuli may elicit and which has to be interpreted as the sign of a synchronous deactivation of large populations of neurons in the underlying cortex. The interesting feature in the present context is the high correlation of its amplitude with the unexpectedness of the stimuli concerned. Other event-related potentials too, have shown sensitivity to the novelty of a stimulus.

However, the P3 wave is by no means a simple phenomenon and lively controversies still surround it. The reader will find a good account and cross-section of rival views in Verleger (1988), Dolchin & Coles (1988), and in their Open Peer Commentaries. (For references see also Grossberg 1980.)

Yet it appears to be generally agreed that, although task relevance and the direction of attention are important, the phenomenon relates to 'strategic' rather than 'tactical' changes in the brain, i.e. to the updating of the brain's internal model rather than merely to the task at hand. The phenomenon is found in the frontal, frontocentral and parietal areas, and in view of the different role these cortical areas would appear to play in the brains' internal model of the world (cf. Chapter 6), it cannot surprise that the P3 manifestations in these areas differ in character (for example that the parietal wave appears to be more closely related to effort than do the other waves). It is also interesting to note that when the stimulus event first has to be categorized by the subject, the P3 waves shows a correspondingly increased latency.

My next section will be entirely devoted to the brain's reactions to the unexpected.

3. The orienting reactions

Violated expectancies cause surprise reactions. These include the brain's responses to novel stimuli and comprise a range of reactions which, as I have mentioned, are known as *orienting reactions*. The most commonly observed ones have both specific components (e.g specific shifts of attention) and non- specific components (e.g. an enhanced state of arousal, vigilance or alertness). They must be of special interest to us since failed expectancies are indicators of discrepancies between the brain's model of the world and the realities it actually encounters. The systematic study of the resulting surprise reactions was first undertaken by Pavlov in dogs. He introduced the notion of *orienting reflexes*, and aptly described them as *what-is-it? reflexes*. Apart from behavioural symptoms, orienting reactions may also give rise to physiological symptoms, such as changes in galvanic skin response, heart rate, respiration rate, pupillary dilation, muscle tone, and EEG rhythms.

When upon repetition a novel stimulus loses its novelty, the orienting reactions vanish. For example, the first time an animal is presented with a loud noise N, t seconds after a signal S, it will show a startle response. If this happens a number of times, that response to N will gradually wane. But if N fails to occur after t seconds, the startle response will reappear since this failure will now be the unexpected event. However, this waning of the response does not appear to be due to a rise of some neural threshold, perhaps through exhaustion of transmitter at the synapses. Rather, it appears to be an adaptive reaction consisting of a *dynamic inhibition* of the orienting reactions which occurs as an integral part of the processes of familiarization and habituation. This suggests that orienting reactions are, as it were, the brain's original response to any stimulus-event, disappearing only as the result of inhibitions that come into being during familiarization, viz., as an integral part of the way in which during familiarization the brain adapts to the stimuli concerned and develops responses appropriately related to the specific nature of the stimulus and the context in which it occurs. Learning to respond correctly to any given kind of stimulus is generally a process which involves inhibitory as well as facilitatory changes in the brain. The inhibition of the orienting reactions may thus be viewed as just part of the inhibitory fraction of those changes.

This inhibition of the orienting reaction can be sensitive to virtually any feature of a stimulus sequence. For example, cats can habituate to a series of tones played in ascending order and will then give orienting reactions if the same notes are played in descending order. Animals habituated to a loud beep of a horn will show orienting reactions when a softer beep is substituted. In human subjects, orienting reactions (here measured through finger vasoconstriction) have been found by Unger (1964) to occur when a subject is habituated to a given number sequence, and that sequence is then presented with just one number out of place.

Extensive research in this field has been conducted by Sokolov (1963, 1969), who used the suppression of the alpha rythm in the cortical surface potentials (a sign of alertness) as main indicator of an orienting response. He demonstrated, *inter alia* that orienting reactions may also be released by the non-occurrence of an event. Having habituated subjects to long tones, he then substituted short tones, and noticed the release of orienting reactions at the onset of the unaccustomed silence that followed the short tones. An anecdote frequently quoted in this connection is that of the citizens of a New York district which every night at 3 a.m. was crossed by a fast and noisy express. They soon learnt to sleep through this event. But when one night the train failed to arrive, many of them woke up with a feeling of unease.

Sokolov also saw in the orienting reactions a cogent argument for the assumption that the brain forms an internal model of the world and has special mechanisms for detecting discrepancies between what this model leads the brain to anticipate and what actually comes to pass. And, indeed, one has to concur that the phenomena I have cited are hard to explain except on the assumption that the brain forms internal models of environmental objects, events, and situations - models which will determine what the organism is to expect at a given point in time.

The mobilization of the brain which we call *arousal* is largely mediated by the reticular formation (Fig. 6.2). Sokolov concluded that this formation is activated both by the non-specific aspects of a stimulus-event via collaterals of the sensory afferents and in the case of novel stimuli also by the cortex. With familiarization the cortex withholds its excitation and at the same time blocks the excitatory non- specific excitations from afferent collaterals. According to Moruzzi (1960) this is mediated by inhibitory mechanisms within the reticular formation at the midpontine level (see also Kilmer et al., 1968).

In this connection it is also of interest that specifically novelty-registering cells have been found in the hippocampus - in this case cells which are normally silent but which fire in response to a novel stimulus. Their response ceases if the same stimulus is repeated.

Expectancies, as we know, also enter into the control of *attention*, and shifts of attention tend to be part of the orienting reaction. Indeed, arousal may be regarded as a non- specific, non-directed state of attention. Grossberg (1980) has studied the role of attention in the interaction between sensory inputs and expectancies. From his studies of classical conditioning, notably the so-called phenomenon of overshading, he has concluded that what is conditioned depends on our expectancies, and these in turn regulate the cues to which we pay attention. Thus two competing forces seem to operate here. On the one hand attention is drawn to failed expectancies because they reveal discrepancies between model and reality and that is something that needs attending. On the other hand in uncertain circumstances the brain will tend to focus on that part of the internal model which has proved to be reliable in the given situation, but it will also then be that part in which any subsequently detected mismatch will cause the greatest stir. In this connection Grossberg was lead to the interesting conclusion that the unattended cues are nevertheless registered, albeit in suppressed form, and that when such a mismatch is noted in the attended cues, the brain returns to the previously unattended cues: these will now be amplified and resume a controlling influence.

All of this is obviously relevant in the context of the Conditional Expectancy Hypothesis. But this hypothesis also forces the conclusion that the range of readily observable orienting reactions, such as those illustrated in the above paragraphs, cover only a small part of the ground. For, as we have seen in Section 5.3, expectancies operate at many different levels. And, although we must assume that at all these levels discrepancies can be detected by the brain between what is expected and what actually comes to pass, the corrective reactions triggered by these discrepancies must differ greatly in scale and in the range of brain structures they engage. To take a crass example, we cannot assume that the attentional reactions which draw the eyes to a moving object at the edge of the visual field will engage the brain and its arousal-controlling mechanisms on a scale comparable with, for example, the attentional reactions elicited in the occupant of a room when the doors opens and another person enters. According to the expectancy theory, therefore, the total family of orienting reactions occurring in the normal functioning of the brain covers a very wide spectrum: from small-scale reactions difficult to access by the experimenter to large- scale reactions which offer him quite a variety of readily accessible components. These considerations also apply to a topic discussed earlier, viz. the P3 waves and their relation to the unexpectedness of a stimulus-event. They show the complexity and variability of the brain processes that must be underlying this phenomenon and rule out the very suggestion that the P3 waves may be an invariable and invariant concomitant of the occurrence of every kind of unexpectedness.

4. The role of movement in perception

According to the Conditional Expectancy Hypothesis, many of the conditional expectancies concerned in the processes of visual perception relate to the changes in visual inputs to be expected from such movement options as moving the eyes, head or body. In Section 5.3 I have given an outline of the role of eye-movements in the elementary processes of shape recognition, and also of head- and body- movements in the stabilization of the visual scene. The importance of movement is bound to be most pronounced during the learning stages of visual perception. This conclusion is supported by a variety of studies of the cognitive development of the human infant. The summary I once gave of their findings (Sommerhoff 1974) would still seem to be valid to-day. I quote:

"In the newborn infant the eyes move in unison and it can fixate only momentarily and monocularly. At a later stage monocular fixation alternates between the two eyes and after about eight weeks this alternation has resulted in a teaming of the eyes which enables them to converge simultaneously on an object of interest. Ocular prehension precedes manual prehension: at about 20 weeks the infant can pick up a small object on the table with the eyes but only after another 20 weeks or so can he pick the object up with his fingers. Then begins a period of intense development of hand-eye coordination and active manipulation of objects. The subsequent development of the faculty of vision is dominated by such object-oriented manipulations. The solid shape of objects, therefore, comes to be recognized before the geometric shapes of the projection in the visual plane. It is objects that are handled, not abstract shapes. In consequence infant vision is faithful to objects rather than their retinal projections. Tilted plates are seen as tilted plates and not as ellipses. If geometric shapes such as circles are shown at that stage the infant sees them as suggesting real objects such as balls. It is not surprising that under the age of 6 - 7 years perception is 'syncretic': wholes emerge more readily distinguishable than what in later life we learn to recognize as significant details."

The point made by this passage is an important one, and one that has become well supported by the evidence: seeing is not a passive process. The eye is a prehensile organ and in the infant, ocular prehension precedes manual prehension: infants *grope* with the eyes before they grope with the hands. Thus the development of visual perception has to be seen as an integral part of the development of the child's *behavioural* skills.

What is learnt here is learnt by way of self-induced movement: of eyes at first, of hands thereafter. It results in a set of act-outcome expectancies, and already at the earliest stages the infant will show surprise if these fail to be fulfilled. Bower et al.(1970) have shown that if illusionary objects are dangled before the eyes by means of a stereoscopic shadow-caster, even an infant of only seven days will show surprise reactions if the hands pass through that empty space and feel nothing.

That spatially organized internal representations may assert themselves already at a very basic level, has been shown by studies of Sparks and Mays (1983) on the control of eye movements. Monkeys were trained to look at brief visual targets in a dark room. On some trials, after the visual target disappeared, but before a saccade (rapid eye-movement) to the target could be initiated, the eyes were driven to another position in their orbit by electrical stimulation of the superior colliculus. The resulting reactions then showed that new and correct saccade commands were formed, which were evidently based both on stored information about the location of the retinal image and information about the new position of the eyes.

The blind have to learn to probe the world without the eyes, and here it is worth noting that when a blind person explores a road surface with his cane, he is known to perceive its features clearly *at the tip of the cane*, although the only sensory input comes from the controlling hand and arm. This is interesting evidence not only for the importance of active probing in building up a model of the world, but also for the hierarchical structure of this model: on the basis of his internal representations of what he is holding in his hands, a blind man builds up expectations of what lies before him on the road, and makes these part of his internal model of the world. Mackay (1969) describes all this as "probing with feedforward" - meaning: probing followed by a change of *operational mode* or *set*, just as the present theory would have it.

Related evidence also comes from studies of spatial knowledge in blind young children, which often far exceeds what one might expect in the absence of vision. For example, Landau (1984) has reported of a two-year old congenitally blind girl of average all-round ability: "... she could brush her hair by finding the handle end of the hairbrush, and using the opposite end (bristled) to brush; she used different hand and arm positions to retrieve objects, depending on the spatial structure of the container (e.g. bowl *versus* plate); she could climb up on different pieces of furniture using different motor behaviours; and all these were in one degree or another *anticipatory* reactions."

In all this probing and physical exploration of the world, the brain's internal model of the current body-posture and body-movement and of the whole topology of the body, i.e. its *first order self-awareness,* has a vital role to play. Around the turn of the century, Bonnier labeled this body knowledge the *body schema,* and it is still widely known under this name. In the 1920s Head demonstrated the importance of the integrity of this 'body consciousness' for motor behaviour. Later Schilder (1935) showed that this body schema depends on the integration of proprioceptive, kinesthetic, labyrinthine, tactile, and visual impressions. The parietal lobes of the cortex (Fig. 6.1) appear to be primarily involved in its construction. Effects of lesions in this area will be reviewed in Section 6.5, but two observations should be mentioned here. The first is that parietal lesions may affect the subject's spatial interpretation of the visual inputs, thus demonstrating that this interpretation requires knowledge of body topology and movement as well as purely visual inputs. The second is that a patient with major parietal lesion may fail to recognize parts of the body as belonging to himself although there is no sensory loss (Gerstmann's syndrome). Sometimes, too, a limb may come to suffer total neglect: it has simply ceased to exist in the patient's model of the self-in-the-world. This is worth mentioning, because some of the strongest evidence for the existence of a coherent internal model of the self-in-the world comes precisely from the nature of the gaps that result in this model in consequence of particular brain lesions. For whereas, prima facie, one might expect any such lesion to disrupt or, as it were, scramble the brain's internal model of the body, just as the removal of a set of components in a TV would disrupt or scramble the picture on the screen, it produces, in fact, merely a failure to complete or particularize some part of the model. That little segment of the world just ceases to exist for the patient concerned, without disturbing anything else. This is readily understood on our hypotheses. It means that no expectancies are formed or maintained that would normally have served as a model for that part of the world; and, of course, it follows from the nature of expectancies, that the loss of any of their number need in no way interfere with those that remain.

There is further evidence which supports the notion that perception is a process of constructing an internal model of the world from the available sensori-motor information, while arguing against the idea that visual perception consist simply in the projection of a facsimile of the visual world onto some cortical screen in the inner recesses of the brain. The observations I have in mind relate to the remarkable degree to which visual perception can recover after seemingly vital brain areas have been destroyed. Thus Norton *et al.* (1966) have shown that continued performance on shape discrimination can be maintained by animals retaining only 2% of the fibres in the optic tract, and that in others it can be recovered after total loss of the striate cortex. Whereas these observations are fatal to the naive projection view I have cited above, they are not surprising if one accepts the Conditional Expectancy Hypothesis. For, according to this hypothesis, the perception of patterns remains possible, so long as a scanning of the scene (no matter by how circumscribed a probe), yields enough sensory information and relevant motor information to form a set of conditional expectancies which is capable of characterizing the shapes concerned. Hence the essential parts of the brain are those at which the relevant sensory and motor information can be integrated.

Some well-known experiments with kittens illustrate the importance of movement information also in the primary development of visual skills. Held and Hein (1963) reared these kittens with one of their eyes open only during normal, active movement, and the other open only while they were being passively transported over equivalent areas. Following several months of this procedure, the first eye produced normal visually guided behaviour, whereas the second proved to be functionally 'blind'.

Similar results were obtained with human subjects. Here the subjects were fitted with prismatic spectacles which displaced all objects by, say, 15 degrees to the left. It was then found that they would learn to compensate for this within one hour if they were allowed to walk actively in an optically structured environment; but no such compensation occurred if they were merely moved passively in a wheelchair for the same length of time. These results also support the stress I have placed on the importance of one particular category of conditional expectancies, viz. the act-outcome expectancies.

5. Attentional reactions

At various points I have mentioned *attentional* reactions, i.e. selective reactions which enhance the influence of some particular category of inputs on the ongoing processes in the brain. This occurs, for example, when the eyes are drawn to some unexpected event occurring in the visual field. There are two important remarks to be made in this connection.

Firstly, such reactions need not take the form of enhancing that particular set of stimuli, as if a spotlight were turned on to them, so to speak, or enhancing the brain's responsiveness to the selected category of inputs in absolute terms. It can, and apparently often does, take the form of inhibiting inputs from all but the selected category - thus amounting to a kind of masking (Treisman and Gormican, 1988). In consequence, for example, an attentional reaction in the cortex may show on EEG recordings as a bulk *suppression* of underlying activity.

Secondly, reactions of selective attention may occur at a great variety of levels in the brain's activities. They may range all the way from enhancing the relative effectiveness of some small set of stimuli in the primary stages of the processing of sensory information to enhancing the relative influence on conscious activity of the complex internal representation of some real or imagined object, event or situation. Thus the question one sometimes sees raised, viz., whether we attend to an object before or after we have identified it, seems rather meaningless. Selective attention is not a unitary brain process. At any point in time different kinds of selective enhancements may occur at different levels in the compound operations of the brain.

***

5.5 -THE LEARNING CHANGES INVOLVED

My object in this section is to consider the learning changes through which the brain may come to form its internal representation of the world and comes to respond to this in the manner asserted by the theory. The specific questions I want to ask are the following:

1. Through what internal mechanism could the repeated occurrence of a particular what-leads-to-what *experience* in a given context, lead to the formation in the organism of a corresponding what-leads-to-what *expectancy*? Through what mechanism, for example, would the recurrent experience that an action A has the effect E, cause the formation of the conditional expectancy that if A is performed E will follow?

According to our definition of expectancy, that state of expectancy for E consists of

   i) a disposition to enter a state of readiness for E, e.g. to prepare for the responses the organism has learnt to be required by E in the given circumstances;

   ii) a disposition to respond with orienting reactions if E fails to occur.

The genesis of both will have to be explained, and here, as well in what follows, I shall mainly concentrate on a particular class of what-leads-to-what expectancies, viz., act- outcome expectancies.

2. How do *sets* of such conditional expectancies come to be linked, so that a particular stimulus object may come to elicit the whole set *en bloc*?

3. What are the cognitive effects of such a linkage?

4. How may a set of linked act-outcome expectancies elicited by some stimulus object jointly come to enter into the *determination* of the organism's behavioural response to that object? The question also covers the general question of how voluntary behaviour as whole can become determined by that comprehensive set of conditional expectancies which, according to the theory, constitutes the brain's global model of the world and the self-in-the-world.

5. How may action *strategies* come to be formed on the basis of acquired expectancies? Most voluntary behaviour patterns consist of a *sequence* of steps in which each step prepares the stage for the next, until the final goal is reached: I enter the room, go to the desk, reach for the pen, put pen to paper, etc.

*

1. The first part of the first question was how the repeated experience that an action A is followed by an effect E, comes to produce the gradual formation of associations in the brain, which in due course cause the initiation of A to elicit reactions that prepare the organism for the occurrence of E and the responses it calls for.

A plausible answer is here suggested by a process which has long been familiar to all students of learning theory. I refer to the observation that if any response R occurs as part of a frequently recurring sequence of external or internal stimulus events, then in due course R, or parts of R, will tend to advance along that sequence and occur already in response to earlier members of the sequence. In this case R would be part of the responses required by E. It is assumed that this advance happens because R, or the relevant fractions of R, become conditioned by contiguity to the traces left by earlier elements in that sequence, and eventually to the elements themselves.

Prominent among such elements may be the proprioceptive stimuli that accompanied physical movements preceding the given response. This, for example, appears to be how, as we learn new motor skills, the initial slow, deliberate, and disjointed movements gradually become a smooth-running sequence of movements in which the execution of each step is already accompanied by a facilitation of the next step. Typical examples are a child learning to tie its shoe laces, or a student learning to type.

Which parts of R will advance in this fashion and at what point of the sequence of events the advance will be halted, is likely to depend on a number of factors. One obvious factor is the requirement that these advances must not reach a point at which they come to impede progress towards the goal which the responses in question are designed to serve. If they exceed this limit, interfering processes are likely to come into play which will inhibit any further advance. In point of fact, these inhibitory mechanisms may not always succeed straight away. For example, N.E. Miller (1963) has reported that if a rat is systematically rewarded for pressing a bar by the presentation of a food dish, an anticipatory ducking towards this dish may at times be observed to interfere with the standing up needed to press the bar.

Another relevant consideration is that those parts of R are likely to advance furthest along the preceding sequence of stimuli which are of the *most general* preparatory nature in respect of the events that follow. When you play tennis it is obviously important to prepare yourself as soon as possible to meet the opponent's ball at the anticipated point of arrival, but you won't win the game if your preparations are too specific, e.g. if you actually strike before the ball has come within striking distance.

To give another example of such anticipatory reactions. The first time you fire a shotgun, the recoil catches you by surprise. The second time you brace yourself: parts of the reactions required to meet the coming event have advanced to the moment you decide to pull the trigger.

In Section 6.2 I shall cite electrical measurements on the brain which demonstrate such anticipatory reactions. I shall also cite observations with implanted microelectrodes which show that at the cellular level this kind of anticipation of response requirements may occur already after a single trial.

As regards the second part of my first question, I have already explained the view that orienting reactions are the normal response of the organism to any event. Their omission in the case of familiar events occurs as a habituation, in this case by way of a dynamic inhibition. Thus we can regard this inhibition of the orienting reactions as part of the organism's way of preparing itself for an *expected* event, viz., by paving the way for the specific responses that event has proved to require. I shall return to this topic in Section 6.3.

*

2. The second question was: How do sets of acquired conditional expectancies come to be linked, so that a particular stimulus object may come to elicit the whole set *en bloc*?

Let me answer this with a simple and idealized example. Since the eyes look at the world from different points, the images of nearby objects will not fall on the same locations of the retina when the eyes gaze straight ahead. This binocular disparity is detected in the occipital cortex and causes the eyes to converge if a nearby object needs to be fixated. Let us assume for the sake of the argument that this is the first thing the organism learns about the distance variable. It will then also in due course learn that if a stimulus object is a nearby one (as indicated by the degree of this binocular parallax), then a movement of the head in any direction will cause the stimulus object to move in the opposite direction against the background of the scene - and the nearer the object the greater this movement. Since this is a regular experience we can assume that in due course the organism will learn this connection between binocular parallax and movement parallax. In other words, in due course the occurrence of any such binocular parallax will cause the organism to expect the object-shifts resulting from given head-movements and it will come to elicit this particular set of act-outcome expectancies *en bloc*. Conversely, the perception of a movement parallax will elicit expectancies of the ocular convergence required to fixate the stimulus object concerned.

*

3. As regards the cognitive effects of such a linkage, I shall cite two.

i) The experience of binocular parallax of a certain degree and the detection of movement parallax of a certain degree have now become *equivalent* stimuli as cues to distance. *Either* can serve the organism as a relevant cue when it comes to reaching out for the stimulus object in nearby space, for example.

ii) Nearby objects now elicit a categorizing reaction to which two sets of expectancies are attached: those relating to binocular- and movement parallax respectively. Clearly, this reaction will be more broadly based in the brain, and also more closely representative of the intrinsic nature of the distance variable. In other words, the organism has learnt a bit more about the nature of three-dimensional space.

*

4. How may a set of linked act-outcome expectancies elicited by some stimulus object jointly enter into the determination of the organism's behavioural response to that object?

This question clearly boils down to the question of how a set of act-outcome expectancies can be effective *before* any of the acts are initiated. Or, in more general terms, how can a set of conditional expectancies become effective in circumstances in which none of the conditionals are yet realized?

Here I must begin by bringing to the mind some general facts about neural activities in the brain.

i) Behavioural responses can become conditioned to any stimulus or combination of stimuli (including internal stimuli), the underlying reality being that the brain's final outputs depend on the balance of excitatory over inhibitory impulses across the bulk of the brain's neural networks. This dependence of a stimulus-response relation on the background of activities of the brain as a whole, is often described as its dependence on the current set of the brain.

ii) Even an undisturbed brain is never truly quiescent. There is a constant traffic of activity, consisting of a firing rate of neurons which tends to lie between 3/sec to 30/sec, averaging about 10/sec. These discharges are commonly called 'spontaneous' activity, though 'spontaneous' here means no more than that their causes are unknown.

iii) In the lightly anaesthetized animal the average level of excitation in the cortex remains remarkably constant over considerable periods of time, even in the disturbed brain. Thus the general effect of afferent sensory stimuli would seem to be a *redistribution* of cell discharges in the receptive cortical areas concerned, rather than a *net addition* to the area's total activity. This redistribution has temporal as well as spatial components. In the undisturbed brain the 'spontaneous' discharges show a measure of synchrony which manifests itself in rhythmic changes of the summated electrical surface potential of the cortex, as shown in the records obtained from large scalp electrodes. Disturbances in the monitored areas of the cortex show up in these EEG recordings as disturbances in these rhythmic changes.

iv) The initiation of voluntary motor activity appears to be preceded in the cortex by extensive preparatory reactions, as I have already mentioned in Section 5.4. This is called the *readiness set* of the motor activity concerned. It manifests itself in surface recordings over the motor or premotor areas of the cortex as potential changes which have come to be known as *readiness potentials*. Microelectrode recordings taken inside these cortical areas confirm the reality of such preparatory reactions at the cellular level. Examples which I shall cite in Section 6.2 show that they occur already in the execution of comparatively simple tasks. These preparatory reactions tend to be unconscious. Libet et al. (1983) have shown that in any voluntary act they tend to precede the subject's recallable awareness of having taken the decision to act.

Now to return to our main question: how sets of act-outcome expectancies elicited by a particular stimulus object (or the global world model as a whole) can become effective determinants of behaviour, i.e. become effective *before* any acts have been initiated that would elicit the respective outcome expectancies.

The above remarks suggest a plausible answer. If voluntary activities are preceded by certain preparatory activities or action precursors (as evidenced in the readiness sets), one can envisage that act-outcome expectancies, including expectancies of need-satisfaction, may through conditioning advance as far as these action precursors. Next, one can envisage that prior to the initiation of any action the brain scans or samples alternative action precursors in a search for one that elicits the expectancy of need satisfaction. In other words, prior to the action decision the brain tests alternative intentions, so to speak, for their expected consequences, activating them at a level below the threshold level at which the corresponding action would be triggered, finally energizing the one which elicits outcome expectancies of need satisfaction. With further experience these tests may later drop out again, since the behavioural responses they produce may gradually come to be conditioned to the general circumstances and set of the brain that initiated the tests. The result would be smoother and more automatic responses.

*

5. My last question was how action *strategies* may come to be formed on the basis of acquired expectancies.

So far I have only discussed the case of single actions determined by some given set of act-outcome expectancies in accordance with the expectancies of need-satisfaction they may contain individually or collectively. Many behaviour patterns, though, consist of a *sequence* of actions, each concerned with the completion of some particular step along the path towards the final goal. For example, if I want to go to London, I shall embark upon a sequence which is punctuated by such steps as phoning for a taxi, directing the driver to take me to the station, buying the right ticket, waiting for the right train, etc. Each step in this sequence carries the expectancy of need satisfaction only in the context of my ultimate desire to get to London. That desire has lead to the formation of an *action strategy* prior to my deciding on the first step to take.

I shall look at these processes under a separate heading in Section 6.4.

***

5.6 - FURTHER COMMENTS ON VISUAL PERCEPTION

Let me return to the basic categories of conditional expectancies that we can suppose to be involved in the visual perception of a three-dimensional world, beginning with the perception of (flat-form) shapes.

As has been said, the characteristics of the contours of a perceived shape can be adequately represented by conditional expectancies relating to the sequence of eye movements required to follow that contour and/or to jump from one salient point of the shape to another. In the early developmental stages of vision, these movements will have to be actually executed to build up the required sets of conditional expectancies, but in due course the learnt associations may cause them to be released by the retinal contour impression directly. Comparative records of actual eye-movements executed by a child gazing at a shape, have shown that, whereas in early infancy jumps from the centroid of the shape to its salient points prevail, at 5 or 6 years contour following prevails (Zaporozhets, 1965).

The perception of major and minor axes, and of the orientation of these axes to the vertical or horizontal (direction of gravity, orientation of horizon), is also known to play a part. The same applies to the perception of symmetries. The expectancy theory has no problems with these elements in the processes of perception. If a certain category of shapes is regularly seen in one particular orientation, human faces for example, it is plausible to assume that the eye-movements involved in exploring the features of the shape tend to be registered in the brain in relation to the vertical axis. This would explain why we have such difficulty recognizing such shapes if they are presented in unusual orientations, e.g. upside down (cf. Corballis, 1988).

The phenomenal (flat-form) shape of any one object-surface is covered by the kind of expectancies I have mentioned above, while the phenomenal spatial relations between the different visible surfaces of the object, are covered by such expectancies as those relating to the jumps the line of fixation must execute to move from (say) the centroid of the one to that of the other. Representations of the object's solid shape would rest on additional expectancies, e.g. expectancies relating to the way the phenomenal shape will transform as the perceiver moves head or body - sideways, up or down, or in the line of the object (movement parallax). In the latter case these would include expectancies about the way the phenomenal shape will expand or shrink as the perceiver approaches or retreats. Such expectancies would suffice to add the dimensions of depth which establish the slant of the surfaces and the distance from the observer of the object itself. But they may be supplemented by expectations that cover such experiences as the sharpening of focus through ocular accommodation and changes in binocular disparity through ocular convergence.

Through the effect of learnt associations, of course, these movement related expectancies may also come to be elicited directly by appropriate static cues, e.g.

binocular disparity, texture gradients, perspective convergence of parallel lines, and others. When a familiar object is occluded by the interposition of another, the brain's internal representation of what is seen will be further supplemented by expectations relating to what would be seen if either the occluded object were removed or the observer moved close enough to the occluded object to obtain a full view.

So far as the constancies are concerned, *size constancy* is readily explained in terms of expectancies. Although the image of a distant man is small, it yet elicits the internal representation of a full-sized man. We can explain this on the assumption that, using a variety of cues, the brain represents the true size of a distant object in terms of expectancies relating to the size it would be found to have if approached closely, touched, handled etc.

In the absence of relevant distance cues, a different set of expectancies may take charge. For example, when an illuminated oversize playing card is suspended in a dark room without any distance cues, the observer will tend to see it as a card of normal (i.e. expected) size suspended at a distance that would yield a retinal image of the received size. Indeed, Wallach and Bode (1972) have found that under certain conditions familiar size may serve as sole indicator of true distance.

I have already touched upon *position constancy*, i.e., the ability of the brain to form a stable representation of the perceived scene despite the instability of the retinal projections as the eyes move in their sockets. Thus to compensate for the movements of the eyes in their sockets, the perceptual system requires the formation of expectancies covering the shifts in retinal images that follow the movements of the eyes. It is interesting to note in this connection that the brain here appears to rely upon anticipatory information drawn from the centres governing voluntary eye-movements, i.e. on signals reflecting intended eye-movements. It is known that signals about eye-movements can be recorded in various brain areas about 100 msecs. before the movements occur. In the absence of such signals, for example if the eyes are moved passively, e.g. by pressing with the thumb, the position constancy fails: the perceived scene jumps about. Again, in patients with oculomotor palsy the scene is known to shift when eye-movements are attempted: the intentional signals are formed, but the movement fails to occur. On the other hand, when the brain has to compensate for head- and body- movements, it seems that it avails itself also of contextual information, not just motor signals. This is shown by the observation that compensation also occurs if the subject is moved passively, e.g. in a wheelchair.

The adaptive nature of the whole perceptual process, and hence the role of learning in vision, becomes apparent when the optical conditions are changed. To return to an earlier example: when prisms are fitted to the eyes, then at first the scene will appear to shift whenever the subject moves, but in due course the brain adapts and vision returns to normal: an appropriate set of new conditional expectancies has been formed. It follows from this plasticity and the expectancy theory, that if the images on the retina

5.6 - Further comments on visual perception

were artificially stabilized by some optical means, the very possibility of vision would cease, since there would now be no input changes that could either confirm or disconfirm existing expectancies, and, indeed, vision is known to vanish in such circumstances: after a short while all becomes just a misty grey.

Sometimes the internal representation, i.e. the linked set of elicited expectancies, attributes more features to an event than are in fact warranted, but no correction occurs because this part of the representation involves no expectancies that come to be violated in consequence of the discrepancy. Stroboscopic vision is a case in point. Our normal experience of a line which is first perceived at position A and a moment later at position B, is generally due to a movement of the line from A to B. If merely the initial and terminal positions of the line are projected on a screen in quick succession, the elicited internal representation tends still to be that of a line which has moved from A to B. And, unless the interval is so great that we would normally have become aware of the intermediate position of the line (an independent set of elicited expectancies), this movement-assuming interpretation produces no violated expectancies.

On occasions, when the brain has formed a false interpretation of the perceived object and expectancies are disconfirmed, an alternative interpretation may be available which attributes this experience to a movement of the object, rather than to an error in interpretation. For example, a monocularly viewed wire cube against a neutral background can be seen by a stationary observer in reverse (nearest edges seen as furthest edges). Its retinal projection has a genuine ambiguity in this respect. But when the observer then moves the head, his expectancies of the resulting transformations of the perceived form will be disconfirmed. However, instead of correcting them it seems that in this case the brain tends to cling to them and to attribute the discrepancies to a movement of the cube - there being insufficient information available to rule out this alternative.

I must add here a general remark on the subject of optical illusions. Alternative theories of perception are often judged according to their success in explaining that whole range of optical illusions which clever experimenters have been able to produce in the laboratory. The expectancy theory is not in quite the same position, because many of the explanations currently suggested for those illusions tend to be formulated in terms of theories about the choice of optical cues on which the mechanisms of perception rely, rather than about the structure of these mechanism themselves. The expectancy theory is clearly irrelevant to these explanations and under no compulsion to look for alternative ones. It can accept that the brain uses the suggested cues as main anchor-points for the relevant set of conditional expectancies.

The conditional expectancies we have discussed in connection with visual perception depend on a prior ability of the visual system to respond to such elementary pictorial elements as shading edges, lines, angled lines, tilted lines, extended blotches etc. And so we must expect there to exist neural mechanisms designed to detect, make explicit, or give adequate prominence to such basic pictorial elements. Moreover, since this is virtually a *precondition* for adequate vision regardless of circumstance (e..g the wearing of glasses, prisms etc.), we may well expect these to be either innate structures, or structures that in early postnatal days mature with only a minimum of actual exercise. This has been amply confirmed. For example, it has been know for some time that the retina is backed by neural structures which enhance contrast. We are dealing here with mechanisms whose outputs supply the brain with an improved set of stimuli to which the expectancies that form the ultimate substance of the brain's internal representations of the outer world, can become attached - and against which they can be tested.

One of the most important discoveries made in this connection was that of Hubel and Wiesel (1965), who managed to isolate neurons in the visual cortex of the cat that were selectively responsive to lines and slits in particular orientations, bars stopped at one end, angled bars, and to the location of an optical stimulus inside the receptive field of the cell concerned. Since those first discoveries, the work has been greatly expanded. Cells are now known, for example, that respond selectively to movement of visual inputs in different directions, others that respond to the extension of blotches, and yet others that respond maximally to specific degrees of binocular disparity. Similar results have been obtained in primates, including man.

The general picture that has since emerged attributes to the primary visual cortex, the *striate* cortex (area 17, Fig. 6.1) the role of acting as a primary receptive area which not only integrates some of the information received but also distributes information to a number of *prestriate* visual areas (mainly area 18, Fig. 6.1), each of which maps the visual field and appears to specialize in processing specific kinds of information, e.g. information relating to orientation, movement, or colour respectively. According to Zeki (1987) there are at least six such directly involved prestriate areas. In other words, these areas do not analyse the same features of the visual scene at an increasing level of complexity, but analyse different kinds of basic features.

Since we are here dealing with basic pictorial elements which have proved to be key elements in visual perception throughout evolution of the species, these results are perhaps not surprising. But in view of the ever changing variety of shapes that need to be recognized and of their common complexity, it would be naive to expect that we shall eventually also find cells that are wired in to be selectively responsive to complete shapes as such, e.g. the retinal projection of a bicycle or of one's grandmother - nor multiple input/single output modules yielding the same result. Of course, it may well be the case that some investigated cells are found to fire only if particular shapes are being perceived, and such results are reported from time to time. In other words, cells which appear to be uniquely involved in a unique experience and the associated categorizing reactions (cf. Section 5.5). But such observations do not demonstrate that we have here traced the output of any *modular analyzer* for the shapes concerned - a fact only too often overlooked by those who pin their faith on modular analyzers. It would merely demonstrate that we have here penetrated an area of the brain containing cells whose level of excitation depends on the end- products of the perceptual process, and that there happens to be one which happens to be maximally engaged by one particular category of shapes.

Rock (1983) has argued that shape perception cannot be the product of experience, because, if this were the case, no entities would be perceived at the outset, and perceptual and perceptual learning could never begin. But in the innate pre- processing I have discussed, we do have a beginning. The neonate could soon learn, for example, how the location in the visual field an innately recognizable line element will change in consequence of different eye-movements and develop the corresponding expectancies. Similarly, the neonate could learn to form internal representations of the spatial relations between salient points of a shape in terms of expectancies relating to the eye-movement required to move from fixation on one to fixation on another.

We live in a three-dimensional and constantly changing world whose images are projected onto the retina with the added movement caused by the movements of the eye. It is no more than plausible to suggest that the representations which the brain forms of that world have a dynamic basis of the kind I have suggested.

For the sake of completeness I must mention one constancy that does not fall within the orbit of the dynamic relationships I have discussed in this section. That is the colour constancy: the plants' leaves look to us green no matter in what light they are seen, be it the glaring light of the noonday sun, the rosy light at sunset, or the diffused light of a cloudy sky. But the phenomenon is nevertheless of interest because it illustrates once again the degree of integration that goes into the perceptual powers of the brain. For the now prevalent view appears to be that colour constancy is achieved by way of brain registering the *spectral reflectance* of the perceived objects and comparing this with the spectral reflectance of the surrounding objects.

***

5.7 - CONTRASTS WITH OTHER THEORIES OF PERCEPTION

Theories of perception are especially relevant to the study of consciousness, because, as has been said, perception is the process through which the sense organs add to our consciousness of the world. And visual perception leads the field in this respect. The theory at which we have arrived is entirely formulated in a physical language, and in this respect it can claim to remedy what must be seen as a major weakness of the main theories that dominate the contemporary literature. To complete this chapter, therefore, I shall try to give a brief summary of these theories.

By far the greater part of cognitive studies to-day is devoted to the mechanisms of vision and much of this work has been very rigorous, both at an observational and at an elementary theoretical level. For example, a great deal of experimentation has been concerned with the investigation of the optical cues on which the mechanisms of vision rely in performing their task, e.g. in forming representations of the position, size, distance, and shape of the objects in the visual field. Experiments may be devised in which the eyes are deprived of particular categories of information and the resulting visual defects or illusions are studied in detail. Thus a considerable number of cues suggest themselves as cues for the perception of depth and distance, such as: binocular convergence, ocular accommodation, binocular disparity, movement parallax, texture gradients, perspective, interposed objects, apparent size of familiar objects and other, more subtle ones. The actual use by the brain of any of these can be studied systematically. Again, experiments may be devised to determine how much of an organism's visual competence is innate and how much acquired postnatally by learning. There is generally little to fault in the logical reasoning found in this kind of detective work. In a proper scientific manner, hypotheses are formulated about the significance of particular cues in particular perceptual contexts, predictions are made, and experiments devised to check the predictions. Equally sound, scientifically, is the exploration of neural unit responses in the visual cortex or other brain locations when the subject is given specific tasks to perform.

The weakness to which I have referred appears only when one turns to the more abstract theoretical concepts which investigators tend to introduce in order to explain their findings, and when they come to consider the nature of the internal representations that are the end-products of perception. Sometimes these concepts are left wholly undefined; sometimes they are defined in woolly or mentalistic terms; or they rest on metaphors of questionable validity, e.g. analogies with computer operations of one kind or another, or analogies with the processes of rational thought.

*

The following are the five major strands of thought which one can discern in the best-known theories of perception:-

1. The view of the behaviourists and S-R (stimulus-response) theorists. The notion of consciousness, as such, does not figure in these theories at all. Instead, the end-product of the processes of visual perception is taken to be simply a state of the subject in which the visual stimuli produced by external objects elicit responses which are conditioned to these optical cues. No intervening structures are claimed to be involved. Since I have already discussed the shortcomings of the S-R theorists, I need say no more at this point.

2. Theories which take the end-result of the processes of visual perception to consist of the collective reports of a family or hierarchy of 'wired-in' feature detectors, i.e, neural networks which are selectively sensitive to the presence of some particular feature in a perceived shape. These theories were given a substantial boost through the discoveries of Hubel and Wiesel which I have mentioned above. However, whereas it is plausible to assume that the visual system is especially wired to detect contrasts, contour features and other elements which are relevant in the recognition of *any shape whatsoever*, attempts to generalize this modular conception of feature detection soon run into trouble.

Unless one makes the obviously absurd assumption that we have wired-in detectors for every possible shape and perceptible feature, *however novel* from an evolutionary point of view, e.g. for the shape of a bicycle or Bugatti racecar, this type of modular theory has to assume that complex shapes are recognized by reference to property lists. Yet whereas discrete feature-detectors coupled to property lists of one kind or another, are acceptable strategies for solving certain narrowly defined problems in AI, they won't do for solving the complex problem of human vision, i.e., the problem of how our sense organs add to our consciousness of the world. In AI, for example, they can be used in devices designed to recognize printed alphanumeric characters and other standard shapes. But, as Barsalou and Bower (1984) have stressed, human beings do not recognize a table or a car by running in the mind through the galaxy of the possible properties which any object in front of our eyes could possibly possess. Moreover, in some cases the *absence* of a particular feature can be as critical as might be its *presence* in a different case. The theory also has great difficulty with dynamic, complex relational and general contextual cues, and with the problem of shapes seen at different angles.

The proper place for such wired-in feature detectors lies in the preconscious, knowledge free, and initial phase in the processing of the visual inputs. I mean the phase at which those primitive features are picked out from the flow of physical stimuli that play a vital role in the recognition of any shape whatsoever. One of the best-known theoretical studies in this field are those of Marr (1982). They centred on a search for possible neural networks that could single out such items as shading edges, extended edges, points, blobs and bars that might vary in fuzziness, contrast, lightness, position, orientation, size or terminal points. Such networks would yield what he calls the 'primal sketch'. According to Marr, this sketch is then analysed by processes that are capable of grouping lines, blobs, bars and points together in various ways, resulting eventually in a separation of figure from ground. The action of these particular networks would be followed by processes that would use binocular disparities to yield a depth dimensions and thus produce what he called a 'two-and-a-half-dimensional primal sketch'. This, he believes, is then 'read' by higher level processes, asking such questions as "Is this a tree?".

I believe an approach of this kind to be valid in respect of those initial input-processing mechanisms for visual information which pick out the picture elements on which the brain has come to rely in the recognition of any kind of shape. Since they are needed for any visual input whatever, we can expect these mechanisms to be wired in. But this, I would think, is as far as it goes. Binocular vision, for example, is only a minor factor in our overall perception of depth in the three-dimensional world right up to the horizon. Here quite different factors enter, such as movement parallax, texture gradients, size constancy etc. And, more tellingly, it is hard to envisage any veridical distance detection in the brain until that higher level of representation has been reached at which the brain has constructed the representation of a stable visual scene from the succession of images that jump across the retina as the eyes move about in their sockets - that higher level of representation, for example, at which there exist error-correcting processes which can cope with such problems as adapting the visual system to a new set of spectacles.

Finally, as regards the question "Is this a tree?", it seems to me that far more categories are involved here than those relating merely to optical shape. For a tree is much more than just that.

3. Theories which see perception as a process in which the visual inputs are compared with templates stored in the memory, the end-result taking the form of an activated template, plus, in some theories, a report on the degree of fit.

Apart from failing to give a plausible account of how such a comparison with a template might be effected, these theories also show a remarkable divergence of opinion on how these templates themselves are to be conceived. The most commonly used name for such templates is *schemata*, but here is little agreement about their nature, and frequently this is described in mental terms or metaphors only. Thus Boden's schemata consist of "knowledge of relations between sensory inputs and its interpretation". Thorndyke and Hayes-Roth (1982) regard them as "configurations of concepts" and "mental abstractions" of a prototype kind (e.g. a typical face). The individual instantiation is then classed as a prototype + distortion. Fodor (1975), on the other hand, sees schemata as "formal systems of symbols". However, since notions like 'knowledge', 'concepts', 'mental abstractions', and 'symbols', are generally left unclear in physical terms, these theories still beg the question of the true nature of the brain's internal representation of external objects.

Not all theories fall into this category, though. Piaget's notion of schemata, for example, is rather more physical and physiological (Piaget 1977). On his view the processes of perception are assisted by 'schemata' which are "sensori-motor plans" derived from the motor explorations through which in infancy we first discover spatial relationships in the surrounding world. With its emphasis on the role of movement, this theory comes a little closer to our own.

Still closer to our theory are those theorists who think of schemata in terms of expectancies. According to Hochberg (1968), for example, a 'schematic map' is a "program of possible samplings of an extended scene and of contingent expectancies of what will be seen as the result of these samples". Goodman (1980) talks about "knowledge structures formed by interrelated real world expectancies"; while Minsky (1975) conceives of the end-products of the mechanisms of vision as 'frames' which are "data-structures containing information about how to use the frame, what can be expected to happen next and what to do if the expectations are not confirmed". Neisser (1976) postulates that visual explorations are directed by "anticipatory schemata" which are "plans for perceptual action as well as readiness for particular kinds of optical structure". Perception is "the interaction of schema and available information". But, by contrast with the theory at which we have arrived, none of these theories offers an adequate analysis of what in physical terms constitutes a state of expectancy or anticipation in the brain. From our point of view this is their main weakness, since it prevents one from drawing inferences about the neural correlates of these expectancies.

4. According to another school to be mentioned, the processes of perception are taken to be based on intuitive inferences. The idea was introduced by Helmholtz already in 1867. Perception, as he saw it, is a process in which the brain draws *unconscious inferences* about the outside world from the stimuli it receives. These inferences are based on past experiences, and result in judgements about what is present in the visual scene. These judgements are not judgements on a perception, they are the perception.

This line of thought has been followed up by Bruner (1957), who sees the processes of perception mainly as processes in which the subject infers the categorical identity of the perceived objects on the basis of past experiences, thus going "beyond the information given". His whole emphasis, therefore, is on acts of categorization.

With all of this we must be in broad agreement However, there is a danger of thinking about these processes of *inference* in unduly and confusingly mentalistic terms. Rock (1983), for example, also operates with the notion of unconscious inference, but then he introduces such descriptions of these inferences as "deductions made from premisses based on rules". Indeed, his theory depicts the whole process of perception as an attempt by the brain to find an intelligent solution to the problem of what entities in the world the given array of optical stimuli represent. According to the criteria we have set ourselves, the main weakness of this kind of theory lies in its heavy reliance on analogies with the processes of rational thought, thus bringing us no closer to an explanation of the processes of perception in physical and, eventually, neural terms.

5. In stark contrast with all these *indirect* theories of vision stands the *direct* theory of James and Eleanor Gibson. Perception is here seen as a *single* act. According to this theory every scene we perceive has an underlying invariant structure which undergoes disturbances of various kinds, and according to James Gibson (1976)

> "perceiving is a registering of certain definite dimensions of invariance in the stimulus flux together with definite parameters of disturbance. The invariants are invariants of structure and the disturbances are disturbances of structure. The structure, for vision, is that of the ambient optical array."

The perceptual system is said to *resonate* to these invariances - the visual system hunting and exploring until all invariants are extracted. For example, if we walk around a table, all the eyes can see of the tabletop is a number of trapezoid forms. But the unchanging relations between the four angles and the invariant proportions over the set uniquely specify the rectangular surface. The main invariants of the territorial environment, its persistent features, are the layout of its surfaces and the reflectances of these surfaces - so the theory claims.

Rather confusingly, Gibson uses the term *information* to denote the specification of the observer's environment. Hence the process of the perceptual system 'resonating' to the invariances is described as a *pick-up* of information. This pick-up is held to depend on the input-output loop of the perceptual system.

Finally, what the environment offers the animal, what it *provides* or *furnishes*, either for good or for ill (to use his expressions), is here dubbed the *affordance* of the environment, and the end-result of the process of perceiving an object is an awareness of its 'affordance'.

Once again, however, the whole theory is uncomfortably vague and although the Gibsons acknowledge that it is incomplete, it leaves far too many crucial questions unanswered. Their notions of 'resonating' and 'information pick-up' merely cloak the things that really need explaining, while the whole teleological context of any act of perception is cloaked by the single notion of 'affordance'.

*

In the present chapter I have given in physical terms a general account of the nature of the brain's internal representations of the world and the self-in-the-world. It was a *physical* account in the sense that all our key concepts, e.g. the concept of *expectancy*, had earlier on been defined in physical terms (Chapter 3). I have stressed throughout that we need a physical account of the nature of the brain's internal representations in order to be able to draw inferences about the way these representations may come to be realized in the neural networks of the brain. In view of the critical role which the brain's internal representations play in my account of the nature of consciousness (Chapter 4) and mental events generally (Chapter 7), these neural correlates are of special importance. Indeed, we must see in them the bridge between mind and matter. They form the subject of the next chapter.

***

Chapter 6

THE NEURAL CORRELATES OF THE INTERNAL REPRESENTATIONS

SUMMARY

To reach the neural correlates of human consciousness, we must search for the neural correlates of the categories of internal representations in terms of which I have described it (Chapter 4). To begin with, though, I shall confine myself again to just one of these: the internal representation of the outside world and of the body in relation to that world. The others will be left to Chapter 7.

According to the Conditional Expectancy Hypothesis, these internal representations consist in the main of states of conditional expectancy, especially act-outcome expectancies. I shall begin, therefore with a recapitulation of the physical definition of a state of expectancy given in Section 3.4. According to this definition, states of expectancy are states of readiness for an event which are followed by adaptive modifications if the event fails to occur. These modifications are initiated by the 'orienting reactions', which include, for example, shifts of attention to the part of the field in which the expectancies have been violated. After a brief recapitulation, I shall therefore proceed by considering first the neural correlates of states of readiness or anticipation in the brain (Section 6.2), and then the nature of the orienting reactions and the mechanisms that control them (Section 6.3). In both cases I shall rely on published evidence, mainly evidence based on microelectrode recordings.

In Section 6.4 I shall then turn to the general organization of behaviour and the way in which the brain's internal representations enter here. In particular I shall consider how the goals of intended actions come to be represented in the brain, and how action strategies come to be formed that match these goals. The contribution which different brain structures make to the totality of these processes will then be reviewed briefly in Section 6.5 with an eye to questions relating to the the seat of consciousness in the brain.

Finally, I shall briefly discuss some conflicting attitudes found in the literature of brain research (Section 6.6.).

***

6.1 - RECAPITULATION

As our first step, we must turn to the question of the neural correlates of states of conditional expectancy which we take to constitute the brain's internal representations of the world. In everyday use 'expectation' is a *mental* term: to 'expect ' something to happen is to entertain certain beliefs about coming events. Thus to talk about 'expectations' is to engage in 'mind talk'. In our work, on the other hand, we need a prescriptive definition of a *state of expectancy* which is couched in strictly physical terms. This has been given in Section 3.4, and I shall repeat it here.

We began with the concept of *readiness* (or *preparedness*, since I treat these as synonyms). This was defined as follows:-

DEFINITION:.

> *A physical system is in a state of **readiness** for an event X, if it is in a state which will facilitate the making of an appropriate response to X should X occur.*

Using this as our starting point, we then defined *expectancy* in the following objective terms:-

DEFINITION:

> *An organism is in a state of **expectancy** for an event, if (i) it is in a state of readiness for that event, and (ii) this state undergoes adaptive modifications if the event in question fails to occur (or fails to occur at some specific point in time).*

The reader will also recall that the notions of *appropriate* responses and *adaptive* changes, which occur in these definitions, had at that point already been clarified in Chapter 3.

The two definitions here recalled settle the *technical* sense in which the phrases *states of readiness* and *states of expectancy* are used in the present work. They were not designed to cover the various connotations these terms tend to have in everyday use. Nevertheless, they are close to those connotation and they reflect the main distinction which we draw in everyday life between *preparing* oneself for a contingency, and *expecting* that contingency. When I go abroad I may *prepare* myself for a variety of contingencies, for example, by packing pills against indigestion and having inoculations against tropical diseases. But I do not necessarily *expect* these things to happen: I shall not be surprised if I fail to get indigestion or to contract a contagious disease. By contrast, if I expect to meet my friend at the airport, I shall be surprised if he fails to turn up, I shall try to find out what happened, and then adjust my plans accordingly.

Our definitions also accord with the following difference between the common notions of *readiness* and *expectancy*: one can simultaneously be in a state of *readiness* for two mutually exclusive events: when I go on holiday I can pack clothes both for sunny and rainy days. But one cannot simultaneously be in a state of expectancy for two mutually exclusive events: I cannot both expect my friend to arrive *and* not to arrive at the airport. However, I *can* expect his arrival with varying *degrees of confidence*. Viewed objectively, this will be reflected in the relative extent to which I prepare myself for the possibility of his failing to arrive, while yet expecting his arrival.

States of readiness may differ in many ways. They may be more or less active or passive: forewarned of a possible burglary, the night-watchman may respond either by patrolling the building or simply by listening more attentively. They may be more or less absorbing: a parachutist preparing for the moment of touchdown has little latitude at that time for preparing himself physically for anything else. Again, states of readiness may be for something more or less specific: I can prepare for a very specific event like catching a ball, or for a very general condition, like bad weather. And they can be more or less time-bound: I can prepare myself for having to phone a friend at some specified hour, or just phoning him when convenient.

***

6.2 - THE NEURAL CORRELATES OF STATES OF READINESS

To get some idea of the neural correlates of the states of expectancy with which we are concerned in the representational functions of the brain, I shall begin with their main component, viz. the corresponding states of readiness for coming events and response requirements.

The brain can prepare itself for coming events in a variety of ways. Evidence for some of these has already been cited in section 5.4. As example I cited the expectancy waves which Walter detected in the brain's surface potentials when the subject expects a click to be followed by a second click after a specific interval of time. This is hard evidence for the brain's capacity to enter into states of perceptual readiness. A common way of measuring the degree of readiness for a coming event is to measure the subject's reaction-time to that event. Such experiments have also shown that states of readiness for an event can be linked to the occurrence of that event at some specific point in time, e.g. a specific number of seconds after a given signal. If, after the initial training period, the event is made to follow the signal either before or after the accustomed point of time, the reaction-time generally proves to be longer. Different types of orienting reactions can also be used to indicate the occurrence of an event unexpected by the subject, e.g. the galvanic skin response.

For direct evidence about the neural correlates of states of readiness we must turn to microelectrode recordings of the activity of individual neurons.

Turning first to the kind of expectancies which the Conditional Expectancy Hypothesis states to be the foundation of the brain's internal representations of the world, I have described in Section 5.3 how these may be arranged in ascending orders according to a variety of criteria, e.g. the degree of complexity of the action to whose outcomes the expectancies relate.

At the lowest level here we have expectancies relating to the consequences of simple eye-movements, e.g. expectancies concerning the shifts of retinal images that follow the saccades. These expectancies are important, for example, in helping the brain to decide whether any particular shift of the retinal image is due to a movement of the eyes or to a movement of the visual scene itself. This in turn helps the brain in the task of constructing the representation of a stable scene from the images that flit across the retina as the eyes move in their sockets.

At the next higher level we have expectancies relating to the visual consequences of moving the head and body as well as the eyes. And so on.

For the lowest levels of this ascending order a number of observations with microelectrodes have been made in recent years which are of special interest in the present context.

Toyama et al. (1984) have studied the responses of saccade- depressed and saccade-excited cells in the striate (primary visual) cortex, (cf. area 17, Fig 6.1) of the cat. Their findings suggest that an integration of retinal and motor signals occurs here which is based on an 'efference copy' of the eye-movements, i.e., on inputs related, not to proprioceptive afferent information about the eye-movements, but to information about the motor commands issued to the eyes (cf. Section 5.5). They conclude that this makes sense if it helps these areas to prepare for the consequences of the eye- movements concerned, i.e. the resulting shifts of the retinal image. These 'motor efferent copies' illustrate one of the channels of information-flow which the brain can use to prepare for coming events in this field.

6.2 – The neural correlates of states of readiness

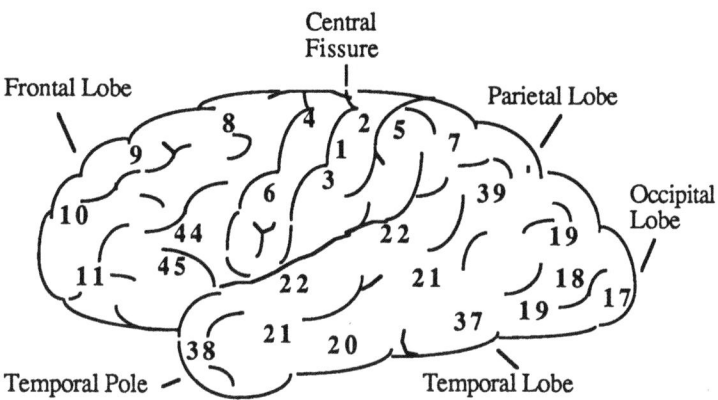

Figure 6.1 –*Lateral surface of the human cortex showing approximate location of some of the cytoarchitectural fields distinguished and numbered by Brodman.* The numbering has remained in use despite the fact that it is based mainly on the order in which they were studied and has no functional significance.

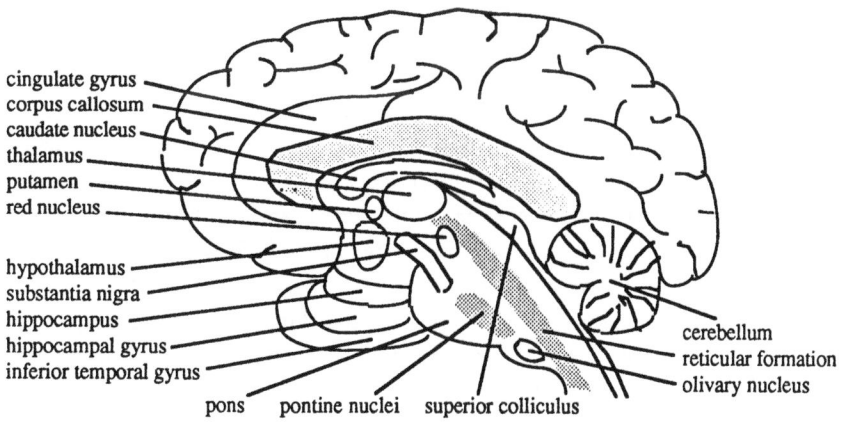

Figure 6.2 –*Schematic median section of the human brain, showing most of the structures that will be mentioned in this chapter.* Although the thalamus is shown as a single structure, it is really a conglomerate of separate but closely interlinked nuclei acting as relay stations of afferent pathways to the cortex, but also with lateral connections and with feedback to each nucleus from the particular cortical area to which it projects.

Evidence for preparatory reactions of a different kind, but equally based on information about eye-movements, has also recently come to light. Here the anticipation is not of coming events so much as of coming response requirements. It takes the form of a facilitation of the responses required in a given situation, by way of an anticipatory sensitization of neurons to the sensory stimuli that have proved to be significant cues in the given context. This has been called the phenomenon of *enhancement*.

For example, Wurtz and his colleagues (1980) trained monkeys to fixate on some point P in the visual field and to change that fixation to a light spot appearing at some other point Q as soon as that light appeared. They then investigated the reaction of cells in the superior colliculus (Fig. 6.2), which were responsive to light at the point Q while fixation still remained on P. It was found that after only one rewarded trial such cells would respond more actively to light at Q even before the change of fixation took place, i.e. the enhancement preceded the eye-movement. This could be detected because, due to the reaction time of the animal, the saccade occurs only .2 secs. after the onset of the stimulus. In short, the brain has learnt the significance in the given situation of light at Q as a target for the coming saccade, and it has increased its sensitivity to Q accordingly.

Similar observations were made in the parietal cortex (Wurtz *et al.* 1982) and certain differences noted (Fig. 6.3). Phenomena of enhancement were later found also in the frontal eye-fields, prestriate cortex, and substantia nigra (Wurtz *et al.* 1984) while Mountcastle (1981, 1984) and his team have detected additional ones in the posterior parietal cortex, and Schlag-Rey and Schlag (1984) in the central thalamus. Some enhancement reactions of this kind have also been found in the striate cortex (see below). Recordings made by Jasper in the visual cortex of intact animals (Eccles, 1966) are also worth noting. These showed that in certain neurons there occurred an enhanced response to a light stimulus after that stimulus had been conditioned to produce a motor response. Once again, therefore, we see here a change in the responsiveness of a neural unit to a visual stimulus which has become a cue for further action.

I have mentioned the parietal cortex as the cortical area most closely implicated in the brain's internal representation of the topology of the body and of body posture or body movement, and I have mentioned its role in the spatial interpretation of visual inputs. The neurons sensitive to visual inputs are found mainly in the posterior areas of this lobe. These parietal visual neurons have been extensively investigated by Mountcastle et al. (1981,1984) and Sakata *et al.* (1983). Some of the examined neurons responded in a manner similar to those of the superior colliculus.[1] Others responded in a less specific manner, for they would also show an enhancement if the external light-stimulus was not used as the target of a saccade but as the target of some other

---

[1] The superior colliculus is a midbrain nucleus which receives inputs from the retina although it does not lie in the main pathway from the retina to the cortex. This pathway passes through the lateral geniculate nucleus of the thalamus (see below), which, however, also projects to the superior colliculus. Together with the frontal eye-fields this structure appears to be implicated in the initiation and control of saccades (see below).

6.2. - The neural correlates of states of readiness

movement, e.g a movement of the arm. In these cases, therefore, the brain appears to have registered the broader significance of the stimulus concerned.

It should be noted in this connection that such sensitizations are really of the nature of an *attentional* reaction. For to react to a stimulus as a *significant* one, is to potentiate it as as a cue for behaviour, and that is also what we mean when we assert that the brain is *attending* to a stimulus. Attentional reactions, in turn, are relevant to the topic of *readiness*, for they contribute to the brain's readiness for the things it has to do next.

The teams of Mountcastle and Sakata also found neurons in the posterior parietal lobes whose enhancement appeared to reflect the learnt significance, both of the target itself *and* of its spatial location: a high percentage of the measured neurons were active only when the animal gazed at a desired object *within reach*. There was also other evidence to suggest that this area is implicated in space vision, at any rate for proximal space. Since the parietal areas are closely linked with the internal representation of body posture and body movement, their involvement in the above manner supports the suggestion that proximate space is (at least partly) learnt by way of active movement.

The same investigators also found units which appeared to be responsive, not to the significance of an object within reach, but to the significance of a particular flow of movements when that object was the target of those movements.

Thus the posterior parietal lobe seems to be mainly involved in representations relating to *where* objects are, as distinct from *what* the objects are. By contrast, recordings in the inferior temporal lobe (Fig. 6.1) suggest that in the brain's interpretation of visual inputs this area is vital to the identification of shapes and objects, to the identification of the *what* rather than the *where*, including the categorizing reactions involved in the discrimination between different shapes and objects (Kosslyn 1987). In line with this, Richmond *et al.* (1983) have shown that the effective stimulus in these areas is not the target of a movement, but the target of a visual discrimination. Again, bilateral lesions in the temporal areas have been found to produce failure to recognize shapes or objects met previously and to demonstrate or describe such objects. I shall return to these areas of the brain in Section 7.5.

Meanwhile it would appear that the *frontal* cortex is heavily implicated in the final synthesis of the brains' global world model and in the evaluation of that model in the light of the subject's needs. Here seem to be decided the subject's ultimate preferences and the action-strategies required for their fulfilment. More about this in Sections 6.4 and 6.5.

In view of their close connections with the frontal areas and the involvement of the latter in the construction of the brain's model of the world, it is not surprising that the *frontal eye-fields* (in area 8, Fig. 6.1) appear to be instrumental in organizing the *voluntary* exploration by the eyes of the visual scene, by contrast with the superior colliculus (see above), which appears to be mainly concerned in the *involuntary* fixation of targets. For we understand by 'voluntary' actions those actions that are determined by the brain's internal model of the world and the self-in-the-world.

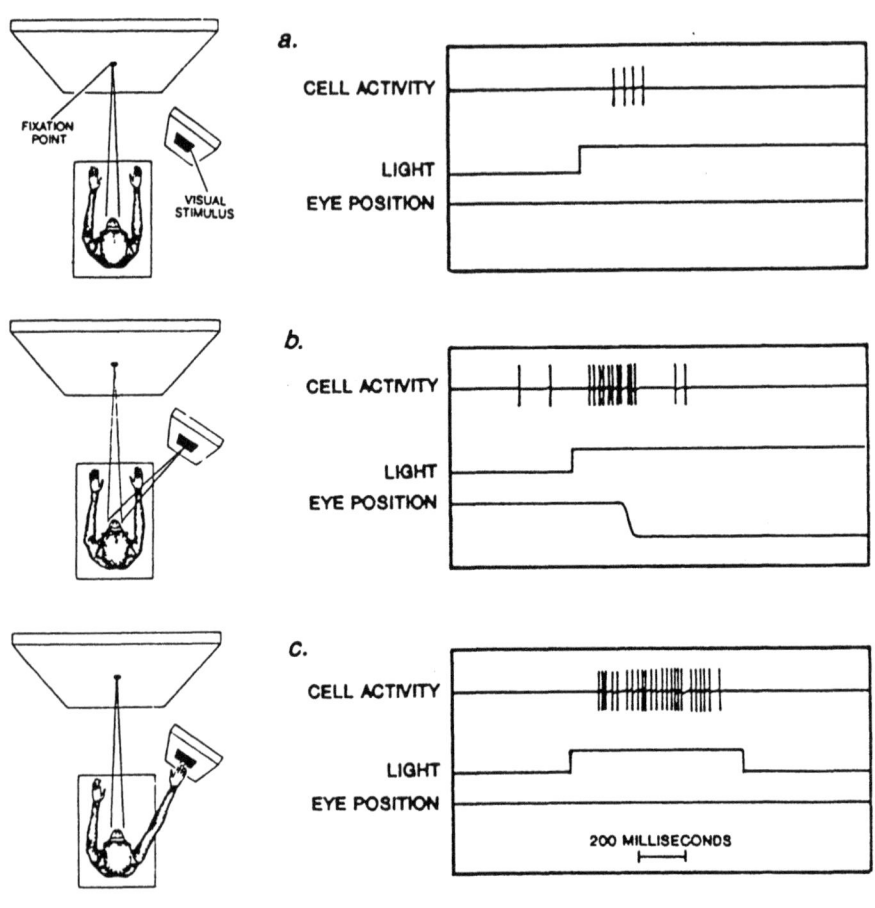

Figure 6.3 – *Enhanced cell responses in the posterior parietal cortex in monkey.*
By contrast with the superior colliculus, striate cortex and frontal eye-fields, the endhanced activity was here found when the monkey attended tø the spot of light regardless of how it attended to it. In a the monkey showed no reaction: a posterior parietal cell responded to the lightspot with a few action potentials. In b the spot was the target for a saccade: the cell's activity was enhanced In c the spot, was the targert of an arm movement and again the cell's activity was enhanced. – From Wurtz (1982), p. 104, by courtesy of Scientific American.

6.2. - The neural correlates of states of readiness

Crowne (1983) has listed evidence from unit responses which suggests that what he calls "model-comparator processes" are represented in afferent properties of neurons in the frontal eye-fields, and that these neurons may also participate in the preparation of motor responses. Thus one of the functions of this area appears to be to potentiate visual stimuli which have proved to be significant in the control of voluntary horizontal eye-movements, given the motivational state of the organism. This conclusion is also supported indirectly by studies which I cited in Section 5.4, and which demonstrated the involvement of the frontal as well as parietal areas in the *readiness sets* relating to difficult visuo-motor tasks.

Evidence from implanted microelectrodes also suggests that neural sensitivities in the frontal eye-fields are modified by the immediately prior history of their inputs. This may be part of the process through which what is anticipated now is shaped by what has gone before.

In Section 5.4 I have already mentioned readiness sets as examples of the brain preparing itself for coming stimuli or response requirements. In the same vein, Tanji and Evarts (1976) have demonstrated preparatory activity in the sensorimotor cortex relating to the direction of an intended movement. Monkeys were instructed to respond to a target stimulus demanding a specific movement but to wait for a specific cue. Recordings in the pre- and postcentral sensorimotor cortex (cf. areas 4,3,2, Fig. 6.1) then revealed instruction-induced changes of neuronal activity in two types of neurons. There were those which responded to target stimuli demanding one kind of movement rather than another, thus appearing to be correlated with the intention to move or response selection. They were quite specific. For example, the changes of neural activity observed might be specific to the direction of the intended movement. Others fired during the waiting period when a movement had been cued but the animal had to wait for a 'go' signal. This suggests that they were participating in a response preparation, possibly involving a presetting of spinal cord reflex excitability specific to the nature of the impending movement.

Assuming similar reactions in the human case, it is evident that they would lie below the level of consciousness. We are not consciously aware of preset changes in spinal motor reflexes. In our terms: the brain does not project these changes into its internal model of the current state of the self. By contrast, Roland (1978) has concluded from the effect of certain brain lesions that *conscious* feelings of the expected extent of a limb-movement and of the muscular force required may accompany descending motor signals from the cerebral cortex at the initiation of a voluntary contraction.

The examples I have given go to show that *getting ready* for some stimulus or some required activity is a universal process throughout the brain - although that readiness is always conditional on the context. Where the relevant contextual information is supplied by the visual inputs, it must, of course, be made available to the areas involved in the preparatory responses. This consideration sheds some light on the significance of recent discoveries about the multiple projection areas of visual stimuli. Up to a few decades ago it used to be thought that there were only two visual projection areas in the cortex, viz., the tip of the occipital lobe and the frontal eye fields. To-day the picture has changed dramatically and, indeed, very much in line with what the present theory would lead us to expect. More and more visual projection areas have been discovered in the cerebral cortex - as many as twelve in the monkey.

According to the general picture which has gradually emerged the primary visual cortex (the primary cortical recipient of visual stimuli, discharges certain selective functions in respect of primary optical elements such as the detection of contours and their orientation, and the integration of information coming from both eyes, while at the same time distributing the processed outputs to the prestriate areas (18 and 19) for further processing and analysis according to different criteria. These, in turn, will distribute outputs to areas further afield, such as the adjacent parietal and temporal areas. For even at the prestriate level the processing of the visual information is far from complete.

Indeed, it now seems that up to 60% of the neocortex may become involved in this way. It has also been found that most of these areas have extensive connections to central motor centres. Indeed, in all the cortical activities I have mentioned, the subcortical centres, too, remain intimately involved, as will be illustrated in Section 6.4. While some neurophysiologists still find all this mystifying, the Conditional Expectancy Hypothesis enables us to see all this in a new perspective. For it sees the internal representations formed as the end-products of perception as a vast network of learnt associations between possible actions and the consequent changes to be anticipated in the sensory inputs - and not as something based merely on sensory-sensory associations formed in uncommitted areas of the cortex, as was once thought to be the case. It is of interest to note in this connection that a number of neurophysiologists have now come to see the process of visual perception as a dynamic one, and one which largely rests on powers of the cortex to detect covariances. (Phillips *et al.*, 1984) - on our assumptions: covariances of the act-outcome type.

*

So much then for the neural correlates of the main component of states of conditional expectancy, viz. states of conditional readiness. Although at present we have mere glimpses of how these states of readiness come to be realized in the brain, the involvement of the brain as a total system, albeit with a regional distribution of particular categories of readiness, has become clearly apparent. We have also arrived at some idea of the direction in which future research needs to look if the picture is to gain in detail and clarity.

From states of *readiness* for an event we must now turn to states of *expectancy*, bearing in mind that, according to our definitions, the latter differ from the former in that they are open to revision if the anticipated events fail to occur. Next, therefore, we must turn to the way in which the brain detects discrepancies between its internal model of the world and the realities it comes up against, i.e. discrepancies between what is expected and what actually occurs. And we must consider how it reacts to such discrepancies.

***

6.3 - THE NEURAL CORRELATES OF STATES OF EXPECTANCY

Errors in the brain's internal model of the world reveal themselves in the form of violated expectancies. They reveal themselves also in failed behavioural responses, but we can treat this under the same heading. Violated expectancies initiate corrective activities in the brain and we have already seen in the orienting reactions the form in which this initiation occurs (Section 5.4).

According to our definitions, a state of expectancy is a state of readiness for an event which will become modified if that event fails to occur. Since we have now dealt with the neural correlates of states of readiness we must next turn to the neural correlates of the processes that elicit this modification. In other words, we must turn to the neural correlates of the brain's way of detecting a failed expectancy and of triggering the corresponding orienting reactions.

Let me begin with the fact that upon repetition of a novel stimulus the orienting reactions will cease. Sokolov (1969) attributed this to the brain now having updated its model of the world so that the discrepancies which triggered the orienting reaction have vanished.

In answer to the question how the brain detects discrepancies between model and reality, Sokolov (1969) postulated the existence of comparator neurons, or comparator networks of neurons, which compare stimuli with the brain's model and excite the orienting reactions if a mismatch is found. However, his investigations showed that during a continuously applied or repeated stimulus inhibitory potentials build up in specific neural structures. For example, he demonstrated the existence in the cortex of neurons in which this inhibitory effect can be observed as a depression of the neuron's spontaneous discharges. This lead him to the conclusion that this is a second channel through which orienting reactions are controlled - this particular influence being an inhibitory one. In other words, the mechanisms controlling the orienting reactions receive excitatory inputs from the comparator networks and inhibitory inputs along a more direct channel. He also produced evidence suggesting that both the formation of the model of the stimulus event and the discrepancy-detecting mechanisms reside in the cortex, but that transcortical connections play no part in this. And, in his view, the lower brain structures, such as the reticular formation, appeared to be involved merely in the execution of the orienting reactions.

The main point to be made in this connection is that on the Conditional Expectancy Hypothesis we can conceive of a considerable simplification in this theoretical scheme. Because on this hypothesis there is really no need to assume that the brain must contain comparator neurons or comparator networks specifically designed to detect a mismatch between model and reality. Such comparator units are difficult to conceive in detail and the need felt by believers in internal models to postulate the existence of such units has always been a source of difficulties.

A more economic theory would be to assume that the mere occurrence of an orienting reaction constitutes the required mismatch signal. In other words, that orienting reactions are the brain's original response to *any* stimulus-event, but they come to be dynamically inhibited upon repetition of a novel stimulus as an integral part of the general learning processes through which the brain upon repetition learns to meet a stimulus with the specific response its nature requires under the given circumstances if that response is to be a rewarding one.

Thus my answer would be: discrepancies between model and reality are not explicitly detected by special comparator mechanisms. Rather, they come to be responded to in an appropriate fashion simply by virtue of the fact that

a) an expected event is an event which is part of a familiar situation or familiar sequence of events;
b) the brain is predisposed to meet any stimulus-event with an orienting reaction, but as the brain becomes familiarized with the type of event (or sequence of events) in question, the orienting reactions come to be inhibited as part of the general processes of facilitation and inhibition through which the brain during familiarization comes to optimize its responses to the events concerned;
c) unexpected events, therefore, will elicit orienting reactions simply because no familiarization has as yet occurred;
d) the orienting reactions predispose the brain to acquire improved expectancies by way of attentional shifts which focus activity on the sensory area at which the existing expectancies have been violated.

According to these assumptions, then, a state of expectancy for an event E may be described as a state of readiness for E which is coupled with a conditional inhibition of the orienting reactions (the conditional here being the occurrence of E).

It should be mentioned in passing that habituation (familiarization) can be a very rapid process. A variety of animal experiments have revealed circumstances in which an analog of habituation is observed after just one trial. This comes under the heading of 'one-trial learning', a phenomenon that has brought much of Skinner's work into prominence.

If the above assumptions are correct, it follows that to understand the neurology of the error detecting devices for the brain's internal model of the world, one has to look at those of the brain's structures and processes that appear to govern the orienting reactions and their inhibition, especially the mechanisms that govern shifts of attention.

Some observations of incidental interest have been made by Kornheiser (1976). He investigated what happens when some experimental set-up (in this case fitting prisms to the eyes) causes the expectancies relating to one sensory modality to clash with those relating to another modality. Here the clash was between what the subject was expecting to see with what he was expecting to feel proprioceptively. It appears that in these cases that modality will with practice adapt, which is the lesser one normally used by the subject in his attempts to perform tasks of the required kind.

*

So much then for the moment about the way in which states of readiness and states of expectancy come to be realized in the neural networks of the brain as constituents of the brain's internal model of the outside world - and about the way in which discrepancies between model and reality manifest themselves and the required corrections come to be initiated. The picture is obviously still a very fragmentary one. Nevertheless, it suggests the general nature of the beast and points to the directions in which future research should look.

***

6.4 - INTENTIONS AND THEIR EXECUTION

Our discussion of the neural correlates of states of conditional expectancy, and of way in which the brain detects discrepancies between expectancies and reality, has given us some idea of the brickwork of which the brain's internal representation of the world is constructed. In this section I want to look at intentional behaviour, i.e. at the way in which the brain may operate when the individual sets up goals, and how it may organize their pursuit in the light of its internal model of the world. So far I have only touched upon this subject when I discussed how activities come to be selected according to expectancies of need satisfaction. But 'need satisfaction' is a very broad concept and we shall have to be more specific.

Let the desired outcome of some human action be called the *target situation* of that action. To function as a determinant of behaviour, this target situation has to be represented internally. In Section 4.5 I have distinguished between three categories of internal representations, and clearly the internal representation of a target situation belongs to the second of those categories, for it is the representation of a not-yet-existing situation, i.e. the representation of a *possibility* not *actuality*.

The internal representation of a target situation may or may not be a representation of which the subject is conscious. That depends on whether its occurrence is covered by what I have described as second-order self-awareness, i.e. whether its occurrence is registered in the brain's internal model of the current state of the self. If the internal representation of a target situation is a conscious one, I shall call it an *intention*. I feel that this comes closest to the use of that word in everyday life. However, I shall not confine myself to this conscious case and propose to consider nature of internal representations of target situations and how they enter into the organization of behaviour, regardless of whether they are conscious or unconscious representations.

The structure and neural correlates of the internal representations of target situations will be covered in some detail in Chapter 7 when we turn to the subject of the imagination and the structure (as well as neural correlates) of internal representations of category 2 (cf. Section 4.5). The suggestion I shall make there is that the representations occurring in mental imagery consist of mere states of conditional *readiness*, by contrast with the states of conditional *expectancy* that form the basis of representations of category 1. That is to say, they are distinguished from the latter by the absence of the processes of *reality testing* and control over the orienting reactions. In the present section, therefore, I shall say no more about this aspect of the matter and merely discuss the way in which internal representations of target situations come to be formed and how they enter into the general organization of purposeful behaviour.

Although the cerebral cortex is indispensable for the higher brain functions it would be a fallacy to think of it as enclosing the seat of those functions, and to expect it to be neatly divided into discrete modules serving them. Rather, the general picture to which one has to adjust one's thinking is that the brain as a whole is engaged in adapting the organism's responses to the demands of the world. In this process the subcortical structures effect, as it were, the coarse and tonic adjustments, while the outputs of the cortex supplement, articulate or modulate these in the manner of a fine and phasic adjustment and elaboration of detail. The same applies to the brain's internal representations of the world and the phasic- as well as tonic changes these representations have to undergo in the light of current sensory experience. However, as has been said, the brain's model of the world and of the self- in-the-world has to be a state of the brain which can accomodate a very high degree of variety and which must also be instantly modifiable. We can have little doubt that the main contributions to this high dimensionality and plasticity come from the cortex. In my LOGIC OF THE LIVING

6.4 - Intentions and their execution

BRAIN I have suggested which of those local intracortical networks that seem to repeat themselves massively throughout the cortex, would appear to be especially suited for this phasic plasticity. I also sketched an idealized model of the relevant networks, which I called the 'lambda system'. In his theory of cerebral learning, Eccles (1978) has suggested an alternative model for such localized networks, achieving broadly the same effects. But in the matter of phasic plasticity his conclusions agree with mine.

The interaction between the cortex and subcortical structures is illustrated by the two-way communication between different cortical areas and their associated nuclei in the *thalamus* (Fig. 6.2). Some of these nuclei act as relay stations of specific sensory afferents on the way to the primary cortical receiving areas, such as area 17 for the visual perceptions and areas 1,2,3 for the body perceptions. Others are actuated by diffuse and multisynaptic afferent pathways and/or inputs derived from within the thalamus. These nuclei project to correspondingly 'uncommitted' areas of the cortex. All nuclei in turn receive inputs from the cortical areas associated with them.

Thus the large neuronal mass of the thalamus is more than just a set of relay stations. Within its domain fibres are regrouped and there are many opportunities for the interaction of different modalities. Fibres cross all its nuclear boundaries. To a greater or lesser extent, therefore, these nuclei also appear to act as dedicated integrating stations.

Motor control by cortical areas, predominantly the so-called *motor cortex* (areas 4 and 6, Fig.6.2), operates mainly through the descending relay- and integrating nuclei of the *basal ganglia,* which lie adjacent to the thalamus and interact with it. These comprise the caudate nucleus and the putamen and globus pallidus (not shown). The scope of their integrating functions is indicated by the fact that they receive cortical projections from the frontal pole of the cortex, from the precentral motor areas, the parietal lobes, the temporal lobes and the occipital (visual) cortex (Kemp and Powell, 1970).

The basal ganglia, in turn, discharge to motor nuclei further down the system, such as the 'red nucleus', the substantia nigra and the all-important *reticular formation* (Fig.6.2). The latter comprises a mesh of neurons and embedded nuclei, which stretches from the brainstem up to the thalamic region, with massive convergent fibres from all sensory modalities. It is involved in various regulatory physiological activities, and in settling the 'tone' or 'general feeling' of any sensory stimulus. It also has tonic motor functions. In addition it has extensive projections to the higher cortical centres, through which it governs *inter alia* the state of general *arousal* of the brain. (Arousal, too, is a preparatory reaction, albeit a non-specific one. For, whereas *attention* is a potentiating function operating on specific categories of inputs, *arousal* denotes a general state of sensitization or mobilization which primes the brain to the fact that some task has to be undertaken, some problem to be faced.)

The motor system described above is known as the *extra- pyramidal* motor system, to distinguish it from the *pyramidal* motor system. This consists of a tract which originates in the motor areas of the cortex (mainly area 4) and, by-passing all these intermediate structures, exerts its influence on muscular movement by projecting directly to the spinal motor nuclei.

Three examples may help to illustrate the contribution of the cortex: if the connections are severed that relay traumatic stimuli from the thalamus to the somato-sensory region of the cortex (areas 1,2,3), pain is still felt, but the patient cannot locate it; if the so-called motor area of the cortex (area 4) is removed, there is permanent loss of fine movements, particularly of hand and foot, but the patient can still walk quite well; finally, if cortical incisions are made which isolate a specific sensory reception area from the neighbouring cortical areas, these do not seem to impair the accuracy of sensory discriminations in that modality, nor does the isolation of motor area (4) impair the fine control of voluntary movements which is held to be one of its normal functions. This last example suggests that the main integration between processes in different modalities occurs at the subcortical level.

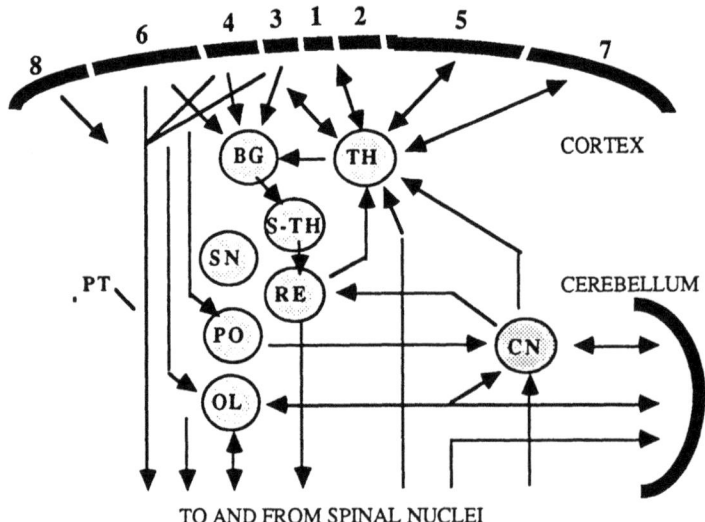

Figure 6.4 - *Some pathways and nuclei of pyramidal- and extrapyramidal motor system, illustrating scope for feedback at all levels.* **Legend:** BG, basal ganglia; CN, cerebellar nuclei; OL, olivary nucleus; PO, pontine nucleus; PT, pyramidal tract; RE, red nucleus; SN, substantia nigra; S-TH, subthalamic nucleus. Double-headed arrows symbolize information-flow in both directions, though not necessarily along parallel pathways.

6.4 - Intentions and their execution

An important point to note is that the resulting finely tuned responses of the cortex supplement, modulate, trigger, or inhibit, but do not replace the coarse responses of the subcortical structures. *Both together* determine the end result. For example, the subcortical motor nuclei may contain pairs of mutually opposing reflex systems, and the cortical output may decide the balance between them. Again, if the link between cortex and lower centres formed by the basal ganglia is damaged, compulsive behaviour patterns may result (Rapoport, 1989).

The integral activity of the whole motor system rests on a variety of feedback loops. Opportunities for such loops exist in the spinal motor nuclei and at all intermediate motor nuclei. One additional, and extremely important loop is the cortico-cerebellar loop. For one of the main functions of the cerebellum (Fig. 6.2) is to take over from the cortex in a semi-automatic fashion the correctly timed initiation of the sequences of movements required in the performance and progressive perfection of individual motor skills. The schematic presentation of Fig. 6.4 illustrates the scope for such feedback loops.

Next, a word must be said about the origin of the incentive to act at all, and about the way in which the brain sets up the goals of our actions - at least, how we might possibly conceive all this, for no clear picture has so far emerged in this respect. My own suggestions, therefore, are entirely speculative.

The question is how a desired situation - what I have called the *target situation* - may be represented in the brain, and subsequently result in action strategies that are appropriately related to the difference between that target situation and the actual situation prevailing at the time. For example, how the desire to go to London elicits some such strategy as: go to the telephone, order a taxi, take it to the station, buy a ticket, and catch a London train.

The roots of our motivational forces are generally held to lie in a set of structures known as the *limbic system* . This is closely linked to (and partly embraces) the main physiological drive centres of the brain, and projects extensively to the thalamus and to the cortex, both directly and indirectly (e.g. via the thalamus). These structures include the cingulate gyrus (which has projections to vast areas of the cortex), the hippocampus, hippocampal gyrus, septum and amygdala, all of which project to the dorsal nucleus of the thalamus, and the hypothalamus, which comprises some of the main nuclear masses controlling drive and emotional behaviour. Some of these are closely related to receptors in the reticular formation which monitor the body's metabolic and endocrine functions. Hence the lymbic system is generally held to be closely concerned in the elaboration of emotional behaviour and, at the cortical level, in the motivational evaluation of (as we must see it) the brain's current model of the world. These hypotheses have been supported, *inter alia*, by the discovery of two types of zones in the limbic system which have been dubbed 'pleasure centres' or 'go centres', and 'punishment centres' or 'no go' centres, respectively. They appear to play a key role in the networks that cause rewards and punishments to become effective. For, in 1954 Olds and Milner, working with rats, discovered that there are regions in the limbic system of the rat where the application of a small electrical stimulus would act as reinforcer to any activity to which this stimulus is added. If the stimulus is applied to such a 'pleasure centre', whenever the rat goes to a particular corner in its box, the animal appears to experience a pleasurable sensation, and soon develops a strong tendency to return to that corner for more of the same. If the rat is given the opportunity to produce itself this pleasurable electrical stimulus by pressing a lever, it would take to pressing this lever several thousand times an hour without any sign of satiation. Even a starved rat would prefer this self-stimulation to any food that might be on offer. In the same year, Delgado, Roberts and Miller (1954) discovered other zones in the brain which acted in the opposite way: the 'punishment centres'. Electrical stimulation in any of these zones produced aversive reactions and manifestations of fright, frustration or pain. Similar zones have been found in the human brain, where stimulation was found to produce feelings of comfort and discomfort respectively, though not with the compulsive consequences observed in rats.

To arrive at some idea of the form in which, under influence of the limbic system, the brain sets up complex goals for our activities, I propose to begin by looking again at the paradigm of a servomechanism based on negative feedback control, as illustrated in Fig. 3.1. Typically a servomechanism maintains some situational variable at a value determined by the command signal, by comparing its actual value with the commanded value, setting up appropriate errors signals and feeding these to the motor output system.

I want to borrow two notions from this paradigm: that of the command signal as a reference signal which determines the goal of the activity, and the notion of an intermediate output which depends on a comparison of that reference signal with a representation of the current actual state of the situation to be controlled.

In the organization of purposeful behaviour the internal representation of a target situation clearly take the place of what is here called a 'reference' signal. In the following paragraphs therefore, I shall for brevity call such a representation the *reference state* of the ensuing action.

In most human behaviour purposeful activity tends to consist of a sequence of actions, each with its own feedback loops, e.g. the sequence of steps I would follow if I wanted to go to London, beginning with a phone call to the taxi rank. In general, therefore, the ultimate goal of an action generates a sequence of sub-goals, and the resulting action follows a pattern in which the realization of each sub-goal triggers an activity directed at the next sub-goal in the sequence. In our terms: the ultimate reference state (ultimately desired state of affairs) of a behaviour sequence generates a set of *subreference* states in the brain.

One can logically conceive of this process as one in which the brain seeks representations of a sequence of intermediate situations which are such that (a) the act-outcome expectancies attached to each member of the sequence include among possible

6.4 - Intentions and their execution

outcomes the situation which is next in line, and (b) the act- outcome expectancies of the last member in the sequence including expectancies of need satisfaction. Such a sequence may be called the internal representation of an action *strategy*. The act-outcome expectancies mentioned will in the main be the product of previous act-outcome experiences, although, of course, they may also be based on other sources of information.

If that same ultimate goal is repeatedly pursued from the same initial situation, the relevant act-outcome expectancies will tend to advance along the line, and thus the sequence will shorten. For example, in a sequence A,B,C of subreference states B will drop out as soon as C comes to figure already among the act-outcome expectancies attached to A. If B is retained at all, it is likely to be merely as a check on ongoing activities. Thus if the goal is a novel situation, the search I mentioned above will be a genuine search, whereas in the pursuit of a familiar goal the brain will come up with the right movement commands more or less straight away.

The idealized model I have sketched above will suffice to show in general terms what types of representations and linking processes must be expected to participate in the organization of purposeful behaviour. The following is a brief summary of the key processes, such as I have sketched them:-

1. It all begins after birth with what-leads-to-what experiences, mainly of the act-outcome type. These result in what-leads-to-what expectancies which expand with further experience and gradually amount to an internal model of the world and the self-in-the-world (beginning with proximate space and gradually spreading beyond).

2. States of expectancy for an event E are states of readiness for E, where the readiness includes inter alia a state of conditional inhibition of the orienting reactions, the condition being the occurrence of E. Should E fail to occur, the now still operative orienting reactions amount in effect to error detectors, i.e. detectors of discrepancies between the brain's internal model and the reality encountered. At the same time these reactions initiate the appropriate responses to such discrepancies (e.g. arousal and shifts of attention). I have called these processes the *reality testing* processes.

3. Voluntary actions are defined as actions which flow from the brain's internal model of the world and the self-in-the- world. This flow is directed by the act-outcome expectancies, viz., expectancies of outcomes which in turn elicit expectancies of need satisfaction - where the latter will be governed by the current motivational state of the organism as determined by the limbic system and its need detectors.

4. The brain also has a power to form internal representations of as-yet-non-existing situations (representations of category 2). These are acts of the imagination. If any such

imagined situation carries expectancies of need satisfaction it may become a guide to voluntary activity. That is to say, it may act as an internal *reference state* in the sense that the resulting activity will be based on the discrepancy between the existing environmental situation and the desired one. This may be likened to the *command-* or *reference* input to a servo-mechanism.

5. If the act-outcome expectancies attached to the existing environmental situation do not cover the desired situation, the brain may (prior to any action) search for an internal representation of a sequence of intermediate situations which is such that the act-outcome expectancies attached to each member cover the situation which is next in line, while the last member of the sequence consists of the ultimately desired situation. This set of sequenced subreference states defines the action's strategy.

6. An appropriate sequence of subreference states having been found, the action is initiated. Each member of the sequence assumes control as soon as the preceding member of the sequence has been reached and a new set of conditional expectancies thus comes into force. I have suggested that, if the internal representation of the final target situation of any act is a *conscious* one, it corresponds to what in ordinary discourse we would call the *intention* of the act.

7. What has here been depicted as a step by step process in the genesis of a purposeful behaviour sequence, will with repetition and practice tend to become one single flowing movement through various degrees of *automation*, e.g. the early anticipation at any stage of what is required for the next (the familiar advance of response components under repetition). Automation is especially prevalent in the execution of the movements of body and limbs that routinely occur in any normal action sequence, such as moving an arm to some predetermined position. According to Bullock and Grossberg (1988) such a movement is governed by two independent commands: a target positioning command specifying where the arm has to move and a velocity determining command. Automatic processes then convert this information into an arm trajectory and a velocity profile whose global shape has proved to be remarkably invariant.

*

As I have said, the above is an idealized account, and as such it is bound to be an oversimplification. Its main merit lies in the fact that it suggests some idea of what is going on in the genesis and execution of intentional behaviour and that, unlike other theories, it does so exclusively in terms which have been carefully defined in a physical language, a precondition for progress in the search for the neural correlates of the processes concerned. As to the areas of the cortex specifically implicated in these processes, the evidence points to the frontal cortex (cf. Section 6.5).

***

6.5 - THE SEAT OF CONSCIOUSNESS AND THE EFFECT OF LESIONS

On the present theory the question of the seat of consciousness revolves around the question of the brain locations involved in the brain's internal model of the world and of the self-in-the-world, since I have suggested this to be the basis of consciousness. For reasons already explained we have to look here especially at the cortical contributions to this model.

Although the neurosciences have so far uncovered no more than a tiny fraction of the relevant facts, there are a number of pointers worth mentioning, and I shall confine myself here mainly to the brain's interpretation of its visual inputs. Even so, the material is so complex and the sources are so varied, that in a short space I can attempt no more than cite some of the main data and sketch the interpretations they suggest. The relevant evidence comes mainly from the observed effects of cortical lesions and of electric stimuli applied briefly to selected cortical loci. Unfortunately, such evidence is rarely unequivocal, since it is virtually impossible to determine either the exact extent of a lesion, or the ultimate destination of electrical impulses traveling along the axons of the stimulated neurons. Moreover, the functional development of the cortex can show significant variations from individual to individual.

According to our analysis (see especially Section 5.5), the brain's internal model of the world and its application rest on the following groups of reactions or processes :-

1. *Categorizing reactions,* i.e. discriminations amounting to a partitioning of the world of stimulus objects into those that evoke the reaction and those that do not. These reactions may be purely internal and more or less plastic, i.e. modifiable by experience. At the lowest levels they may be produced by way of wired-in neural analyzers which are dedicated, for example, to the detection in the visual field of contrasts and contours, of their orientation, direction of movement etc..

2. *What-leads-to-what expectancies,* mainly of the act-outcome kind, which are based on past what-leads-to-what experiences. These become attached to categorizing reactions in the following sense. The organism will learn from experience that a stimulus object eliciting a categorizing reaction A will produce results K,L, or M, if acted on by actions R,S, or T respectively. In due course, then, the reaction A automatically comes to elicit the conditional expectancies K -> R, L -> S and M -> T. This expanded reaction then amounts to a new categorizing (partitioning) reaction which will be representative of properties of the stimulus object not previously covered. We may also say that A has now acquired a new and deeper meaning or significance for the organism.

3. As the result of further learning, still higher levels of categorizing reactions may be formed along analogous lines as the organism learns that particular sets or combinations of the categorizing reactions elicited by a stimulus object, will give certain results if acted on in certain ways. These, then, will add still further to the properties or features of the outside world of which the brain now has an internal representation.

4. The resulting internal representations of objects, their properties and mutual relationships, of the body and of the current state of the self jointly amount to a coherent and comprehensive internal representation of the world and the self-in-the-world.

5. The occurrence of overt activities which are based on this global internal model. We identify these with the so-called *voluntary* activities.

6. The formation of action *strategies*, leading to action *sequences*, in which some distant goal is pursued in a stepwise fashion, organized in the manner I have suggested in Section 6.4.

*

It follows that the deleterious effects of brain lesions should be interpretable in terms of their interference with

i) the acquired categorizing reactions, e.g. by way of depriving them of the inputs which they would have partitioned in the undamaged brain;

ii) the what-leads-to what expectancies which such categorizing reactions have come to elicit in consequence of past what-leads-to-what experiences, and which thus give rise to categorizing reactions of a higher order;

iii) the still higher levels of categorizing reactions mentioned under (3) above.

iv) the ensemble of categorizing reactions which cause an object to be fully perceived and fitted into the brain's global world model;

v) the behavioural responses that have been learnt as appropriate reactions to the brain's internal representation of the world and the organism's current needs;

vi) the power to form action strategies in the manner I have tentatively described in Section 6.4.

6.5 - The seat of consciousness and the effect of lesions

In addition to these main categories of interference, one other may be mentioned here:

vii) interference with the dynamic inhibition of the orienting reactions when an occurrence is an expected one (cf. Section 6.3).

This does not, of course, exhaust the list of possible effects. There are other important ones, for example, interference with the conscious recall of past events and interference with speech functions. I shall consider these when we reach those particular functions of the brain in Chapters 7 and 8.

Broadly speaking, we must expect (i) to produce loss of discriminating powers - perhaps even loss of the power to discriminate important environmental variables, e.g. loss of movement detection in the environment, as reported by Zihl et al (1983).

Defects (ii) and (iii) must be expected to produce loss of meaning at different levels of what is discriminated, hence to produce agnosias of one kind or another. Defects (iv) may produce failure to add external stimulus objects to the brain's global world model, even though the sensory stimuli they produce may still have effects below the level of consciousness. An example is the phenomenon of blindsight: patients blinded by severe lesions of the visual cortex may still prove able to point at some bright external object, though they are unable to perceive it consciously.

From defects (v) we must expect loss of previously learnt responses to perceived situations, while defects (vi) will produce failed tests on familiarity: objects will appear strange when they should be familiar or vice versa..

At the same time, though, one has to realize how very difficult is bound to be the design of experiments that can separate out the categories of interference I have listed. How, for example, is one to discover whether a particular lesion has prevented a particular internal categorizing reaction from being formed, or has merely prevented it from exercizing any manifest influence? And if it failed to be formed, which of the required confluent channels of information for the conditional expectancies concerned might have been the impaired one?

*

In the visual modality, we must look for interference with the lowest levels of categorizing reactions at the cortical level to lesions in the *striate cortex* (area 17, Fig. 6.1), whose analysing and distributive functions have already been mentioned. In monkey, removal of striate cortex abolishes the capacity for perceiving spatially organized visual inputs and leaves only an abnormal capacity to respond to total luminous flux and to such 'textural' features as speckledness.

Area 17 projects to the surrounding *parastriate* areas (areas 18 and 19). These show responses not unlike those of area 17, but with a relative preponderance of neurons selectively responsive to the more complex rather than simpler features of lines and contours. These areas also seem to be involved in the perception of distance. Thus in monkey cells have been found in area 18 which responded only when the stimulus object was a certain distance. Possibly this may rest on binocular parallax, since many cells in the visual cortex receive inputs from both eyes and are therefore able to detect discrepancies in the retinal location of a nearby stimulus object before the eyes have been made to converge on that object (but see also below).

All these areas are involved in the control of eye movements, and in view of our hypotheses about the importance of act- outcome expectancies, it is relevant to note that they all receive movement information. Thus Noda *et al.* (1972) have reported cells in area 17 which respond only during saccadic eye movements, and areas 18 and 19 are known to receive projections from the frontal eye fields (adjacent to area 8), areas which appear to be closely involved in *voluntary* eye movements.

In man, lesions in areas 18 or 19 may produce severe impairment of object recognition. The patient may indeed see the object, but attach no meaning to it: he may walk around a ladder, thus showing that he sees it, but be unable to recognize it for what it is. Clearly we have here a loss in the act-outcome expectancies attached to the sight of the object, viz. those that define the potential use of a ladder. However, it is not to be inferred that these particular conditional expectancies themselves are formed in the areas now under consideration. For, the informed opinion seems to be that areas 17,18 and 19 supply only cues to the interpretation of visual inputs, e.g. in the inferior temporal lobe (see below).

Stimulation in these areas may produce sensations like flashes, stars, stripes and wheels, but not complete images (by contrast with stimulation of the inferior temporal lobe to which these areas project amongst others - see below).

It is interesting to note in this connection that when a lesion blinds the centre of the visual field, and a figure is presented to the eyes whose outlines lie entirely outside the blind area, say a square or circle, that shape can still be recognized without difficulty. This refutes the theory that shape recognition depends on neural activities within the boundaries of the shape, while supporting the idea that it is based on contours and on the spatial relation of salient points of the contour (relations which we take to be registered in terms of act-outcome expectancies in respect of the eye movements they invite).

The role of the *parietal lobes* (areas 5 and 7) in the formation of a coherent representation of the topography of the body and of posture and movement has already been mentioned (Section 5.4), as have some telling effects of major lesions in this area. Proprioceptive information plays a large part in this body knowledge and the whole area seems to stand under its influence. Thus it suffices for a lesion to sever the proprioceptive afferents from an arm to cause that arm to be experienced by the patient as a foreign object.

In addition, area 7 contains neurons showing responses correlated to eye position and the fixation of target objects, such as extraretinal signals relating to direction of gaze, convergence and accommodation. as well as signals relating to the target object as such and its surroundings (Sakata *et al*, 1980).

Since the body forms the frame of reference upon which our knowledge of nearby space is based, it is plausible to assume that these areas also participate in forming representations of the spatial location of nearby targets in relation to the body - representations of the *where* rather than the *what* (Section 5.4). Patients with bilateral parieto-occipital lesions tend to have difficulties knowing where an object is without any difficulty identifying the object for what it is (DeRenzi, 1982). According to Mountcastle *et al* (1984) these areas also contain command structures relating to actions in nearby space. It follows that if parietal regions are damaged the patient may lose representations which are vital in the execution of his intended action strategies: he may know what to do but cannot mobilize the kinetic formula needed to do it.

Thus tasks relating to nearby space are frequently impaired by parietal lesions. By contrast, actions relating to distant space tend to be impaired by lesions in the frontal (premotor) cortex. These may leave intact tasks relating to nearby space (Rizzolatti *et al*, 1983). This makes sense in our theory in view of the fact that the brain has various kinds of rather direct act-outcome expectancies available for the internal representation of nearby space which are not available for the representation of distant space (e.g. expectancies relating to ocular convergence and to what can be reached, touched or manipulated. It is also worth reflecting here that the infant's first knowledge of nearby space comes from the movements it has to execute to reach nearby targets, and that knowledge of these movements must, therefore, guide the development of its interpretation of the visual inputs. It also makes sense that lesions in the posterior parietal regions may interfere with shape recognition and with the recognition of the weight, consistency, and structure of an object. In some lesions these particular faculties may be spared, but the integration of relevant perceptions is impaired: the patient may feel a key placed on his hand, he may feel it to be heavy and cold; but he cannot bring these and other facts together and arrive at a concept of a key in the absence of visual stimuli. However, on the evidence I shall cite below, it may well be that this failure results from a failure to supply certain frontal regions with the cues they need.

Whereas parietal lesions tend to result mainly in tactile rather than visual deficits, the opposite is the case in the inferotemporal lobe. Various pointers suggest that the awareness of body and spatial relations achieved with the aid of the parietal cortex comes to be supplemented in the *temporal lobes* by categorizing reactions and associated what-leads-to-what expectancies which complete the brain's representation of external stimulus objects as entities with discernible solid shapes, and other features or properties, as well as spatial relationships. Hence the infero-temporal cortex is often dubbed the *interpretative cortex*. By contrast with the parietal areas it deals with the *what* rather than the *where*. In connection with what I have to say in Section 7.3 about mental imagery, it is also of interest to note that in this domain, too, a *what-where* dissociation may occur. Thus Levine *et al* (1985) have reported patients with parietal lesions who could imagine shape but not location and, conversely, patients with temporal lesions who could imagine location but not shape.

In the absence of parietal lesions, complex visual discriminations, especially learnt discriminations, appear to be mainly dependent on the temporal areas. That these areas may be the site at which the conditional expectancies are formed on which the discrimination of patterns depends, is suggested by the fact that lesions here may impair the postoperative retention and learning of such discriminations. Although lesions may not impair simple visual tasks, like tracking a moving object with the eyes, they do tend to impair visual processing abilities as a whole. Not surprisingly patients may also fail on tests for familiarity. A variety of agnosias may result. For example, a patient may recognize his wife but deny that he is married. Memory may be impaired, too - as may be the ability to judge the temporal order of experienced events.

It is in these areas, too, that cells are found which respond maximally only in the presence of some particular, and particularly significant, shape in the perceptual field - for example, to the perception of a familiar face. The latter category of responses has been extensively studied in recent years. Cells in the upper bank of the superior temporal sulcus have been identified which could acquire a remarkably specific response to the faces of particular individuals. Perception of the face concerned could double or treble the discharge rate which the cells showed from other causes in the absence of that stimulus object. Moreover, this response might occur regardless of such variables as the position of the face in the cell's receptive field, the distance of the face (hence size of the retinal image, the orientation of the face (upright, horizontal) or changes in the strength, colour or intensity of illumination. Other cells have proved to be responsive to facial expressions rather than facial identity. (Perrett *et al*, 1987). All of which suggests that we are here dealing with cells under the predominant influence of categorizing reactions of a very high level of sophistication, involving contributions from numerous categorizing reactions of a lesser order and their associated sets of conditional expectancies. And that some cells are maximally involved with just one of the tested sets of categorizing reactions.

The importance of both the temporal cortex and the parietal cortex in the construction of a stable scene from the images that dart across the retina, is suggested by the fact that stimulation in either area may produce feelings of dizziness. However, other cortical areas also seem to be involved. Lesions in the frontal cortex, for example, may

6.5 - The seat of consciousness and the effect of lesions

result in a failure to distinguish between changes in sensory inputs due to the subject's own self-induced movements and changes due to actual movements in the environment.

Stimulation in the temporal areas may result in a variety of interpretative illusions: objects may appear larger than they are or to recede when they are not. Objects may also seem either familiar when they should not, or strange when they should not. Clearly we have here an interference with the mechanisms controlling the appropriate inhibition of orienting reactions.

Stimulation may also result in a general spatial disorientation. This suggests that the stimulation activates categorizing reactions and their attached expectancies out of their proper context. Psychic response may also occur, such as hallucinations (a breakdown of reality testing?). Also an uninvited recall of some past visual (or auditory) experience. According to the observations of Penfield (1966), the experiences recalled are always visual or auditory experiences attended to at the time of their original occurrence, and the recall ceases as soon as the electrical stimulus is terminated. He never found recall of emotional experiences, nor of pain or body-functions or introspective self-awareness.

However, it has to be pointed out that the two hemispheres do not contribute equally to the particular brain functions I have mentioned above, since a large portion of the posterior temporal lobe of one hemisphere (called the *dominant* one) is usurped by speech functions - mainly the areas lying behind the auditory reception area (40). Hence on the dominant side (usually the left) only the frontal pole and part of the inferior temporal lobe are left for the interpretation of visual inputs, whereas the whole of the inferior temporal lobe (areas 20 and 21, Fig. 6.2) is available on the non-dominant side. Generally speaking, disorders in visual perception, e.g. faulty perceptual categorizations, tend to be mainly associated with lesions in the right (non-dominant) hemisphere. By contrast, lesions in the left hemisphere may impair the ability to name or describe an object or event without impairing the ability to draw that object or mime that event.

For the part which the cortex plays in the evaluation of the brain's model of the world in accordance with the subject's motivational state, we must turn to the *frontal areas*, as already mentioned in Section 6.4. Their close connection with the limbic system allows these areas to function as the sites at which needs and motivations are articulated and the brain's internal representation of the current situation is evaluated in the light of those needs. Here, then, seems to occur the differentiation of discrete preferences, the attribution of subjective significance to perceived objects or events and a corresponding degree of control over attention. This is where the goals are staked out for the subject's voluntary activities and the strategies for their attainment (cf. Section 6.4). Thus a patient with frontal lesions may rigidly cling to old behaviour patterns, unable to devise new patterns in response to a novel situation. Control over parietal areas may also be impaired: the patient may feel compelled to grasp or use objects presented in nearby space. Again, patients with frontal lesions may show a severe loss in the power to anticipate; they may fail to structure their intentions and work out appropriate strategies. For example, they may be seen to strike a match against the cigarette box. In the monkey, lesions in these areas may result in the disruption of sequences of actions which depend on each stage being released by the successful completion of the preceding stage (Pribram 1960). The damaged animal also appears to be unconcerned about the positive or negative outcome of its actions. There is even indifference to injury: a damaged animal may repeatedly grasp a flaming match, for example. It also seems that the frontal lobe of the dominant hemisphere is most closely involved in the programming of voluntary actions, while that of the other hemisphere is more closely involved in monitoring the temporal sequence of externally ordered events.

Although decrements in the ability to form mental images are mainly associated with lesions in the parietal, occipital and temporal areas (Farah 1984), it is not surprising that a frontal lesion may also have an effect if the images relate to action sequences. Thus a patient with frontal lesions may be unable to demonstrate the use of a hammer without holding a hammer in his hands; or he may be able to drink out of a full glass but not out of an empty one.

As is also to be expected in the light of our analysis, the frontal areas exercise considerable control over the organism's orienting reactions. In monkey, resection of the frontal pole can abolish all physiological components of the orienting reactions, bar the changes observed in the surface potentials. Apart from the latter, only behavioural reactions appear to remain (Pribram 1971).

The frontal areas appear to exert their cortical effects on motor outputs mainly in conjunction with the adjacent *supplementary motor area* (8) and *premotor area* (6), which in turn borders on the *motor area* (4). As already mentioned, area 4 has (*inter alia*) direct control over mainly distal muscles via the pyramidal tract. This serves as a fine control which is superimposed on the framework of movement control established by the extrapyramidal motor system. Results of stimulation have suggested that it is movements rather than individual muscles which are represented here. Its role in voluntary movement (i.e. movement controlled by the subject's global world model) is illustrated by the fact that lesions in this area may disrupt the capacity to smile intentionally, but (so long as area 8 is intact) the patient may still smile unintentionally when he hears a joke. And Penfield has reported that, whereas stimulation of the pyramidal tract caused movements which are experienced by the subject as *involuntary* movements, stimulation in some areas of the sensorimotor cortex produced movements experienced as *willed*: the patient will say "I moved because I wanted to" - a good pointer to the fact that voluntary behaviour is based on the same brain levels as those that we take to be mainly responsible for the brain's detailed representation of the world and the self-in-the-world.

Blood flow studies have suggested that, in the absence of concomitant voluntary movements, the supplementary motor areas remain silent during purely somatosensory discriminations of shapes and objects. However, they appear to be heavily implicated in the programming of subroutines in sequential behaviour patterns, and in forming a queue

6.5 - The seat of consciousness and the effect of lesions

of time-ordered motor commands before voluntary movements are executed by way of the primary motor area (Roland et al, 1980).

The postcentral areas 1,2, and 3 are designated as the somesthetic areas, since they receive via their thalamic relays, extensive projections from the interoceptors for touch, temperature and pain, as well as proprioception. However they also project to subcortical motor centres. Conversely, sensory afferents are known to reach the precentral motor area. All this makes sense in view of our assumptions of the importance of sensorimotor associations in the formation of the brain's act-outcome expectancies. In view of these facts, and also because the morphological boundary between the two regions is not a strictly functional one, it has become common practice to speak of these areas jointly as the *sensorimotor* cortex.

By contrast with area 4, areas 6 and 8 project mainly to subcortical nuclei, e.g. basal ganglia, and, upon stimulation, show complex and more highly organized motor effects. Area 8 is credited with the organization of associative movements, i.e. the semi-automatic movement routines involved in the execution of familiar motor tasks. Some of these are independent of conscious control (see the example cited above), presumably because they have become automated through the action of the cerebellum, to which this area loops. Indeed, the cerebellum appears to play a major role in this process of automation, thus leaving to the cortical areas merely the task of setting in train sequences of steps whose members have become linked elsewhere in the brain.

That area 6 plays an important part in the realization of intentions is illustrated by the observation that lesions here may cause the patient to behave as if he performed a complex movement for the first time. But acquired kinetic formulae remain intact: as in area 4, they can still be mobilized in an automatic context.

A revealing light on the differential roles which the motor areas play, has been shed by studies of cerebral blood flow. In simple motor tasks, like keeping a spring compressed between thumb and index finger, blood flow increased in area 4 but not in 6 or 8. But complex sequencing tasks, involving all digits, increased blood flow in these areas as well. On the other hand, a purely mental *rehearsal* of these tasks increased blood flow only in area 8. It seems then that this area plays a particularly significant role in the elaboration and/or sampling of intentions in this kind of task, the final outputs being decided by the act-outcome expectancies which each sampled intention elicits. This is the search process I discussed in Section 6.4

All these reflections bear on the question of the seat of consciousness - the very heading I used for this section. They show that consciousness has no seat in the brain in the sense of a specific site. The brain's internal model of the world and of the self-in-the-world, and its application in the control of behaviour in accordance with the subject's needs, implicates virtually the whole of the brain. Thus we must see in consciousness an operational mode of the brain in which the cortex plays a crucial but not an exhaustive role. Subcortical centres can exercise a critical control over this operational mode, causing it to be lost or to recover as the case may be. The sleep controlling centres known to exist in the reticular formation are a case in point. Again, localized lesions in the thalamus may result in sensory disturbances, disturbances of the body schema, agnosias of various kinds, affective and psychotic disorders and other disturbances of consciousness (Wallesh et.al, 1983).

***

6.6 - CONFLICTING BELIEFS IN BRAIN RESEARCH

Among those who are directly involved in the study of the brain and its anatomy or physiology, we find contrasting views about the relevance of their work to the problem of the nature of consciousness and mental events generally. It is worth looking at some of these divisions in the light of our conclusions.

For example, most neuroscientists believe, as I do, that mental events have physical correlates in the brain, in the sense that the physical events in the brain are held to be a sufficient condition for mental events, and not merely a necessary condition. But there are others, though only a minority, who believe that the mind is a unique and non- material entity which is required in addition to the physical apparatus of the brain to account for our mental experiences. On this view, therefore, the translation of mental into non-mental terms (as I am carrying out) cannot succeed. The best way to refute them, of course, is to show that it can succeed.

As regards the question of where this elusive entity is supposed to interact with the physical brain, these scientists differ from the most famous proponent of this dualist theory, Descartes, who saw the locus of this interaction in the pineal gland (which he singled out on the grounds that he believed it to be the highest *unitary* structure below the cerebral hemispheres). The distinguished neurophysiologist Eccles (1977), for example, subscribes to a dualist theory of mind and matter, which postulates the existence in the brain of some entity which he calls the 'Self-conscious Mind'. This entity is supposed to draw information from the brain, to digest it in ways that have no physical correlates, and then to direct brain-events in accordance with the results of its deliberations by effecting small deviations in the physiological processes that occur in the synaptic clefts of the nerve cells - the brain's final outputs then being controlled through the bulk effect of these small deviations.

The opponents of this view would claim (rightly, in my view) that by postulating such a ghostly entity one explains nothing. Such attempts merely create the semblance of an explanation by gathering together the unexplained phenomena, e.g. the influence of thought on action, and then invent a convenient entity - here the 'Self-conscious Mind' - whose properties are defined in terms of some magical power to produce these effects.

6.6 - Conflicting beliefs in brain research.

One might as well explain a rainbow by postulating the existence of a sun-energized 'rainbowizer' acting on the raindrops. Such hypotheses evade, but do not solve, the real problem. It is ironic that in the same volume in which (much to the surprise of some of his admirers) Karl Popper supports Eccles in this dualistic stance he should in a different context have quoted Beloff's remark that "a doctrine which sustains itself only by elaborate evasions is little better than humbug".

Of course, scientists themselves frequently postulate the existence of some new entity to account for some observed and hitherto unexplained effects. 'Black holes' are a good example. But they would not claim to have explained those effects in terms of that suggested entity until they had succeeded in elaborating their idea to the point at which it has become part of a single coherent theory embracing both causes and effects, while at the same time satisfying the stringent criteria, semantic as well as methodological, by which the natural sciences judge the respectability of their theories.

However, this dualist view is not the only way in which the real problems are sometimes evaded. I am thinking in particular of such simplistic and reductionist, but not uncommon, formulae as those that equate consciousness and the 'inner life' with total brain activity, self-awareness with receptivity to impulses originating in the body, or the 'self' with the totality of neural activities since birth - all of which may even to-day be found in the publications of distinguished neuroscientists.

The majority of neuroscientists fall into neither of these traps. But they can still hold very different views about the relevance of contemporary work in neuroanatomy and neurophysiology to our understanding of the higher brain functions. There are those who would claim that in modern times the neurosciences have made great strides towards an understanding of the physical counterpart of mental events: and in support of this claim they would cite the impressive volume of facts which have come to light about such things as the mental effects of particular drugs or drug deficiencies, the composition of neurotransmitters, the effects of hormones - not to mention the mental effects of brain lesions and electrical stimulations I have briefly surveyed in the last section.

These claims are criticized (and again, I think, rightly) on the grounds that, however valuable these discoveries may be in their own right, they often reveal no more than some *necessary condition* for healthy operation of the brain mechanisms concerned. They do not reveal the way the mechanisms operate. To take a familiar analogy: the discovery that your TV set will fail if a particular resistor is removed, does indeed reveal a necessary condition for the set's proper operation; but it is unlikely to bring you any closer to proper understanding of how the set works. Thus the outcome of many of the experiments I have mentioned tells us no more than that the physical counterpart of mental events is of a kind that is *affected* by the kind of chemical, electrical or anatomical changes in question - and in some cases critically so.

Not unrelated to these considerations are the two contrasting views which people may hold about the best way to approach the problem in hand: the *bottom-up* approach and the *top-down* or *systems* approach. Neurobiologists of the first school believe that in studying the physiology and anatomical connections of the nerve cells of the brain, one is in effect studying all that needs to be studied to arrive eventually at a complete understanding of the brain's mental functions. By contrast, the opposition believes, as I do, that in all conscious activities and mental operations the brain is involved as an integral dynamic whole, whose mode of operation can only be understood if one looks at it as an *integral system* whose properties and operational modes need to be analyzed, defined with accuracy and then explained in what amounts in effect to a *top-down* approach.

This is not to say, of course, that what is discovered at the level of individual neurons cannot in some respect affect our views about the brain as a system. It invariably will. The discoveries of Hubel and Wiesel about unit responses in the visual cortex are a typical example. And, of course, any prediction a top-down theory such as ours makes about the neural correlates of mental events needs to be tested against observed unit responses or categories thereof, as indeed I have tried to do. The two approaches have to go hand in hand. The sadness, though, is that those who are committed to the bottom-up approach often tend to shun the effort of rising to the more abstract levels of accurate thought and analysis demanded by a methodical top-down approach.

Conflicting beliefs are also found about the value for brain research of computer stimulations of the kind effected in the field of AI. I shall return to that topic in Chapter 9.

***

Chapter 7

SELF-AWARENESS, IMAGINATION AND MEMORY

RECAPITULATION AND SUMMARY

In the last chapter we have dealt with the physical counterpart of the brain's internal representation of the outer world. This forms one part of the brain's global world model which I have taken to be the basis of consciousness. We must now turn to the other part, the part that relates to the self. This, in turn, consists of two parts, defined in Chapter 4 as *first-order* and *second-order* self-awareness respectively. As will be recalled, the former consists of body-awareness, while the latter relates to currently occurring internal representations of the brain and amounts to an awareness that they are part of the current state of the self. This is the equivalent of what Locke described as "the perception of what passes in a man's own mind".

I shall also look at the often expressed fear that any assumption of the brain observing itself, as it were, must logically lead to an infinite regress.

To cover the main facets of our mental life, other kinds of internal representations have to be examined as well, foremost those occurring in the exercise of the imagination (Sections 7.2 and 7.3) and in episodic memory-recall (Sections 7.4 and 7.5). Each of these may have either a quasi-pictorial or verbal (propositional) form. One can imagine a story pictorially, i.e. in terms of visual images, or verbally: "once upon a time....". In this chapter I shall only deal with the quasi- pictorial form, leaving the verbal form to Chapter 8.

In the faculty of the imagination we are dealing with a capacity of the brain to form internal representations, not only of *actualities*, but also of mere *possibilities*. Once again we shall look at the requirements of dimensionality of this new class of representations, and we shall draw further clues from the fact that they are exempt from control by current sensory inputs. In particular, they are exempt from reality testing, and thus divorced from the brain-mechanisms controlling the orienting reactions. This will lead us to the conclusion that when one imagines an object, the brain enters certain states of *readiness* for particular sensory experiences, but without also entering corresponding states of *expectancy*. Hence no surprise and orienting reactions follow the non-occurrence of the sensory experiences concerned

This conclusion will also bear on our final topic, viz. the nature of memory in general and of episodic memory recall in particular: the recall of events experienced or witnessed in the past. We shall come to see this as a case of the brain entering a state akin to that entered when an event is imagined, except that this state has now come to be accepted by the brain as the representation of a past event - thus adding to the historic dimension of the brain's global world model. To get some insight into the processes involved here, I shall try to trace what happens to the internal representation of a current event as that event gradually recedes into the past.

***

7.1 - SELF-AWARENESS

I. AWARENESS OF THE BODY (First-order self-awareness)

By self-awareness I mean that part of the brain's internal model of the world and of the self-in-the-world which relates to the current state or condition of the self. As has been said above, this representation of the self can be divided into two parts. In the present section I shall consider the first of these, viz. representations which cover the body's topology, posture, movements, physiological needs, sites of injury etc.It is the physical equivalent of what may be called body-knowledge.

In speaking of our primary body-awareness as a first-order *self*-awareness I am using the term 'self' in the same sense as the one in which it occurs in such common phrases as *self*-regulation and *self*-reproduction. This sense differs from the philosophical sense, in which the 'self' is taken to denote the object of introspection, i.e. the self as represented in the brain by both first-order and second-order self-awareness.

From the standpoint of our analysis, one's primary awareness of the body presents no special problems. It is the so-called *body schema* we discussed in Section 5.4: the internal representation of the body's topology, current posture, movement, and certain major physiological variables. As we have seen, the parietal lobes of the cortex appear to be heavily involved in this process. These internal representations draw their information from two distinct sets of sources: one *exteroceptive*, the other *interoceptive*. The first comprises the visual and tactile perception of the body and limbs. It yields a representation of the body as a physical object situated in space among a multitude of other physical objects, and develops as an integral part of the internal model of the world which the brain forms on the basis of its peripheral sensory inputs and active exploration of the world. According to Piaget (1977) this internal representation achieves full articulation only in the second year of life as an awareness of individual existence. Until then the body and the rest of the physical world appear to be one. They separate only gradually as the infant's world model progressively gains in scope and articulation. Indeed, Piaget believes that the discovery of one's own body may play an important part in developing concepts of the external world and relations between objects.

The interoceptive sources of information are provided by such sensors as those placed in the muscles, tendons, joints and vestibular structures.

In the mature organism these two sources of information will be fully integrated. Yet there is evidence to suggest that the internal sensors play a dominant role in this partnership. For example, the reader will recall that if the proprioceptive afferents of an arm are severed, the arm will feel to the patient as if it were a foreign object even though it is still *seen* as part of the body.

It is noteworthy in this connection that the interoceptive impulses which contribute to the brain's internal model of its current body-posture and -movement, do not individually enter consciousness. They are not individually projected into the brain's internal model of the self as part of the current state of the self. Nor, therefore, are they evaluated at the level of consciousness. This is understandable, since no conscious action normally has to be taken in respect of these stimuli individually. Indeed, owing to the time required for such a projection, up to half a second according to Libet (1978), it would considerably slow our reactions if these individual inputs were to be evaluated at a conscious level.

Stimuli resulting from many kinds of injury and from physiological needs like hunger and thirst, on the other hand, do enter consciousness, and, on our assumptions, do so by way of being projected into the brain's global world model as being part of the current state of the self.

As to the physical nature of the brain's first-order representations of the self, we can make the same assumption as we have made for its representation of external objects, viz. that the current state of the body is represented by sets of conditional expectancies of the what-leads-to-what kind, based on past what-leads-to-what experiences, mainly of an act- outcome type. Examples would be: expectancies relating to what will be the interoceptive consequences of an intended movement; what change of arm position will be perceived visually if the arm is moved under proprioceptive control; what movement of the hands will bring them into touch; what movements will restore the balance of the body if it is disturbed, etc. Speaking generally, an acquired set of what- leads-to-what expectancies is really the brain's way of mapping a set of causal relationships. And the set of causal relationships which the brain maps in the first-order self- awareness is of the kind I have illustrated.

*

II. SELF-CONSCIOUSNESS (Second-order self-awareness)

Matters appear to become a little more complicated when we turn to the brain's power to form internal representations not only of what we perceive, but also of the fact that we perceive what we perceive: In the brain's global world model this representation adds the *I perceive it*, *I think it* and *I remember it* to our perceptions, thoughts and memories. This awareness of what passes in the mind, or, as we have called it, *second-order self-awareness*, is always with us, even though it may not be attended to. It finds expression in everyday sentences like "I am hungry", "I saw an accident", and "I think I took the wrong turning". Whatever I consciously experience is, as William James put it, experienced in a special way as belonging to myself. Thus the 'I' in "I think", "I feel", etc. denotes for the speaker the self as an entity which is the possessor of a body, the perceiver of the outside world and the haver of thoughts and feelings. As the reader will recall, this faculty of second-order self-awareness is also what we have decided to mean by *self-consciousness*, while by *introspection* we mean the process of attending to these internal representations. The latter forms the basis of what is commonly described as the 'inner' world of subjective experience, or the 'inner life'. For Kierkegaard and the Existentialists, but also for Blake among the poets, real truth lies only in this world of complete subjectivity.

How is the structure of these representations of category 2 (second-order self-awareness) to be conceived? On the face of it, one may feel that we are dealing here with an exceptionally sophisticated faculty of the brain, perhaps one that can only be expected to exist in the very highest orders of life. But I shall suggest otherwise.

To particularize: let A denote the internal representation of some object, event or situation, real or imaginary, past or present. Second-order self-awareness in this respect then amounts to a further representation, call it $B(A)$, which, in the brain's model of the world, represents the occurrence of A as being part of the current state of the self. According to our definition of internal representations, $B(A)$ will be a state or 'set' of the brain which, in an appropriate manner, adds the occurrence of A to the factors determining responses that need to be appropriately related to the current state of the self as a whole. ('Appropriateness' being here again understood in the technical sense defined in Chapter 3).

On the Conditional Expectancy Hypothesis A will consist of a set of what-leads-to-what expectancies which characterizes the nature of the object represented, and which have been formed in consequence of past what-leads-to-what experiences with similar objects. On the same hypothesis we can conceive of $B(A)$ as consisting of a set of what-leads-to-what expectancies which characterize the fact that A is part of the current state of the self. What kind of expectancies would these be? I suggest that whatever are the consequences for the organism of A being part of the current state of the self (rather than the state of some other object) can be mapped by a corresponding set of conditional

7.1 –Self-awareness

expectancies. And these can be acquired by experience in the normal manner. An example would be the expectancy that A is selectable by voluntary shifts of attention.

I can see no reason why this category of expectancies should in principle be more difficult for the brain to form than any other category of expectancies based on past experience. And if this view is correct, then it follows that our earlier decision to make second-order self-awareness one of the necessary conditions for consciousness in the proposed technical definition of consciousness, does not significantly prejudice the possibility of infra-human species having capacities satisfying that definition. I have mentioned this conclusion already in Chapter 1 in reply to those who feel that self-consciousness may be an exclusively human faculty.

Finally let me add that the account of self-awareness I have given should dispose of a not uncommon misconception, viz. that in self-awareness the brain models *itself*. Richard Dawkins, the Oxford zoologist, for example, believes that as the brain's simulation of the world becomes complete it is driven to model itself, and Johnson-Laird in Cambridge seems to share a similar view. Indeed it is difficult to see in what precise sense 'modelling itself' is here to be understood.

III. THE QUESTION OF THE INFINITE REGRESS.

Since second-order self-awareness includes representations of the fact that one has representations, it might at first seem that this idea of being aware of one's own awareness, so to speak, leads us into an infinite regress of a kind that might broadly be expressed as "I am aware that I am aware that I am aware........". However, it should be clear from our definition of second-order self-awareness that no such infinite regress follows.

What people have in mind when they fear that the notion of the brain observing itself may lead to an infinite regress, may be illustrated by pointing a video camera at a monitor displaying its own output (Fig. 7). With low ambient illumination and small aperture the effect is the same as looking at yourself in a hall of mirrors, in which you will see an infinite train of progressively diminishing images of yourself receding into the distance. In the video case you see a stack of diminishing monitor frames around a central bright and blurred area - Fig. 7.(a). On the other hand, if the aperture is opened up and a small source of light is introduced into the field (such as a burning match), the recurrent positive reinforcement of that initial stimulus creates a blindingly bright central patch of unpredictable shape and movement – Fig.7 (b).This is one of the many phenomena that have come to be studied by the modern mathematics of chaos. The effect illustrates the well-known power of positive feedback to undermine the stability of a system. Clearly, neither happens in humans. We experience neither a hall-of-mirrors-effect nor the latter kind of feedback-induced instability.

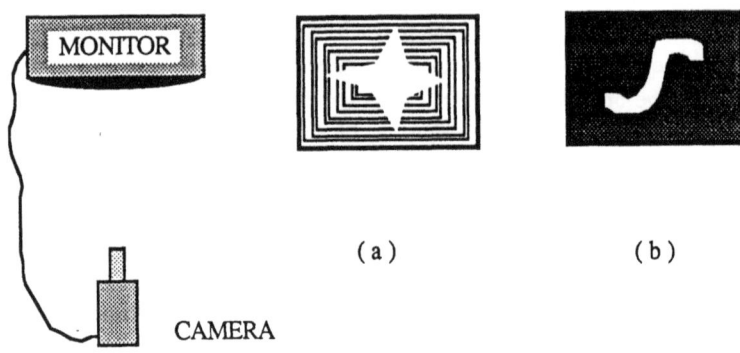

Figure 7 – *Instability through positive feedback.*

It is not uncommon for writers to attempt to avoid running into an infinite regress in their treatment of self-awareness, by postulating that the self of which we become aware comprises only our past thoughts, feelings, volitions etc. Rose (1973), for example, asserts that the "*I am thinking about me thinking about me... ..*" regress denotes a "*pattern of events ordered on a time based sequence*". In other words, each "*thinking*" in this sequence refers to a point in time just before the point in time to which the previous "thinking" refers.

I find this unacceptable. Awareness that a particular thought is part of the current state of the self *accompanies* a thought. It does not *follow* the thought. Only the attempt to cast it into words would follow the thought (provided the latter is temporarily fixed in the memory).

On our analysis there is no need at all to resort to such strategems, for there is no danger of running into infinite regress. On our assumptions, second-order self-awareness merely adds to a representation A of category 1 or 2, a representation $B(A)$ which represents the occurrence of A as being a part of the current state of the self. We would run into an infinite regress only if we were to assume that to every second-order $B(A)$ representation the brain would automatically add a third- order representation $C(B(A))$ representing the occurrence of $B(A)$ as being part of the current state of the self, and that to each such third-order representation it would automatically add an analogous fourth-order representation, and so on. But there is no need to assume any of this. Firstly, our theory does not require that assumption: we do not need it to explain what we want to explain. Secondly, I know of no empirical evidence that would force it.

***

7.2 - PROPERTIES OF THE IMAGINATION

The nature of mental imagery has occupied philosophers long before the birth of cognitive psychology. Hume, for example, believed mental imagery to be a top-down activation of perceptual representations, and this is still held to be true to-day by some psychologists. Others, by contrast, believe mental images to be a categorically different kind of internal representation.

Here, too, the approach we have been following suggests new answers to these and related questions. Moreover, our answers will be given in a strictly physical language. It also suggests that some of the questions commonly asked, or concepts applied to the problem, rest on tacit assumptions whose validity has to be denied. Others will have to be dismissed as too vague - for example the question whether visual imagery actually involves the visual system.

In one form or another, most of our thoughts seem to revolve around imagined objects, events, or situations, including imagined speech. In everyday life, most of our thoughts turn on the possible consequences of our actions, the goals we want to achieve, the dangers we have to avoid, the things we might have said or done instead of what we actually said or did, etc. All of these are representations of *mere possibilities* rather than *actualities*.

Often the two go hand in hand. The classical division of thought has been into imagining and reasoning. At the same time it has generally been realized that the latter may heavily depend on the former. Thus Kant and Hume viewed the imagination as the general instrument that bridged the gap between sensation and intelligible thought. But whereas for Hume the imagination was exclusively the ability to form an image of an object not present in reality ("the cognition of the absent" Price has called it), Kant also assigned to the imagination a role in the formation of abstract concepts: in the concept of a perfect triangles, for example. In his CONCEPT OF THE MIND, Ryle (1949) gave the notion an egocentric twist by describing the imagining of an object as a fancying that one is seeing the object - a description to which he later gave a distinctly behaviourist slant by translating 'fancying that one sees' an object into 'seeming to perceive' an object.

By contrast, Piaget has treated mental images as a kind of internalized imitation. They are an attempt by the brain to reproduce in schematic form the internal representations which are the product of the normal processes of perception. It is a faculty which, he believes, develops after 18 – 24 months.

From a behavioural point of view, the particular significance of the imagination may be seen in the part it plays as a guide to action by translating our needs into images of situations from which the satisfaction of those needs can be expected. It thus forms the basis of what we experience as conscious desires and purposes. For to have a conscious *desire* is to have the internal representation of a situation, as yet non-existent, which elicits expectations of need satisfaction; while to have a conscious *purpose* is to deliberately work towards the realization of such a situation. Moreover, in working towards that goal, the imagination will also assist by way of representations of the path that has to be followed to reach the goal and of the hazards that may be encountered. Then again, we use imagery to anticipate what would happen if an object or person were to behave in some particular way. Of obvious importance, too, is the role of the imagination in all creative activity, in planning for the future, in enabling us to escape into a world of phantasy and in the setting up of ideals. And, of course, imagery occurs in memory recall, e.g. the recall of events witnessed in the past.

It is sometimes said that to imagine is not the same as to conceive: that we can conceive a thousand-sided polygon, but we cannot imagine it. This is true if we confine the notion of the imagination to internal representations of a quasi-pictorial kind, or, more generally, of a quasi-sensory type. That is undoubtedly the primary sense in which the term is commonly understood and also the primary sense commonly given in the dictionaries. By contrast, to conceive of a thousand-sided figure is to conceive of a set of descriptive propositions which may subsequently be examined for their logical implications. In the present section I shall confine myself to the visual imagination. It will become obvious that what I have to say here can readily be extended to the other sensory modalities.

What then, are the most general characteristics of the visual imagination.? Opinions differ, but a search through the literature suggests the following list as representative of what most people could agree to.

1. We can imagine objects, events, or situations which we may never have experienced before, but the components of these imaginings and their optical attributes and mutual relations are always based on objects, events or situations we have experienced before. That is to say, the imagination draws its material from past perceptual experiences. However novel may be its creations, the elements of which these creations are fashioned all derive from past perceptual experiences. Although my vision of a unicorn is pure fantasy, it is put together from elements which are not. This 'putting together' or synthesis is not a purely additive summation. It is a linking together of the respective elements in some kind of fitting relationship, the criteria for which will again be based on our experience of relationships encountered in the real world.

2. The creations of the visual imagination are bound by the limits of seeing also in one other respect: as in seeing, they are tied to a standpoint. We cannot imagine what we could not perceive at a glance: we cannot imagine the inside and outside of a house simultaneously. What is imagined is always imagined as seen from a single point of view: what the visual imagination conjures up is not a possible object of a certain kind, but the possible perception of an object of a certain kind.

3. The products of the imagination tend to be abstractive in character. Termini can be left open. I can imagine, put cannot paint, a leopard without a precise assignment of the number and location of its spots, or a face without settling the colour of the eyes.

4. The products of the imagination are informative only in the sense that mental images can help us to see implications of a concept of which we had not been aware. Thus by imagining the operation of a planned machine the engineer may become aware of the fact that, as the design stands, one of its levers will foul another. Yet no new details can be extracted from a mental image by scanning it. In fact what we 'see in the mind's eye' is not generally something seen before some inner eye in a way that enables it to be scanned like a real scene in search of new particulars. All one can seek to do is to shift attention to selected points and seek to elaborate the image at those points.

5. Mental images can be transformed, expanded, rotated, etc., and the results of impressive experiments conducted by Shepard and Metzler (1971) suggest that these operations happen in real time. They found that the time taken to rotate a mental image is linearly related to the angle through which it has to be rotated. For example, if people are shown two stacks of objects in different orientations, and asked whether the two stacks are the same, the time they take to answer proved to be proportional to the degree of rotation that would be needed to turn either stack until it matches the orientation of the other. In more recent work Shepard (1984) has been able to relate image rotation and other image transformations to the mechanisms that underlie the perception of motion. In an elegant series of experiments Georgopolous *et al.* (1989) have made analogous observations in rhesus monkey.

6. There is no interference between what is imagined and what is seen, although selective attention to either process may impair the other. Segal and Fuzella (1970) have shown that in perceptual and immediate memory tasks, subjects are poorer at detecting visual stimuli when imagining pictures, and poorer at detecting auditory stimuli when imagining sounds. Conversely, mental images can be suppressed by current mental tasks (Brooks, 1967).

7. What we imagine is not only less detailed, but also less vivid than what we see. Visual imagery also appears to have a lower resolution than direct vision. For example, Fiske and Kosslyn (1980) have shown that imagined points can only be brought so close before they fuse.

8. The imagination posits its object as *not being,* to use Satre's phrase. In other words, we are aware of the non- reality of what we imagine. Yet, as Warnocke (1976) has

pointed out, images are not separate from our interpretation of the world. This accords with our account of the part which the imagination can play in the expectancies that constitute the brain's model of the world. For example, our internal representations of the properties of actual objects may in part take the conscious form picturing the way they would respond in certain imagined situations.

9. The contents of our mental images can be categorized, judged, and appraised, and they can arouse an impulse to bring about what is here projected in an imaginary form.

***

7.3 - THE NEURAL CORRELATES OF THE IMAGINATION

How then, are we to conceive the physical substrate of the brain's internal representations of mere *possibilities* (representations of category 2 - cf. Section 4.5) as distinct from its internal representations of *actualities* (representations of category 1)? My answer flows logically from two considerations, both of which follow from what has gone before.

> i) Since we can imagine every kind of object, event, or situation that we could also perceive, it follows that this physical substrate must be able to accommodate the full dimensionality of our perception of the real world. Thus, although imagined objects have no clear location in space relative to the subject (other than that they tend to be projected into the centre of the visual field), they can be imagined to have clear spatial relations to each other. They can be imagined to have colour, to be large or small, hard or soft, stationary or moving, noisy or silent.
>
> ii) The mental images that are products of the imagination are obviously removed from the reality testing to which the brain's internal representations of the actual world are constantly being subjected: we do not expect them to be confirmed by our sensory inputs. This is a truism, since by definition they consist of representations which are not elicited by the sensory inputs nor controlled by them. We are therefore dealing here with representations which are divorced from the mechanisms causing orienting reactions: no surprise reactions result from any discordance between what we imagine and what we actually see. On the contrary, surprise reactions would occur if we *did* suddenly see in front of us something we had just imagined, since these sensory inputs would have been unexpected under the circumstances.

Of course, the absence of reality testing does not mean that the creations of the imagination escape any tests at all. They can be subject to very stringent tests, but they

7.3 - The neural correlates of the imagination

are tests of a different kind. Mental images tend to arise in response to some particular need, and they will be subject to tests dictated by that need - for example, when one tries to imagine a mechanical device that would perform a certain task.

Bearing these matters in mind let us contemplate the difference between *seeing* a chair and *imagining* a chair.

The end-product of the visual perception of a chair is an internal representation which is part of the brain's model of the outside world and has its proper place in this model. According to the Conditional Expectancy Hypothesis the features and properties of that chair are characterized in this representation by sets of what-leads-to-what expectancies based on assimilated what-leads-to-what experiences at a multitude of different levels (cf. Section 5.3). At one of the lowest levels, for example, the linearity and orientation of any one of the chair's legs may be represented by expectancies relating to the movement the eye would have to execute to follow it from one end to the other. At a slightly higher level, the spatial relationships in the visual plane of the parts of the chair are represented by expectancies relating to the movements that would take the eye from one part or salient point to another. Still higher up, the distance- and depth-relationships would be represented by expectancies relating to the way the image would transform in consequence of movements of the head or body. The same applies to the internal representation of spatial relationships between the chair and its environs, and the distance of the chair from the viewer. Again, the weight and rigidity of the chair would be represented by conditional expectancies relating to the effort required to lift it or distort its shape respectively. And so on.

All these expectancies will be subject to *reality testing:* if they are violated they will elicit orienting reactions at the respective act-outcome level and tend to undergo appropriate modifications in consequence. For, as the reader will recall, we have defined a state of *expectancy* for an event E as a state of *readiness* for E which will undergo modifications if E fails to occur.

Now, when I am merely *imagining* a chair, the representation that has now entered consciousness will still be one characterizing some or all of the features of a chair, but it will

    i) not be elicited by current visual inputs;
    ii) not be tested for its veracity by current visual inputs; and
    iii) not be part of the brain's internal model of the outside world (although its occurrence would be registered by the brain as being part of the current state of the self).

We can accommodate these differences between a chair seen and a chair imagined, if we assume that in the latter case

i) the representation, though based on past visual experiences, is elicited by internal causes other than current visual experience;

ii) the features and properties of the chair are characterized not by states of conditional *expectancy* of the kind I have illustrated above, but merely by the corresponding states of conditional *readiness*, i.e. states of conditional anticipation dissociated from the mechanisms controlling orienting reactions;

iii) the brain does not respond to these states of conditional anticipation in the manner we have defined in Section 4.4 as the manner that makes a state of the brain function as an internal representation of an external object.

I believe this to be the correct answer, not only because simple logic suggests it, but also because it satisfies the features of the imagination listed in Section 7.2. It explains, for example, why the imagination is bound by the limits of seeing in the sense I have described. For the states of perceptual readiness to which our theory refers derive from the states of conditional expectancy that are the end-product of the perception of actual objects, events or situations. Thus the theory explains why we can no more see a house simultaneously from the inside and from the outside in the imagination than we can in real life.

One can also see several reasons why mental images are less vivid than actual percepts:

a) they are distillates in which the impact of numerous details may be missing;

b) what we imagine is of no immediate consequence to us in the sense in which what we see is of immediate consequence to us;

c) the whole arousal element is missing that accompanies the detection of discrepancies between our internal model of the world and the reality we encounter;

d) also missing is the active exploratory engagement of the eyes with the surrounding world that is part and parcel of the process of seeing: and,

e) gaps in the mental images (like the omission of eye- colour in an imagined face) cannot become the source of unexpected happenings. Hence their absence is no cause of worry, and there is no inherent tendency to complete details other than those relevant in a given context. This may account in part for often rather sparse nature of mental images.

7.3 - The neural correlates of the imagination

Meanwhile the fact that we are conscious of the non-reality of mental images, can be interpreted to mean no more than that we are aware of some of the facts listed above, for example the fact cited under (b).

Our assumptions also explain why what we imagine does not interfere with what we see. In the first place, and as has been explained in Chapter 3, the brain can simultaneously be in a state of readiness for two or more mutually inconsistent events. Hence in principle the states of perceptual readiness that correspond to a mental image need not interfere with the states of perceptual readiness that are implied by the expectancies formed in the visual perception of the world. In one sense it is also true to say that the representations occurring in visual imagery are outside the visual system, viz., if by 'visual system' we mean the pathways through which current visual inputs come to excite states of act-outcome expectancy to which the brain responds in the manner I have defined in Chapter 4 as the manner which causes these states to act effectively as *representations* of the features or properties of external objects.

In another sense, though, they *do* lie within the visual system. For the states of conditional *readiness* concerned have their origin as parts of the states of conditional *expectancy* elicited in direct vision. Hence it cannot surprise that according to modern research-findings tasks requiring visual imagery engage the same cortical areas as does direct vision. Thus studies of the patterns of blood flow in visual imagery tasks show a massive activation of the posterior cortical areas compared with the resting state. Similarly, studies of ERP (event-related potentials) have demonstrated occipital activity during the processing of visual imagery. (For a summary of these and related results, see Farah, 1988.)

Our conclusions are in broad agreement with the view independently arrived at by Neisser (1976), namely that *to imagine is to be ready to perceive:*

> "To imagine something that you know to be unreal", Neisser says, "it is only necessary to detach your visual readiness from your general notions of what will really happen and embed them in a schema of a different sort. When you have an image of a unicorn at your elbow - while quite certain that unicorns are purely mythical animals - you are *making ready* to pick up the visual information that the unicorn will provide, despite being fully aware that your preparations are in vain" (my italics).

Since people are conscious of the mental images occurring in the exercize of the imagination, it follows from our assumptions that we are dealing here with representations which are accompanied by the second- order awareness that their occurrence is part of the current state of the self. (cf. Section 7.1). On the other hand, the processes responsible for the production of any mental image, need not themselves be of a kind of which we are conscious. In the exercise of the imagination we can in general do no more than intentionally initiate the search for a mental image that will satisfy some particular set of criteria, but whether this search will be successful depends on factors beyond our conscious control. These subconscious process form part of what we commonly call our *intuition*.

The picture I have drawn above of the nature of mental imagery casts doubt on the appositeness of a variety of notions that one sees applied in this problem area. For example, the idea that the processes of perception and, for that matter, image formation, can be viewed as computational processes in which a variety of specialized subsystems participate. The states of conditional expectancy that are elicited as the end-product of an act of visual perception and constitute the brain's internal representation of the perceived object, form a spectrum stretching from the simplest act-outcome expectancy (e.g. shift in retinal image following a saccade) to the most sophisticated (e.g. transformation of received retinal images in consequence of moving head and body). That spectrum can be conceptually subdivided in many different ways and according quite arbitrary criteria. Hence it seems inadvisable to think of any such possible set of subdivisions as a discrete set of subsystems to each of which there must correspond a specialized operational mechanism or module. Kosslyn (1987), for example, distinguishes here 12 such subsystems which he regards as *processing* subsystems of a modular nature. This writer also postulates a special storage medium for visual images, which he calls a *buffer* in analogy with certain types of temporary storage spaces provided in computers and their printers. But the nature of this 'buffer' is left undefined. We can be rather more specific. For, according to our interpretation of the nature of mental images, the medium in which these occur is formed by states of conditional readiness in the brain which are based on past visual experiences and which in the act of image generation come to be potentiated and to influence the ongoing activities of the brain in a particular way, viz., in a way which satisfies the functional definition I have given in Section 4.4 of internal representations of category 2.

Our conclusions are also out of sympathy with the well-known view of Pylyshyn (1981) that the imagination is not a faculty, but merely a collection of epiphenomena which emerge when other processes, e.g. processes of memory and reasoning, interact in particular ways. As we have come to see it, the capacity of forming internal representations of category 2 is clearly a contingent capacity in its own right which has very definite tasks to fulfil and can be disrupted or lost in brain damage - in other words, a faculty.

Again, the picture I have drawn shows certain questions to be less than felicitous in their formulation. For example one frequently encounters the question whether visual imagery 'uses the mechanisms' of direct vision. According to our picture the answer must be: in one sense it does, for the basic states of conditional readiness that constitute a mental image are based on past visual experiences; but in another sense it does not, for the mechanisms generating representations of category 2 in any given situation do not overlap with the mechanisms generating representations of category 1 at that point in time (even though both involve the same areas of the cortex).

***

7.4 - ASPECTS OF MEMORY

The subject of memory has to be approached with caution, since we are dealing here with a portmanteau concept which covers a variety of ways in which the current behaviour of an organism may come to be influenced by past experiences. Any enduring change which an individual organism undergoes in consequence of some individual experience and which is capable of affecting subsequent behaviour, tends to be called a memory phenomenon. On this understanding, for example, any form of learning amounts to a memory phenomenon. And memory, when understood in this broad sense, is not a unitary function. Moreover, the scientists' preoccupation with such phenomena as the learning of skills or learning by rote has tended to push into the background one of the most important human memory functions, viz. *episodic memory recall*, the ability to bring back to mind past experiences, plans, intentions, or events one has witnessed. On this important subject contemporary memory theory has had very little to say. This may be partly due to the difficulty of studying this faculty in non-human species, and partly to the fact that for a proper understanding of episodic memory recall one needs to have some clear conception of the nature of consciousness, because we are dealing here with what, at least, *appears* to be the re-entry into consciousness of events once lost from it. For that very reason, of course, this class of memory phenomena is of particular interest in this book. For the theory I have advanced up to this point will enable us to see this class in an entirely new perspective. This will be the sole topic of Section 7.5. In the present section I shall confine myself to aspects of only two kinds of memory phenomena, viz. learning and recognition.

I. THE TELEOLOGICAL ASPECT OF LEARNING

Although learning is sometimes defined as any change in a subject's behaviour in a given situation which is brought about by his previous experiences in the same or similar situations, one has to qualify this in one respect. Traumatic changes of behaviour, e.g. deficits brought about by injury, are typically excluded, as are changes brought about by wear and tear and other chronic deficiencies that may be produced by sustained activities of one kind or another. In fact, most learning theorists would probably agree that only *adaptive* changes should count as learning changes, though they may be hard pressed to define the meaning of 'adaptive' in this context. We have no such problem. For, in Chapter 3, we have defined an *adaptive* change as a change that is *directively correlated* to the circumstances in which it occurs in respect of some desirable outcome. This, then, is also the sense in which learning will be understood in the present chapter.

II. LEARNING AND THE CONCEPT OF INFORMATION

In view of what has been said above about the varieties of effects which are commonly regarded as memory phenomena, it is hard to understand the persistently recurring attempts to treat memory as a unitary function which can throughout be conceived in terms of three component processes, viz., information acquisition, information storage and information retrieval. But owing to the vagueness of what in concrete terms constitutes information in any given context, these abstractions may not in fact be helpful when one looks as the physiological processes involved in learning and in the brain's construction of internal representations of the world. This conceptual uncertainty is reflected in the great diversity of models suggested in the literature, each employing different symbol systems, different subdivisions and categories. While many of these models may be valued as interesting exercises when applied in the field of AI, I do not believe that any unitary theory can cover all the different ways in which the past is brought to bear on the present in the brain's outputs.

The informational approach fails us already at the elementary level of the common learning processes, such as those occurring in classical or instrumental conditioning and in the perfection of motor skills through practice. For if, in a series of training sessions, an animal successfully learns to react to some given stimulus S with some particular response R, the resulting accomplishment is the *cumulative* product of the small individual learning changes produced by each of the preceding trials. And from this cumulative effect the contributions made by each individual trial can no more be recovered than can the contributions made by individual rain showers be recovered from the final water level in a rain gauge. So what, then, can be said about these learning changes in *informational* terms? No more than that in the course of these trials the animal has registered, and learnt to apply, the information that R will be rewarded if it is produced in response to S. There is little to be gained from this terminology. It is much simpler to say that, in the course of the trials, the animal's behaviour has undergone a certain *adaptive* change. This is also a safer formula, because, as we have seen in Chapter 3, it is easier to give a precise physical definition of *adaptive change* than of *information*. Again, if, at some subsequent point of time, the animal is found to have retained the lesson it has learnt, it is both simpler and safer to say that this adaptive change has persisted, rather than to say that certain categories of information have been stored and subsequently been retrieved. Indeed, the latter description may even be misleading, because 'retrieve' is an action word, and 'retrieval' in the present context suggests therefore, that at this point there occurs some discrete internal action (presumably directed at some memory address), whereas no such assumption may be warranted in fact.

III. THE PHYSIOLOGY OF LEARNING CHANGES

As regards the adaptive physiological changes that occur in the brain during learning, these have been most extensively studied for classical (Pavlovian) conditioning. The most widely accepted theory points at the synapses as the loci at which these changes occur. Studies in rats have suggested that the changes take the form of an enhanced sensitivity to the transmitter substance (here acetylcholine) at the receptive regions, viz. the membrane of the postsynaptic neuron, rather than an increased production of transmitter. In these studies it also seemed that the increased overall conductance in the pathways concerned may be solely due to this effect and not to an increase in the number of synapses (Deutsch 1983). A variety of other studies, though, have suggested that both presynaptic and postsynaptic changes may be involved and that the latter may interact with the main cell body of the postsynaptic cell. Nor have changes in the branching of the dendrites and the number of synapses been ruled out as such. Deutsch concluded that

i) after learning, and up to a certain point, the sensitized postsynaptic membrane may become increasingly more sensitive with time. This conclusion is drawn from the observation that rats which have only partly been trained on a maze, require fewer trials to complete their learning when the trials are resumed after 14 days than after 3 days);

ii) after this point, the sensitivity declines, leading to progressive deficits in performance;

iii) apart from this natural decay, responses can extinguish through *habituation*. This takes the form of learnt *inhibitions*, viz., synaptic sensitizations in inhibitory pathways;

iv) it may be the case that the synapses with slow decay time are paralleled by synapses with fast decay times, the latter serving as a kind of short-term store.

As to the conditions under which all such synaptic facilitations take place, Hebb's (1949) original hypothesis envisaged that a repeated firing of a neuron B by synapses from a neuron A would increase the efficacy of B being fired by discharges from A. However, various studies conducted since have suggested that the synaptic facilitation occurs regardless of the spike activity of the post-synaptic neuron (Hawkins 1983)

But the question of which synapses in what pathways undergo these sensitizations, has remained a perplexing one. This is partly due the wide distribution across the brain of the networks participating in the learning process, and partly due to the difficulty of determining whether any actually detected change in synaptic conductivity is a primary learning change, or merely a secondary concomitant.

The scientific world was shocked into an acute awareness of this wide distribution, when, in the late 1920s, Lashley made his famous claim that when a rat has learnt a maze and parts of its brain are then pared away, the amount or learning retained will be proportional to the amount of cortical tissue remaining, no matter which parts were selected for removal. In the sequel Lashley had to qualify these claims considerably, and his 'Law of Mass Action' is now merely regarded as a statement symbolic of the brain's integral mode of operation, rather than an accurate statement of fact.

Troubled by this apparent distributiveness of the brain's 'memory traces' some scientists have searched strenuously for possible physical analogies of the 'mass action' phenomenon. Even analogies with holograms have been attempted (Pribram, 1971). But these analogies eventually proved to be too remote to justify the hopes that had been pinned on them.

IV. LEARNING TO ANTICIPATE

A simple illustration of learning to anticipate is furnished by the development with practice of particular skills, such as typing, playing the piano, riding a bicycle etc. As has been mentioned, in the acquisition of these skills a great deal of the increased speed and error-free performance is due to the fact that with practice the brain learns to *anticipate* response requirements at different stages of the action. According to the classical explanation, in the learnt stimulus- response sequences proper to these skills, components of the responses required anywhere in the sequence tend to become conditioned to the traces of antecedent stimuli and thus gradually advance along the sequence in an anticipatory fashion. I have discussed this classical paradigm in connection with the question of how with repetition,what-leads- to-what *experiences* lead to what-leads-to-what *expectancies* (Section 5.6). It also bears on the next memory phenomenon to be mentioned, viz. the phenomenon of recognition.

V. RECOGNITION

As I shall understand the word, to *recognize* an object (or class of objects) is to apprehend that it has been met before and to transfer to it some or all of the reactions that had been formed to it at the earlier encounter(s). In both humans and animals, this apprehension can be explored in delayed matching-to-sample tests, i.e. experiments in which the subject has to select from a presented pair of objects the member that is identical to a previously presented sample. How is this ability to be interpreted in terms of the brain's internal representations?

When an object is met for the first time it will elicit *some* categorizing reactions and associated conditional expectancies, e.g., those descriptive of what is actually seen and of the familiar objects it most resembles. But since it is not one of them there will features which are not covered, i.e. *novel* features. These will elicit orienting reactions of one kind or another - the brain's common response to novelty. The resulting scrutiny or exploration of the object will fill these gaps in the brain's internal representation of the object, while at the same time eliminating the orienting reactions. When the object is subsequently met again the acquired expectancies may or may not be elicited, hence transfer. If they do, we have *recognition*. And apprehension of the resulting absence of orienting reactions, may be equated with the feeling of *familiarity*.

Under abnormal circumstances one can have a feeling of familiarity when none is warranted. This is the well-known phenomenon of the *déjà vu*. On the present theory it seems likely that this phenomenon reflects a transitory and circumscribed deficit in orienting reactions: hence the appropriate feelings of novelty fail to occur and the automatic result is a feeling of familiarity. Conversely, feelings of familiarity may be missing when they should occur, perhaps due to a failure to inhibit the orienting reactions that familiarity would have come to inhibit. In certain brain lesions, for example, the patient may treat his doctor as a stranger, but presently ask him to renew his last prescription. Both these defects amount to defects in the mechanisms of 'reality testing', as we have come to understand that expression.

Loss of transfer of acquired reactions to an object or class of objects may also occur in abnormal circumstances. For example, a brain-injured patient may fail to recognize an object as the kind of object it is. Seeing a Christmas tree, for example, he may get no further than categorizing it as a kind of shrub.

Assuming that we are not dealing here merely with a failure to name the object correctly, this would seem to be a case in which the categorizing reactions and associated conditional expectancies that had original become attached to a familiar class of objects (Xmas trees in the present case), have either been lost as the result of the injury, or simply fail to be elicited by the present perception of any such object.

We have seen in Chapter 6 that different neural networks are likely to be involved in the formation of categorizing reactions at different levels of abstraction. It follows, that lesions in cortical areas which are responsible for some particular level of abstraction, may cause loss at these levels in the subject's powers of recognition while sparing other levels. This is born out by the evidence. For example, in certain lesions in the temporal cortex pattern discrimination may be unimpaired, yet the patient may fail to recognize a face or even his own face in a mirror, he may confabulate, show lack of mental cohesion and, though perhaps recognizing his wife, deny that he is married. The fact that he may confabulate suggests that here, too, a breakdown has occurred in the mechanisms of reality testing.

A patient suffering from such impairments may try laboriously to infer logically the missing categorizations of the object or situation concerned. Being shown a picture of a Christmas tree in a room, for example, the patient cannot make sense of it, but will try, though often unsuccessfully: recognizing the tree as a kind of shrub he may conclude that the picture must be about something outside; recognizing a picture on the wall as a flat rectangular object which could be a picture but could also be a window, he may conclude (since he believes it to be an outside scene) that it is the window of an adjacent house, etc.

***

7.5 - EPISODIC MEMORY RECALL

Awareness that an object, event or situation has been met before, i.e. *recognition*, does not imply recollection of the occasion, i.e. *episodic memory recall:* remembering a past experience or an event one has witnessed. The importance for man of this faculty needs hardly be stressed. Yet it is the least understood of memory functions. This must in part be due to the fact that we are dealing here not only with traces left by past experience, but also with the conditions that govern re-entry into consciousness, and with an end product that has all the qualities of an imagined event except that the brain accepts it as part of its internal model of the past, i.e. as part of the historic dimension of its global world model. Hence the study of the physical basis of episodic memory recall was bound to be hampered by a lack of clear perceptions of the nature of consciousness as such, and of the imagination in particular. The matter is further complicated by the fact that short-term and long-term memory may involve different mechanisms: after concussion or other traumatic shocks, all memory may be lost for recent events but not for the distant past.

The first point to take up is one already mentioned, viz., that the brain's global world model also has a *historic* dimension: it includes representations, not only of what is the case but also of what is accepted as *having been* the case at some time in the past. This state of affairs is easily expressed in functional terms along the same lines as the definition of internal representation given in Section 4.2. We can define a representation of an event as a past event as follows:

7.5 - Episodic memory recall

DEFINITION:

A state X of the brain acts as an internal representation of the past occurrence of an event E if the brain is set to behave as if, in order to perform an act which is appropriately related to that past occurrence of E, it would suffice to perform an act which is appropriately related to X.

We must assume that the brain has acquired the power to form such representations because it *needs* them. It needs them because without them it could have no representations of *continuity*, of *change* or of *causal connections* in the world. Again, an individual may be required to respond in a manner that is correctly related to the past occurrence of some event - perhaps because he is required to report on such an occurrence, or because he seeks to explain the present in terms of the past. Moreover, our very sense of identity hinges on our perceptions of our own past life. One qualification has to be added, though: it seems that in general we can recollect about the past only such things as had been attended to at the time.

The subject is of special interest to us because the expectancy theory enables us to approach this memory function from a new angle, and one that does not so far seem to have been explored.

To explain this angle, let me begin with the brain's power to form representations of the *recent* past. Consider this simple case: looking out of the window I perceive a chestnut tree, and presently I turn my eyes away again. On the Expectancy Theory, that perception will now have left a trace in the form of the conditional expectancy that when I turn my eyes back I shall see that tree. If next I walk away from the window, a different and more complex expectancy will be formed: the conditional expectancy that if I first move to the window and then look out, I shall again have that particular perceptual experience, see that chestnut tree. *Both cases imply the existence of a state of conditional readiness to see that tree, but the conditionals have changed.*

All this is a natural part of the brain's currently changing internal model of the world and the self-in-the-world. And the importance of these observations is that they show an interesting way in which perceptions leave a trace. This, by definition, is a *memory* process - here described in terms which have already been fully covered in earlier parts of this work.

The question now suggests itself, whether a theory of episodic memory recall could be derived by following up the above considerations. It seems a possibility worth exploring, since it would enable us to see the nature of this faculty in a new light.

Thus we might posit the following two hypotheses:

1. Under certain circumstances the perceptual readiness to see that chestnut tree (which, in the two cases I have single out above, has come to be conditionally elicited as the result of the original perception of the tree) may subsequently come to be elicited also in the absence of the conditionals concerned - thus giving rise to mental images analogous to those created in the imagination.

2. The factors bringing this about are of a kind that will at the same time cause the brain to treat this representation as part of its internal representation of the history of the self.

What these factors and associated mechanisms are, remains to be explained. There are two questions here:

a) what factors can cause a particular state of perceptual readiness for a complex event to recur under conditions other than the ones under which it was first formed, and

b) on what grounds would the brain come to accept such a state of perceptual readiness for a complex event as part of the historic dimension of its global world model and order it correctly on the time scale of the subject's life?

Although I cannot at present see a clear answer to either of these questions, I shall add some thoughts about them at the end of the present section. Meanwhile, the point will have been made that the possibility of this kind of approach makes us see the problem of episodic memory recall in an entirely new perspective.

Both questions are undoubtedly difficult. On the other hand, no alternative hypothesis known to me makes the matter any easier. The traditional, but essentially facile, assumption that memory recall is no more than the activation of the fading traces, or 'engrams' of stimulus patterns, glosses over the nature of conscious mental images and the conditions that need to be satisfied if an event is to enter consciousness. It also glosses over the fact that, what began as part of the brain's model of the *present*, now returns as part of the brain's internal model of the *past*. Finally, it glosses over a fact which I shall take up presently, viz., that the *imagination* also plays a not inconsiderable part in memory recall. Memory recall is in part a *constructive* process. Our assumptions take care of this, since we have come to see both mental images and the content of episodic memory recall to be events of the same genre, viz., events consisting of the brain entering more or less specific states of readiness (though not, of course, expectancy - cf. Section 7.3) for a particular perceptual experience.

One or two further aspects of the matter call for comments before we return to the two main questions which, I said, the theory still leaves in need of an answer.

7.5 - Episodic memory recall

Like the rest of the brain's model of the world, its representations of past events stand in need of *reality testing*. That is to say, the validity of the brain accepting a certain mental image as a representation of some past event, stands in need of support by whatever means the brain has available. It would be confirmed, for example, if the expectancies that flow from it are confirmed by current actualities. In this respect there is a notable difference between a recall of recent events and the recall of more distant events. For, the opportunities for testing the veracity of representations of past actualities are bound to be become the more tenuous the more remote the past. Representations of the distant past may be confirmable only by eye-witnesses whose testimony proves to be as we expect from our memory recall, or by way of enduring results of the remembered events, such as the current perception of objects we remember buying, the testimony of photographs, diary entries, etc. By contrast, events that happened in the recent past are likely to have rich implications for the interpretation of what is happening in the present, or what would be the consequences of particular actions (as in my example of returning to the sight of that chestnut tree). Context, too, may be a decisive factor. For example, consistency with other things recalled.

Conceivably the meagre possibilities for disconfirmation may account for at least part of the stability of distant memories compared with recent memories. However, the well attested vulnerability of short-term memory to concussion, severe psychological shock and electroconvulsive shocks (ECS), suggest that long-term memories owe their stability to some form of progressive *consolidation*, and this accords with the view now held by many memory theorists. Observed effects of ECS in animal learning support the view that the initial consolidation here depends on some kind of *rehearsal* occurring within seconds of the acquired experience (Schneider and Plough, 1983).

In accordance with the first of the two hypotheses I posited above, it seems plausible to assume that this rehearsal takes the form of the brain activating unconditionally, subconsciously, and for the purpose of reinforcement, any (originally conditional) state of perceptual readiness which a recent perceptual experience has induced. In other words, the brain constantly checks its state of preparedness for the recurrence of any perceptual experience it has recently encountered - a suggestion that would make good biological sense.

Published observations also suggest that consolidation continues slowly over a considerable period of time, and may even involve paradoxical sleep. But whether that long drawnout process of consolidation is also a matter of rehearsal remains an open question.

It should be noted in this connection that any theory of rehearsal fits in very well with the Expectancy Theory. For, rehearsal is a kind of preparatory reaction, and the Expectancy Theory hinges on the assumption that preparatory reactions are a common feature of brain activity. Thus a built-in tendency of the brain to remain prepared to meet again what was once experienced may be part of the answer to the first of the two questions which I cited above as in need of further exploration by the theory.

The second question which our theory left unanswered was how the brain comes to allocate a place in time to a recalled experience. How does it come to place that experience into the right niche in its representation of the history of the self? I doubt if we need to assume that the original experiences are stored straightaway with some explicit time-tag, as has been suggested by Yntema and Trask (1963). The evidence here is equivocal. The results of their experiments could equally be explained on the assumption that no more than a sequential ordering of remembered events occurs, and that this can be a by-product of rehearsal, perhaps coupled with subconscious attempts to fit the events together into a sequence offering continuity and internal coherence.

It could well be that the brain also uses contextual clues in assigning a place in time to particular memories. These clues could be internal to the recalled images: thus images of a nursery with remembered toys, of a classroom with remember companions, of a war-time scene, of a university lecture, etc., all fall into place in the order of our life simply by virtue of their content.

Another notable aspect of episodic memory recall is the following. In practice the mental image of a recalled event tends to be no more than a distillate of the original perception in which numerous details may be missing. It may relate merely to some very general features of that event, or just the main features that had been attended to. And in all cases the features recalled depend on the original interpretation of what was seen. When the memory recall covers not just one single event, but a whole sequence of events, these missing details can amount to very considerable gaps, especially if only the end product of the sequence was of direct concern to the subject. And, presumably to achieve congruency, the brain here seems disposed to use the imagination to fill those gaps. In other words, a *creative* and *constructive* element enters into the final presentation.

Frequently, too, we remember events merely in terms of some familiar category we have applied to their interpretation. For example, a jagged shape may be remembered just as a star. This also manifests itself in the relative ease of recall. For example, we tend to have great difficulty remembering visual details, e.g. details of a machine, unless there already exists in the mind an abstract representation of the kind of object, (type of machine) whose details they are.

Often merely the outcome of a sequence of events is remembered, but not the sequence itself, because only the outcome caused a sufficient degree of arousal and attention to impress itself on the memory. The mind then struggles to fill in the gaps. In my younger days I used to fly radio-controlled model aircraft. Occasionally the system failed and the aircraft would spin into the ground. Even when interrogated immediately afterwards, the bystanders were rarely unanimous as to whether the aircraft had spun in a clockwise or anti-clockwise direction. The unreliability of witnesses has been studied systematically by a number of workers, for example Elizabeth Loftus (1975). But my anecdotal example will suffice to make the point.

That both short-term and long-term memories are often the product of a largely constructive activity rather than a faithfully copying one, was recognized by Bartlett (1932) more than 50 years ago. To-day it has become generally recognized in psychology that a recalled experience is rarely a true copy of the original experience. Piaget accordingly speaks of episodic memory recall as *reconstructive* memory, to be contrasted with the *reproductive* memory that is shown in learning by rote.

All this accords with an assumption which we made earlier on, viz., that the conscious representations which arise in episodic memory recall are of the same general genre as the conscious representations formed by the imagination, and differ from these only in that (subject to the tests I have listed) the brain comes to respond to them in the manner I have defined in functional terms at the beginning of this section.

A brief word must be added about the implications of the present theory (hypotheses 1 and 2 above) on the interpretation of the observed effects of brain lesions. One implication is that recognition and memory recall are two very different processes. This is borne out by the evidence. In some brain lesions recognition may be spared but recall impaired: the patient knows he has met an object before but cannot recall when, even though the encounter happened recently. In others the patient may in some contexts perform better on recall than on recognition.

When discussing memory deficits after brain lesions, the question is often raised whether the lesions have affected memory *storage* or memory *retrieval*. If the memory deficit is only temporary a deficit in retrieval would be the obvious answer. A case in point is the Korsakoff syndrome, an amnesic syndrome first described by K.S.Korsakoff for alcohol-deranged subjects. But on the two hypotheses I have advanced above the matter is not as straightforward as it might seem in these terms. According to this theory there are several points at which lesions might disrupt matters. For example, the recent perception of an event X might not leave the brain with a potential readiness to perceive X again, either conditionally or unconditionally. On the present theory severe deficits of this kind would lead to loss of continuity between one conscious experience and the next. For not even the most recent past could be recalled. This is a known effect of some bilateral lesions in the medial temporal lobes. Other lesions might prevent consolidation, and thus interfere with the laying down of long term memory. Again other lesions might leave all the above processes intact, but prevent a subconsciously activated readiness to perceive X from being accepted by the brain as part of its internal model of the history of the self. This would be the case when a recently perceived object is met again and is recognized as a familiar one, but without the subject being able to say when it was met before. In extreme cases, this, too, could lead to loss of continuity between one experience and the next. Some similar deficit may also be underlying the Korsakoff syndrome - at any rate if those investigators are right who claim that the fundamental deficit of the syndrome lies in a failure to integrate new knowledge with the total personality (in our terms, to adjust the internal model of the self and its history). A direct indication of this may be seen in the fact that Korsakoff patients frequently deny that they are ill or suffer from memory loss. The frequently observed loss of mental cohesion points in the same direction. However, there are also workers who would argue against these conclusions, and I would not be competent to weigh one argument against another. My main object here has been to show that the theory I have discussed, opens up avenues of explanation which may be closed to other theories.

One controversial issue must be briefly mentioned before I leave this section. I shall deal with it at greater length in Section 8.5. This concerns the question of whether in memory recall the recalled material has a pictorial or propositional form. Kosslyn (1980) for example, has argued in favour of the imaging hypothesis, whereas Pylyshyn (1980) has defended an exclusively propositional assumption. He believes the world, both past and present, is primarily modelled by the brain in terms of descriptive propositions which may or may not subsequently be turned into visual images. Meanwhile Anderson (1978) has argued that the issue cannot be rigorously decided since, he claims, the two theories lead to identical behaviour predictions.

Although we are dealing here with a logical possibility, in the absence of compelling evidence it seems a gratuitous complication to assume that whenever an event is remembered in a pictorial form, this is only a reconstruction from some verbal description remembered below the level of consciousness.It does not seem plausible to assume that I remember by nursery only in terms of verbal descriptions which I silently made even as a barely literate infant. And what about the deaf and dumb? Are they not able to visualize how to get from A to B in their house or in their town? On our account the brain's internal representation of the world consist of linked set of conditional expectancies which may have been formed either in consequence of events personally encountered (e.g. visually perceived), or of descriptions received in a verbal (propositional) form. And representations existing in either one of these forms can be converted into the other. Furthermore, in Chapter 8 I shall suggest that internal representations cast into a propositional form are a more sophisticated kind of representation than mental images, and I shall describe the operations which the brain has to perform to transform the latter into the former.

Thus, in the absence of proofs to the contrary, I take the view that the brain's internal representations may have either a quasi-sensory or propositional form, often depending on the form in which the relevant information was received in the first instance. In addition there may exist a purely individual bias to convert either form into the other. As has been said, some people are more visually oriented than others, some more linguistically. Painters, sculptors, designers and engineers tend to have a richly visual inner life, while writers and philosophers are more likely to have a richly linguistic life.

7.5 - Episodic memory recall

Empirical evidence against the assumption that the propositional form is the universal basis of internal representations of the world, past or present, comes from the observed effects of certain brain lesions. As I have mentioned, the temporal lobes seem to be heavily implicated in the brain's construction of an internal model of the world. And the effect of lesions in these areas suggests that *both* verbal and pictorial representations normally occur in long-term memory recall and that these lobes are involved differentially. Injury to the left temporal lobe appears to be associated with verbal memory disorders, whereas corresponding injuries to the right lobe impair learning, recall, and recognition of certain non- verbal visual and auditory patterns (Brenda Miller 1978). Observations made on split-brain patients support the same conclusion (see Trevarthen's review in Gregory (1987) p.740- 747).

***

Chapter 8

LANGUAGE AND RATIONAL THOUGHT

INTRODUCTION AND SUMMARY

In the preceding chapters I have discussed the character of the internal representations involved in some of the main contributions to man's stream of consciousness, but not yet of the special representations required by the faculties of speech and rational thought - the internal representation of logical relationships, for example. These will be the topic of the present chapter. Once again my aim will be to characterize the nature of these internal representations in precise and strictly physical terms.

The human language can serve a variety of functions. It can be used to express emotions, commands, calls for help, etc. But, since this book is not a linguistic treatise I shall confine myself to the two most important uses of the human language:

1. The *descriptive* use: its use for the purpose of telling a story or, more simply, for the purpose of saying something about the attributes or features of the real or imaginary objects, events or situations that occupy the speaker's mind. In the main, therefore, we shall be concerned only with sentences in the the subject-predicate form.

2. The *argumentative* use of language - the foundation of rational thought and the power of reason. This field is so broad that I shall confine myself to the nature of logical deductions of the type met in the most common figures of syllogistic reasoning (cf. Appendix B), and the four types of categorical statements that occur as premisses in these syllogisms.

These two uses of language have one critical element in common: both require a power of the brain to form internal representations of the class-membership relation and of logical relations between classes, such as the relations of inclusion, exclusion, union and intersection (see Appendix B).

In all of this I shall do no more than analyse a handful of concrete examples, selected in order to illustrate the main categories of internal representations they require in addition to the categories I have already examined in the earlier chapters.

The three main aspects of language which, each in its own way, occupy the theoretical linguist are: the *syntactic* which relates to the grammatical structure of a sentence; the *phonological* which relates to the sounds occurring in spoken utterances, and the *semantic* which relates to the *meaning* of what is said or written. In contemporary linguistics the weakest element is undoubtedly the link between the syntactic and the semantic. The whole problem area concerning the link between the structure of a sentence and its meaning, has remained a grey area. One of its causes is the uncertainty of the meaning of meaning. The reason for this is not difficult to see. The meaning of a verbal description of any object, event or situation lies in the link between that description and the speaker's internal representation of that object, event or situation. He will seek to describe an object *as he sees it,* and a fact *as he understands it*. Hence, lacking a clear understanding of the nature of the brain's internal representations of the world, linguists were bound to lack a precise conceptual framework on which to rest their attempts to understand the vital link between syntax and meaning - and, by implication, the link between language and reality. Most of my observations, therefore, will be aimed at clarifying this link in the light of our earlier work.

Every descriptive sentence asserts something about something. Typically it asserts that some particular entity has (had, will have) a particular property or feature, or that it stands (stood, will stand) in some particular relation to some other entity. Formally, however, we need not treat properties (or features) and relations as radically different cases, since both can be treated under the heading of *classes*. Every property or feature defines a class of objects. To say that an object is brittle, is to say that it belongs to a class of objects which tend to disintegrate on impact. And every relation between two (or more) objects defines a class of ordered pairs (triplets, n-tuplets) of objects. For example, the dyadic relation *greater than* can formally be treated as a class of ordered pairs of objects, viz. the class of all pairs of objects in which the first member of each pair is greater than the second.

The complexities of the syntax of our language flows primarily from the fact that if the speaker invariably had a word (let's call it a *name*) available for the object he wants to talk about, and equally for the property or feature he wants to attribute to it, all descriptive statements could have a very simple structure. At the lowest level they could consist of just the two names uttered in an appropriate conjunction, as in

'John ill'.

Introduction and summary

At the next higher level they could include such speech components as those which our language provides to distinguish between the past, the present and the future, or between the singular and the plural, as in

> *'John is ill.'*
> *'Paris was beautiful'.*
> *'Cars are dangerous.'*

But they would still be very simple structures, easily formed.

Unfortunately, the vocabulary of all natural languages is finite, whereas the number of things we may wish to talk about is virtually infinite. Thus in most cases in which we want to make such predications we have no names available. My vocabulary has a name for a particular *class* of tall plants in my garden, namely 'trees', but it has no name for the *individual* tree that happens to stand outside my window. My speech community is not in the habit of naming individual trees. Nor has my vocabulary a name for the attribute of casting-a-shadow-over-my-desk. Hence, instead of a single noun as subject of the sentences in which I want to convey this information I have to use a descriptive noun phrase (*'the tree outside my window'*), and instead of a single verb or adjective as my predicate I have to use here a descriptive verb phrase (*'casts a shadow over my desk'*). Thus the bulk of human communications tends to consist of complex sentences with complex *phrase structures*.

Thus the main complexities of the grammar of the descriptive use of language flow from the need to talk about objects, events, situations, properties, features, or relations for which we have no single word. And the main point I have to make in this connection is that the ability to do so demands certain abstract representational powers of the brain, such as the ones I have cited above.

Even to attribute a *nameable* feature to a *nameable* object already requires one representational power of the brain which we have not yet considered, viz. the ability to form an internal representation of the *class-membership relation:* the relation between an object and the class of which it is asserted to be a member. Thus if John strikes me as ill, the verbal response *'John is ill'* presupposes an internal representation of the fact that it is this particular object which has elicited this particular categorizing reaction in me, i.e. a representation of the class-membership relation.

Still more is required when it comes to communicating of some *nameless* object that it has some particular *nameless* feature, as in the example of the tree outside my window Now some descriptive phrase has to be constructed to particularize that nameless *object* in terms of nameable entities, to be coupled with a descriptive phrase to particularize that *nameless* feature in terms of *nameable* entities – yielding then such constructions as the one I have cited:

'The tree outside my window casts a shadow over my desk'.

The general character of this procedure may be described as follows: suppose O is some nameless object about which some declaration is to be made, then typically the speaker will solve this problem by

a) finding a set of nameable classes of entities, say the classes A,B and C, which is such that an intersection between them yields the target object O, and

b) constructing a (noun-)phrase which is able to convey the logical relationship (here, intersection) between A, B, C that yields O - availing himself of the syntactic conventions that have evolved for this purpose in the grammar of his language.

This procedure thus presupposes a capacity of the brain to form not only internal representations of the auxiliary classes A,B and C, also internal representations of such logical relationships as the intersection of two or more classes. In a variety of circumstances, shorter procedures are also available, e.g. contextual references as in

'This tree casts a shadow over my desk'.

However, these will not be my main concern.

If the feature to be attributed to O is also a nameless feature, call it here the feature F, a similar procedure would have to be applied to particularize it. A set of *nameable* classes would be found which are such that their intersection yields the *nameless* class F, and an appropriate (adjectival-)phrase would be constructed able to convey the relationship between these classes that yields the class F.

In short, a very complex set of representational processes has to occur if the original internal representation of the perception that some nameless object O has some nameless feature F is eventually to issue in a verbal statement in which the identity of O and the nature of F are adequately conveyed. I shall call this set of processes the *cognitive resolution into a propositional form* of the *source perception* of the resulting sentence - here the perception that the nameless object O has the nameless feature F.

In accordance with the outline I have just given I shall begin my analysis in this chapter with a brief discussion of classes and the internal representation of classes (Section 8.1), to be followed in Section 8.2 with a discussion of the internal representation of the class-membership relation and of logical relationships between classes, such as the relationship of intersection mentioned above.

The general processes underlying the ability to describe nameless entities in terms of nameable ones will form the subject of Section 8.3. At this point, therefore, I shall explain in greater detail what I mean by the *cognitive resolution* to which any primary internal representation, e.g. a mental image, has to be subjected in order to acquire a *propositional* form, i.e. a form that can be given a verbal expression in accordance with the lexicon and grammar of the language concerned. This will be followed in Section 8.4 by a diagrammatic representation of the logical relationships involved in cognitive resolutions of this kind and of the way in which they are mirrored in the structure of the resulting sentence. This *logical phrase-marker,* as I shall call it, will then be compared with the *generative phrase-markers* used by linguists in their attempts to depict the grammar of our language as a generative process consisting of a finite number of 're-write rules' (See Appendix D).

In Section 8.5 I shall add some further comments on the question of mental images and their resolution into a propositional form along the lines I have intimated above. Finally, I shall move in Section 8.6 from the internal representation of logical relationships between *classes* to the internal representation of logical relationships between *propositions* - thus reaching the foundations of rational thought.

***

8.1 - CLASSES AND THEIR INTERNAL REPRESENTATION

To recall a topic of Section 5.5 : most human (as well as animal) reactions to the objects of the surrounding world tend to have a general character: they divide the world of stimulus objects into those that, singly or jointly, elicit the reaction and those that do not. In this sense they may be described as *categorizing* reactions. To the extent to which these reactions remain consistent throughout time, they define a distinctive *category* or *class* of objects in the life of the individual concerned, viz., the class of objects over which the reaction generalizes (tables, mountains, pins, friends etc.) The internal representation of such a class consists of the respective categorizing reaction plus the associated set of conditional expectancies which reflect the attributes the subject has learnt to associate with objects eliciting the reaction.

As has been said, the logical concept of *class* has the advantage that it enables us to treat *properties* and *relations* under one single heading, since any property may be taken to define a class of objects, and any relation between two (or more) objects may be taken to define a class of ordered pairs (triplets, n-tuplets) of objects (see Appendix B). Typical examples of a dyadic (two-term) relation are *to the left of, belonging to,* and *larger than.* Thus the relational expression *larger than* can be taken to designate a class of ordered pairs of objects in which the first member of each pair is larger than the second member (cf. Appendix A). In other words, the internal representation of a dyadic relation may be described as a categorizing reaction which generalizes over a class of ordered pairs, plus the associated set of conditional expectancies which reflect the attributes the subject has learnt to associate with the pairs of objects eliciting the reaction concerned. By contrast, a typical example of a class of ordered *triplets* would be the relation *between*, as in '*B lies between A and C*'.

Often our categorizing reactions to an object depend on the object having a particular *combination* of features. These features thus constitute jointly the criteria for membership of the object concerned in the corresponding class of objects. To have an internal representation of these criteria is to have a *concept* of that class.

However, the criteria applied in practice need not be the defining criteria that might be listed in a dictionary. The concepts we use in our everyday life are not generally so sharply defined and explicitly represented in the mind. When we talk about lemons, we do not carry in the head a dictionary definition of this fruit. Rather, we tend to carry the model of a *representative sample* or *stereotype* of the fruit which as children we learnt to call 'lemon'. The only attributes we tend to take note of are simply those needed to distinguish this fruit from any other kinds around: size, colour, squashability, edibility, and taste, for example. Thus when in ordinary conversation mention is made of such things as cars, Scotsmen, soldiers, chairs, etc., these words will not in general conjure up in the mind articulate representations of the defining characteristics, but merely representations of familiar examples of the categories concerned and perhaps some of their salient attributes., i.e. of the stereotypes we have formed in the mind.

As in infancy we learn to name particular objects or classes of objects we become members of a language community bonded by *shared* concepts and language habits. Through the powers of the imagination we can also form internal representations of combinations of features of objects regardless of whether these combinations exist in the real world. If they do not, the terms we use to designate such classes of entities would then be said to have a *connotation* but lacking a *denotation*.

***

8.2 - LOGICAL RELATIONSHIPS BETWEEN CLASSES

Classes may stand in various kinds of logical relationships (cf. Appendix A). For example, a class may be *included* in another class: the class of humans is included in the class of mortals. Again, a class may be formed by the intersection of two (or more) other classes: *canals* may be described as a class formed by the intersection of the class *waterways* with the class *human artefacts*. And, of course, there is the *class- membership relation* itself: the relation between an entity and the class of which it is a member.

As has been explained, to be able to deal with the nature of language, we must turn to the way in which the brain forms internal representation of such logical relationships.

8.2 - Logical relationships between classes

Both the descriptive and argumentative use of language critically depend on these. I shall concentrate on the three which are most relevant to the concrete examples which I shall discuss in detail later: the class-membership relation, the relation of inclusion and the relation of intersection.

*

I. THE INTERNAL REPRESENTATION OF THE CLASS-MEMBERSHIP RELATION.

As a simple example of a statement expressing a class- membership relation consider the simple subject-predicate statement

'John is ill.'

The two key elements here are the object (John) and the attribute (ill), both represented internally. However, a *third* element must also be present, viz. an attended internal representation of the *nexus* between these two. We must assume this to take the form of *an internal reaction which brackets (i.e. generalizes over) attended objects and attended categorizing reactions to those objects, and which is itself attended to*.

This third element, the link between subject and predicate, is verbally represented in our example by copula *'is'*. In theory, it could be expressed merely by an ordered juxtaposition of the labels concerned, as in

'John ill.'

This primitive form is not uncommon in baby talk, It also tends to occur in one's first steps to communicate something in a foreign language. For the use of the copula owes its value to the additional functions it is able to discharge, such as distinguishing between the past, present and future, or between the singular and the plural - functions which may not yet have been mastered in the infant's or traveller's first linguistic efforts.

The composite internal representation which yields in English the sentence *'John is ill'* would yield in German *'John ist krank'*. Though these two sentences differ, the underlying composite representation will be the same. The conventional way of putting this is to say that the two *sentences* express the same *proposition*. Hence a composite internal representation which has the composition and structure required by some descriptive statement - e.g. the three-component representation I have cited above as the one underlying both *'John is ill'* and *'John ist krank'* - will henceforth be called a *representation cast into a propositional form* or, for short, a *propositional representation*.

II. THE INTERNAL REPRESENTATION OF THE LOGICAL RELATIONSHIP OF INCLUSION

Turning now to the *logical relationships between classes*, let us consider first the relationship of inclusion, e.g. the internal (propositional) representation of the fact that all members of a class A of objects are also members of a class B. What form will that representation take? In answer, I would point to the fact that a brain which has absorbed this relationship will be set in a way in which any object eliciting the categorizing reaction A (and associated expectancies) will also elicit the categorizing reaction B (and associated expectancies). Thus *the internal representation of the logical relationship of class-inclusion may be conceived as an attended internal reaction which generalizes over all cases in which a categorizing reaction and a fragment of that reaction are attended to individually.*

As regards the learning processes in the brain which cause this kind of representation to evolve, I see no reason why they should differ from the normal learning processes which cause the brain to react in some specific way to the joint occurrence of two distinct events - as when we come to react in some specific way to a person who both smiles and looks at us.

Verbally, class-inclusion is typically represented by a copula in plural form ('*holidays are fun*'), though generally supplemented by the quantifier '*all*' to distinguish the case of *inclusion* from the case of *intersection*, which would use the quantifier '*some*', as in '*Some holidays are fun*'.

Similar considerations would apply to the internal representation of the relationships of *exclusion*, and *union*. For example, we can conceive of the internal representation of the relationship *exclusion* as similar to that of *inclusion* except that for the subject's *acceptance* of all specific attributes of B as attributes also of A we now have to substitute a *rejection*.

III. THE INTERNAL REPRESENTATION OF THE LOGICAL RELATIONSHIP OF INTERSECTION

The class-relationship of intersection is of particular importance in descriptive statements. It is obviously involved in all simple adjectival phrases such as '*tall trees*', '*happy families*' etc. And it is *doubly* involved in phrases expressing a dyadic relation with its two open termini, such as the relationship *owned by*. For, as has been explained, such a relation may be viewed as a class of ordered pairs - the first member of each pair here consisting of an object owned, the second of the owner. As an example, let O denote the relation of ownership, then the internal representation of the class formed by all cars which are owned by *foreigners* would require representations of (a) the intersection of the class cars with the class formed by the first members of the ordered pairs of O, and (b) the intersection of the class *foreigners* with the class formed by the second members of the ordered pairs of O.

How are such intersections represented internally? Consider a class C formed by the intersection of two classes A and B. The internal representation of this case will take the form of a categorizing reaction C which is conditional on the joint occurrence of categorizing reactions A and B, plus an internal representation of this relationship between A,B, and C. Hence the internal representation of the class-relationship of intersection my be described as *an attended internal reaction which generalizes over all cases in which the occurrence of one attended categorizing reaction is conditional on the occurrence of two or more other attended categorizing reactions.*

As regards the internal learning processes in the brain that cause such reactions to be formed, the same remarks apply as those I made in connection with the class-relationship of inclusion.

In terms of expectancies, the difference between class- intersection and class-inclusion may be expressed as follows. In a brain which has absorbed that *All A are B* there will be the *expectancy* that whatever has the properties of A will also have the properties of B, whereas in the case of *Some A are B* (an intersection) there will be a *preparedness* (but not *expectancy*) to find the properties of B associated with A.

Lastly, we can reach the internal representation of a union of two classes if in the above paragraph we substitute the condition *either A or B* for the condition *both A and B*.

***

8.3 - THE BRAIN'S STRUGGLE WITH NAMELESS OBJECTS AND CLASSES

In the first part of the last section I have described what kind of internal representations and specific reactions are required as preconditions for simple subject/predicate statements like

'John is ill'.

This statement could be simple because we had a single word available both to designate the subject (John) and the attribute (ill). Both were *nameable*. But in practice that tends to be the exception rather than the rule. In most cases the particular object about which we wish to make a predication is a *nameless* one, and so may be the property or feature we wish to communicate.

A word must therefore be said about what I shall understand by a *name* in the remainder of this discussion: I shall take the word *name* to be synonymous with what is sometimes called a *fixed designator*. This denotes in the main

1) *proper names*, i.e. expressions used to designate a particular object, person, town, country etc. (*Mars, Peter Smith, London, England*);

2) nouns denoting the ensemble of members of a class (*tables, Englishmen*);

3) nouns denoting a class in the abstract (*stickiness, colour*);

4) adjectives denoting membership in a certain class (*great, happy*), or their adverbial derivatives (*greatly, happily*);

5) prepositions denoting a relationship (*on, in, below, between.*)

Excluded, therefore, are descriptive phrases used to designate a particular object, person, town, etc. ('*the tallest towerblock in London*', '*the father of Peter Smith*'). This point has to be made because some philosophers also refer to such descriptive phrases as names.

To resume: to specify a single nameless object, a set of *nameable* classes has to be found which are such that their intersection yields a class with only one member, viz. the object in question. Similarly, to specify the *nameless* property or feature of the object concerned, a set of *nameable* classes has to be found whose intersection yields the nameless class in question. This having been done, a string of words has to be found which can convey through its structure and composition the logical relationships between these auxiliary classes that yield the object(s) or feature(s) concerned.

Often certain shortcuts are available here. For example, the English language contains words which permit certain *nameless* objects or classes to be designated by an understood reference to the context in which the utterance is made. Typical examples are: *this* and *that* (e.g. the expression *this tree* designates the member of the class *trees* which has been pointed at or mentioned before). Other examples are: *I* and *my* (reference to speaker), *you* and *yours* (reference to the person addressed), *yesterday* (reference to day preceding day of utterance), etc.

However, in respect of the major issues I want to discuss, these shortcuts are of no particular interest.

Let me take some concrete example, and illustrate in specific detail the process of substituting the intersection of an appropriate set of nameable classes (or classes defined by context references) for the nameless objects or classes one may wish to talk about - the most powerful instrument we have in the *descriptive* use of language.

8.3 - The brain's struggle with nameless objects and classes.

Suppose I want to communicate the fact that the only table in the study of a person called Peter stands on a Persian rug. Now I have a name for the general class to which this object belongs, viz. *tables*, but I have no name for the particular table that stands in Peter's study. Nor do I have a name for the attribute of standing on a Persian rug. If the common vocabulary of English could meet both these needs - if, for example, we were in the habit of naming tables like persons, and that particular table went by the proper name 'Tabletom'; and if all objects situated on a Persian rug were generally known as 'luxuriates' - I would have to say no more than

'Tabletom (is a) luxuriate'

to describe the fact I wish to communicate. And if the vocabulary of our language were so extensive that the same needs could be met in the same way in all other cases in which I wished to make a predication, all sentences could be of this simple and transparent structure - and most grammarians would be out of business. *'Tabletom is a luxuriate'* and *'John bought Tabletom'* would be typical examples of the kind of grammatical competences that would suffice for ordinary descriptive discourse.

But as things are, I have no name available either for the object or the attribute concerned. Thus to specify the object I have in mind I have to use the descriptive noun phrase *'The table in Peter's study'*. This table stands in a certain relationship to the carpet in question, viz. in the dyadic relation here specified by the verb phrase *'stands on'*. And the second terminus of that relation is filled in by the noun phrase *'a Persian table'*. The role of the definite article 'The' at the beginning of the sentence is to convey that the intersection of the classes used to specify the subject is a class with only one member, while the indefinite article '*a*' in the last-mentioned noun phrase completes the picture by adding the information that the class yielded by the intersection of the class *tables* with the class *Persian objects* happens to have more than one member. The point of this example is to show what complex internal representations the brain needs to be able to produce even such a comparative simple predication relating to nameless entities. For, to specify the subject it here needs internal representations of:

i) the object *Peter*;
ii) the class *tables*;
iii) the class of all rooms known as *studies*;
iv) the dyadic relation *belonging to*;
v) the dyadic relation *inside of*;
vi) the class of all classes having only one member;

and to specify the selected attribute of that table, it needs representations of:

vii) the dyadic relation *standing on*;
viii) the class *rugs*;
ix) the class of *Persian objects* (in the context taken to be understood as objects of Persian origin);
x) the class of all classes having more than one member.

Finally, it needs

xi) internal representations of the class-membership relation and of the intersections between the named classes that are used to specify both the object of which the predication is made and the feature predicated.

It does not matter whether the description of this state of affairs in Peter's study relates to something actually being perceived, something remembered or something merely imagined. In each case the *source* of the statement lies in a single categorizing reaction - in the first case to a perception, in the second and third to a mental image. And the list I have just given illustrates what additional internal representations have to be brought into play or created in order to communicate verbally both the nature of the object eliciting that categorizing reaction and the nature of that reaction itself. In other words, it is a highly sophisticated step from the single categorizing reaction that was the source of the descriptive sentence we have just analysed, to the complex set of internal representations into which it has to be resolved in order to be able to communicate it verbally in terms of nameable entities. Elsewhere I have called this step the *cognitive resolution* of the nameless into the nameable which is needed to cast a primary representation or categorizing reaction into a *propositional form* (Sommerhoff 1974); and I shall continue to use these expressions.

***

8.4 - WHAT THE SENTENCE-STRUCTURE CONVEYS

We must now turn to the question of how the structure of a phrase or whole sentence conveys the class-membership relations and class-relationships I have analysed in the above example.

In Fig. 8 I have symbolized this by way of an inverted tree structure not unlike the *phrase-markers* used in modern linguistic theory (cf. Appendix D). To distinguish it from these I shall call it a *logical phrase-marker* and refer to the others as *generative phrase-markers*.

8.4 - What the sentence-structure conveys

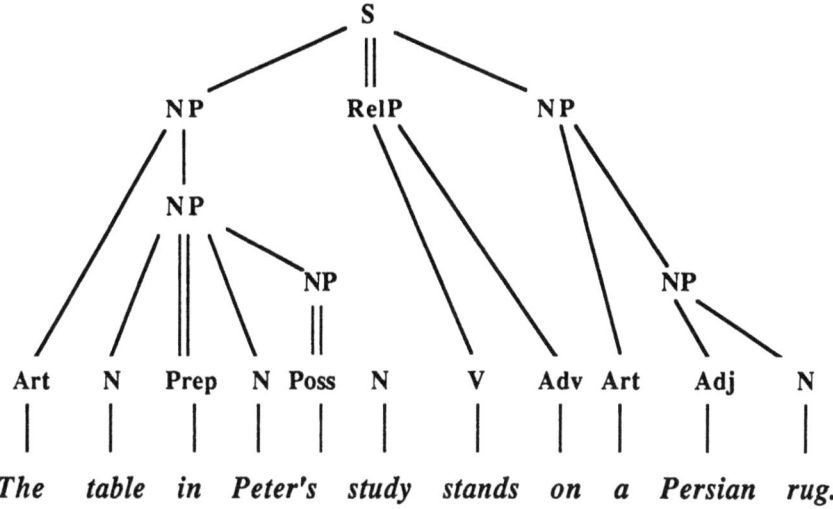

Figure 8 - *Example of a logical phrase marker as defined above.*
Legend: S denotes the content of the sentence as a whole; NP stands for noun-phrase and signifies a phrase (string of words) denoting an object or class of objects; VP stands for verb-phrase and RelP signifies a phrase denoting a relation between (here) two entities. The abbreviations, 'Art', 'N', Poss., Adv., and Adj. speak for themselves.

Simple resolutions of the type

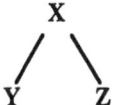

are here intended to symbolize that the entity covered by the phrase X is being specified in terms of the intersection between the two classes of entities designated by Y and Z respectively.

Pairs of such resolutions, combined in the type

are used to symbolize that the entity designated by the phrase **T** is being specified in terms of a particular relation (here designated by **V**) between the two entities designated by **U** and **W** respectively. As I have explained in Section 8.2, the internal representation of a dyadic relation requires internal representations of two intersections - hence the double bar here used to symbolize the link between **T** and **V**.

The logical phrase-marker given in Fig. 8 shows clearly how tidily the sequence of words in the sentence I have illustrated mirrors the logical relationships involved in the cognitive resolution that had to precede the production of the sentence.

We are justified to regard this logical phrase-marker as truly representative of the *deep structure* of the sentence, since it mirrors the structure of the internal representations which form the basis of the sentence and determine its meaning. It would certainly give the phrase *deep structure* a clear meaning in terms of the internal representations underlying the sentence and its construction.

I have remarked that the *generative* grammars constructed by linguists are always intended to be *descriptive* of what are regarded as permissible forms of sentence construction in the language under consideration. They are not *prescriptive*. And their type of phrase-markers is used as an instrument in this approach. By contrast, it follows from our analysis that if one were given the task of constructing a grammar from scratch, and one capable of generating descriptive sentences of the power I have examined, one would have to begin with *logical* phrase-markers of the kind described above as a statement of the linguistic competences the new grammar must be able to confer.

*

Obviously the preceding pages can serve only as a rough sketch of the suggested link between syntax and meaning. Yet they make the point that these are the kind of matters that need to be studied if that link is to be fully understood. The same applies to the link between language and reality, for that link is formed by the brain's internal representations of the world.

Our conclusions are also relevant in connection with certain major issues that still tend to be debated in the field of linguistic theory (cf. Appendix D).

Firstly, the notion of 'deep structure' has not found a uniform definition in different linguistic theories. I would argue that a satisfactory meaning of the term should be one which relates clearly to the processes actually involved in our linguistic competences - as does the interpretation I have suggested above.

Secondly, our analysis has something to say about another controversial issue, viz., the question of the common ground between all natural languages. For, it has brought out at least something that must be common to all natural languages whose syntax permits the nameless to be specified in terms of the nameable, i.e. languages

8.4 - What the sentence-structure conveys

which have truly descriptive powers. Clearly, all such languages need cognitive resolutions into a propositional form along the lines I have illustrated and analysed.

Thirdly, in the dispute between the *Extended Standard* theorists and the *Generative Semantics* theorists (see the final paragraph of Appendix D), my conclusions support the second group. To explain this, let me take the two sentences cited in the appendix as a critical case:

> *'Many pilots do not fly gliders.'*
> *'Gliders are not flown by many pilots.'*

Do these sentences have a common deep structure? Not according to our analysis and our definition of that term. They are based on different internal representations and cognitive resolutions.

In the first sentence we have the internal representation of a class formed by the intersection of the class pilots with the class of *people not flying gliders*. And the sentence states that this class has many members (i.e. is a member of the class of all classes having many members).

In the second sentence we have the representation of a class formed by the intersection of the class of *all gliders flown* with the class of entities *flown by pilots*. And the sentence states that this class has not got many members (i.e. is excluded from the class of all classes having many members).

*

To my mind, it is only through analyses of the kind I have illustrated in the preceding part of this chapter that the science of linguistics can find a sound scientific footing. A great deal of controversy might have been avoided if linguists had realized from the start the importance of tackling the problem of linguistic competence at its source, namely the roots that verbal expressions have in the brain's internal representations of the world, and the demands which the linguistic medium makes on the structure of those internal representations if they are to be capable of being projected into a verbal form. And, as I have said, it is at this level too, that the link between language and reality has to be explained. This link, of course, was one of the major problems that occupied Wittgenstein in the TRACTATUS. But his answers were far from helpful. Propositions, he maintained, are pictures of facts used as vehicles for thought - to which he added the belief that the relation between language and reality cannot meaningfully be discussed in language. Since the present analysis relates language to internal representations and internal representations to reality, we have no problem here. Nor do we have a problem with another issue that troubled Wittgenstein, namely the case of propositions expressing some non-existing, i.e. imagined, state of affairs.

Failure to understand these matters may also be taken to account for major misconceptions about the gap that separates man even from his nearest relations in the animal world, a gap that many devoted researchers have attempted to bridge by teaching higher primates some system of communication that could be called a language system. What they have achieved shows a clear lack in these animals of the power to form cognitive resolutions of the kind I have discussed. At best these animals seem to be able to master only a sign language paralleling the two-term structures of baby talk (*'John ill'*) from which I started in our analysis. Thus the sign-languages which star performers like the chimps Sarah, Lana and Warshoe, or the gorilla Koko, have been able to learn, fall far short of the complex cognitive processes on which, according to our analysis, the linguistic competences of the human species rests. Whether you can call this primitive form of communication, a 'language system' is a matter of definition, but if you do, you may be obscuring categorical differences between the powers of the human brain and those of these primates.

The evidence seems to suggest that in humans, the ability to perform these cognitive resolutions is an ontogenetically acquired one, but one that has to be acquired in infancy if it is to be achieved at all. Hence it seems to depend on genetically transmitted degrees of plasticity in the brain during specific phases of the maturation process. And it seems to be a plasticity whose fruitful development depends on extensive socialization. The suggestion that this ability cannot be acquired in later life, finds support in the famous case of the Wild Boy of Aveyron, a child of 11 or 12 years which was captured in a forest in 1799. Five years of devoted care and training by his mentor, a Dr. Itard, failed to achieve in broad terms more than what, in our time, has been achieved with chimps and gorillas. The boy could not learn to speak, but he could learn to associate signs and written words with objects and with his wants, and to use them as means to obtain satisfaction of his wants. It is also interesting to note in this connection that children who have learnt the use of language may lose it again if isolated for long periods.

***

8.5 - MENTAL IMAGES AND THE PROPOSITIONAL FORM

When we think, do we think in terms of quasi-pictorial images, or in terms of words and sentences, or in terms of something more abstract than sentences, viz., propositions?

Surely, any of these might be the case, depending on one's native or acquired dispositions and the circumstances. The professional engineer depends on his ability to work with diagrams, to picture in the mind some new mechanical device, to visualize it

8.5 - Mental images and the propositional form

going through its paces, and, perhaps, coming to the conclusion that one of its levers is likely to foul one of its gears. At the same time, though, he may recall in a *verbal* form certain mechanical principles, or rehearse in a kind of silent speech how he would explain the advantages of his device to anyone interested. By contrast, a journalist would mainly think in verbal terms.

Thus the individual's reliance on either mode of representation, obviously differs greatly from case to case. I happen to be visually orientated, even my memory is predominantly photographic: at school I could learn irregular French verbs best by visually remembering the relevant page in my French grammar. Other people rely much more heavily on verbal representations and verbal memory. They learn languages easily, but may have great difficulty with scientific diagrams and engineering drawings.

The reality of quasi-pictorial images as a type of internal representation differing radically from the verbal, was dramatically demonstrate by Shepard and Metzler in a series of experiments I have already mentioned: the speed of decisions about whether two geometric figures placed a different angles were actually representations of the same object, was found to be directly related to the angle through which either figure had to be rotated mentally in order to match the other. These findings were later reinforced by Kosslyn, Ball and Reiser (1978) in a series of scanning experiments in which subjects were first asked to memorize a map. Then they were asked to recall the map in their imagination and, on a given signal, to move as rapidly as possible an imaginary black spot from one specified landmark on that map to another specified landmark, and then to press a button. It was found that the time taken to move that spot was a monotonic function of the distance between the two specified landmarks.

Yet, despite such finding, Pylyshyn (1973) has consistently maintained that all our internal representations have a propositional form, and he has continued to defend that position. Put in a nutshell, his contention is that quasi- pictorial images are epiphenomena of beliefs and beliefs he takes to have a propositional form invariably. The argument falls apart if one identifies beliefs, as I shall do in Chapter 9, simply with the brain's internal representations of the world.

Moreover, the precise meaning and implications of 'propositional form' are never adequately clarified in these contexts. For most writers, the 'propositional form', just means the subject-predicate form. But we have seen above what the subject-predicate form presupposes by way of complex cognitive transformations of more basic representational levels. This flatly contradicts the idea that representations in propositional form are the original representations and mental images the derivatives.

Kosslyn's own view appears to be that quasi-pictorial entities occur transiently in active memory, and are accompanied by the subjective experience of having an image, and that these entities are generated from beliefs which are stored in long-term memory in a propositional form. .This receives some support from the well-supported belief that long-term memory depends largely on rehearsal, and no doubt much of that rehearsal may take the form of 'silent speech'. But it is hard to believe, for example, that the laying down of early childhood memories, say of the layout of the nursery, should *depend* on the child rehearsing that layout in silent speech. It seems more likely that it results from the child repeatedly transplanting itself in its pictorial imagination into the environs of its early years. On the other hand, it would not be inconsistent with the present theory to assume that when the brain has learnt to achieve a propositional form it uses this extensively in the rehearsal needed.for memory consolidation. (perhaps because the memory would then be available for recall in the most frequently required form).

Meanwhile Anderson (1978) has argued on logical grounds that, barring decisive physiological data, it will not be possible by any line of experimentation to establish whether a representation is pictorial or propositional. The logic of his arguments has been questioned by Johnson-Laird (1983), although he concedes that it is unlikely that the issue can be settled by psychological experiments. And with that I could concur.

***

8.6 - THE FOUNDATION OF RATIONAL THOUGHT

From the *descriptive* use of language we must now turn to the *argumentative* use. i.e. the brain's power to draw logical inferences in a verbal mode of expression. This, too, forms an exceedingly broad field of enquiry, so that only a few general principles can here be considered in the light of our analysis. Our main object must be to understand the basis on which these powers rest. For purposes of illustration, therefore, I shall confine myself to syllogistic reasoning, and only to the most common forms of the syllogism.

First we must turn to the major premiss of the syllogism and examine its four forms. These are

> All A are B
> No A is B
> Some A are B
> Some A are not B

Hence we must begin with the question of the nature of the internal representation of these alternative major premisses. This subject has already been covered in Section 8.2. The minor premiss, too - e.g. the minor premiss (Socrates is a man) in the classical syllogism

> All men are mortal
> Socrates is a man
> ....
> .: Socrates is mortal -

needs no further comment. But what about the conclusion? Two questions arise here: (1) why does it necessarily follow? and (2) how would we in practice arrive at it?.

To take the first question. The primary relationships on which the cogency of the conclusions of all forms of the syllogism rests, are the relationships between classes - such as the relationships of intersection, inclusion, and union. Venn diagrams (see Appendix A) make that point very clearly. And the rules of syllogistic reasoning are close cousins of the so-called *calculus of classes*. The cogency of their logic derives from the implications of these relationships. It follows simply from the nature of classes that if an object X is a member of a class A of objects, and class A is wholly included in a class B, then X will also be a member of class B. In acquiring a sense of logic, therefore, we are really learning to assimilate the intrinsic nature of the process of classification. Some of that adaptive process already occurs as the brain learns to assimilate that if A can be expected to produce B and B to produce C, then A can also be expected to produce C.

As regards the second question, it is worth noting a conclusion Johnson-Laird (1983) has drawn from a series of systematic studies. These suggest that the human mind does not commonly reason by internally citing the rules of logic, and then applying them to the premisses to which it is attending. In most cases, he asserts, we discover the conclusions that follow from the premisses of a syllogism by imagining simple instantiations of the premisses. For example, in the case of the classical syllogism cited above, a person may well simply picture in the mind a little line of men and opposite each a symbolic coffin or cross. That represents the first premiss. The second premiss may be represented equally imaginatively by attaching the label 'Socrates' to one of the little figures, so that a symbolic Socrates now appears opposite a symbolic coffin or cross. And this, the author suggests, adequately represents the conclusion.

All of this, would of course depend on the particular dispositions of the individual. Someone acquainted with Venn diagrams, might derive a logical inference simply from the mental picture of an appropriate set of such diagrams.

One of the specific examples Johnson-Laird has investigated is particularly telling. Consider the logical argument for a suspect's innocence, given the following premisses:

1. The victim was stabbed to death in a cinema while he was watching the afternoon showing of *Bambi*.

2. The suspect was traveling to Edinburgh on a train when the murder took place.

Any child can see the validity of the suspect's alibi. But how is the conclusion of his innocence arrived at in practice? The replies the author received to his questionnaires, showed clearly that the subjects had arrived at the conclusion by forming images of the premisses, and then searching in vain for any compatible representation of how the murder could have been committed by the suspect.

By way of additional support for this theory the author cites a number of experiments whose outcome suggests that the time taken over a syllogism depends on the number of models it requires, and that children can at first handle only logical arguments which depend on not more than one model. There is nothing in our analysis that conflicts with these conclusions. If this is the way the imagination may in practice enter the process of drawing logical inferences, then this is just one more function to be added to the list of functions I gave in Chapter 7, and it poses no special problems from our point of view.

***

Chapter 9

MIND AND MATTER

SUMMARY

In this chapter I shall begin with a brief review of the functional and physical account of the main categories of mental events given in the earlier chapters. I shall then discuss the bridge between mind and matter which this account has produced, i.e. the main definitions and hypotheses through which it permits statements about mental events to be translated into statements about physical events or physical patterns of relationships (from which, in turn, inferences can be drawn about the neural correlates of those mental events). With this picture before us, I shall then look at a number of important questions frequently debated in connection with the mind/matter relationship: questions relating to the freedom of the will, the unity of the self in split-brain subjects, the precise meaning that can be given to terms like *knowledge* and *belief*, and the conditions AI would have to satisfy to achieve genuine 'mind modelling'.

***

9.1 - CATEGORIES OF MENTAL EVENTS

Mental events appear in our account as internal representations of actual or possible objects, events or situations, which have become part of conscious experience through what we have defined in Chapter 4 as second-order self-awareness (our account of what Locke called "the perception of what passes in a man's own mind"). The greater part of that account concerned the physical nature of these different kinds of internal representations.

As regards the different kinds of mental events, an account of the physical basis of the *mental images* that occur in thought and memory recall was given in Chapter 7, while in Chapter 8 we discussed the internal operations needed for such representations to be converted into the *propositional form* required by verbal utterances. *Desires* were treated as a subcategory of mental images, viz. mental images which have acquired a motivational quality by virtue of the positive expectancies of need satisfaction which they elicit. *Volitions*, in turn, were taken to be desires which are actively being pursued, generally in the form of behaviour designed to diminish the relevant discrepancies between the actual situation and the desired one. Meanwhile, *sensations* and *feelings* were taken to be peripheral or internal stimuli or conditions (e.g. physiological needs) whose occurrence has been added in second-order self-awareness to the brain's global world model as being part of the current state of the self.

According to this account, our *everyday thoughts* are a stream of of attended internal representations of both actual and merely possible objects, events or situations. These representations may or may not be cast into a propositional form, for one can think both in terms of quasi-pictorial images (as when one wonders how to fit a carpet into a room), and in terms of silent speech (as when one reasons what to do next in a given situation). For the former to be transformed into the latter, the brain requires the capacity to form internal representations both of the class-membership relation and of logical relations between classes (inclusion, intersection etc.) - as explained in Chapter 8.

By confining these remarks to 'everyday' thoughts, I wish to indicate that they apply only to the basic kind of thoughts I have illustrated in the chapters concerned. Since our main aim has been merely to understand the fundamentals, it is beyond the scope of this work to go further. To cover the subject of human thought comprehensively, one would have to cover, for example, such sophisticated forms of abstract thought as the manipulation of mathematical symbols. This would take us much further afield than I am qualified to go. The logical foundations of mathematical thought are still in dispute among the professionals.

It has sometimes been argued that thoughts can have no neural counterpart, because, unlike definable physical events, they have no beginning and no end, and unlike concrete physical objects and processes, one cannot have half a thought or three quarters of a thought. But if thoughts are seen as we have come to see them, namely as representations which are part of the continuous flow in the brain of internal representations of actual and merely possible objects, events or situations, then there is no way of timing them introspectively. Because the very attempt to do so would replace these thoughts by other thoughts. Nor can they be timed by an outside observer, except via verbal reports by the thinker, which would have the same effect. Moreover, since these internal representations are operational modes of the brain as a whole, their physical counterpart is clearly not divisible either. You cannot have half an operational mode of the brain or the operational mode of half a brain. You can have an internal representation of half an object, or a deficient representation of a whole object, but you cannot literally have half an internal representation of a whole object.

***

9.2 - THE BRIDGE BETWEEN MIND AND MATTER

Along the pathway I have followed, we have in effect constructed a bridge between mind and matter. For we have formulated a set of definitions and hypotheses which enable us in general terms to translate statements about different categories of *mental* events into statements about different categories of *physical* events or patterns of *physical* relationships. The crucial formulae are contained in the following six steps, here listed in the order in which they would be applied in such a translation:-

1. We have defined *consciousness* and the main categories of mental events in terms of *internal representations* of three major categories (Chapters 4 and 7).

2. We have given a functional *definition* of what is to be meant by an *internal representation*. The definition was formulated in objective terms, viz. in terms of the way in which the brain responds to the internal states concerned (Chapter 4).

3. As regards the *structure* of these representations, we have postulated that the brain's internal representations of *actualities* consist of states of conditional *expectancy* (Chapter 5), while its internal representations of mere *possibilities* consist of states of unconditional *readiness* (but not expectancy), viz. readiness for particular perceptual experiences (Chapter 7). The neural correlates of the relevant states of conditional expectancy or conditional readiness respectively, were discussed in Chapter 6.

4. Prior to this, we had defined *expectancies* in terms of states of *readiness* for particular events coupled with a disposition of the brain to modify these states if the brain detects discrepancies between what is expected and what actually occurs. (Chapter 3). How it detects these discrepancies and then responds to them was discussed in Chapter 6.

5. We had defined states of *readiness* for an event as states which facilitate *appropriate* responses to the event in question should this event occur (Chapter 3).

6. We had defined *appropriate* responses in terms of the concept of *directive correlation* (Chapter 3).

7. And we had defined *directive correlation* in as a particular pattern of *physical* relationships (Chapter 3). This concept, therefore, furnishes the final link in the conceptual bridge we have forged between the mental and the physical.

9.3 - THE UNITY OF THE SELF IN SPLIT-BRAIN SUBJECTS

The account I have given of consciousness and self-awareness enables us to sort out some of the conceptual confusions that tend to occur when questions are asked about the unity of the self in split-brain subjects.

The main communication between the two cerebral hemispheres occurs via the commissural fibres (800 million in all), mainly those of the corpus callosum (Fig. 6.2). When these commissural fibres are severed by a midline section, the brain functions in a kind of Y-configuration, the stem of the 'Y' being formed by the sub-cortical centres in which bilaterality remains intact.

The consequences of such commissurotomy flow mainly from the fact that the two hemispheres differ substantially in some of the categories of sensory inputs they receive and motor outputs they control. Many sensory inputs project to both hemispheres. These include a variety of sensory inputs relating to the body and its posture, all auditory inputs, visceral sensations such as hunger and thirst, and sensations of temperature, pressure and crude pain. Bilateral motor controls are also extensively present and, for example, exploratory movements of the eyes can provide bilateral representations of the perceived objects and scene. (Sperry, 1974, 1984).

On the other hand, each hemisphere receives only inputs from the contralateral half of the visual field of both eyes. Similarly, sensory representations of the right hand, arm and leg are received only in the left hemisphere and the converse applies to the right hemisphere. Thus the *left* hemisphere apprehends, as it were, what is held in the *right* hand and also controls motor adjustments in that hand, while the *right* hemisphere deals with the *left* hand. Moreover, in 95 % of the population the main constructive language functions are primarily linked to the left hemisphere, whereas the right hemisphere tends to specialize in non-verbal functions such as hand-eye coordination and visual imagery.

Typically, therefore, split-brain subjects are unable to recognize by sight any object which has just been perceived in the right half of the visual field when that object is shifted to the left half. Similarly, objects which have been explored and identified tactually with the right hand while out of sight, cannot subsequently be recognized tactually by the left hand. Nor can they be recognized visually if subsequently presented in the left half of the visual field. Typically, too, split-brain subjects cannot report verbally on objects felt with the left hand or seen in the left half of the visual field.

During recovery from the operation the severed connections do not reform. Superficially, therefore, it may seem surprising that, after recovery these subjects are found to behave so normally that only expert tests can reveal their deficiencies. But it cannot be a surprise once one has come to see that the main cognitive function of the hemispheres lies in assisting the construction and up-dating of the brain's internal model of the world and of the self-in-the-world.

According to our analysis, the main body of this internal model consists of states of conditional expectancy relating to the way in which inputs will transform in consequence of the outputs open to the organism. The severance of the commissural fibres will result in absence of certain conditional expectancies that would have formed under normal circumstances, e.g. conditional expectancies requiring cross-hemispherical associations between, for example, information about a self-induced movement and information about the sensory consequences of that movement. Thus no expectancies may be able to form about the sensory consequences of an act if the act can only be registered in the left hemisphere and the consequences only in the right hemisphere. For example, if the subject maintains a fixed gaze which keeps the left half of the table in front of him firmly in the left half of the visual field, no expectancies will be formed of what the right hand will feel if it moves across the left half of the table. Again, perception of an object in the right half of the visual field will fail to trigger expectancies about the sensory inputs that will result if the object is shifted to the left half of the field, and so on. But since in practice the eyes would always scan the scene and since, anyhow, the conditional expectancies that constitute the subject's internal model of the world have a high degree of redundancy, these gaps are unlikely to make a noticeable impact on the observable behaviour of the split-brain subject. At the worst, they may result in *neglect* of certain features of the world.

On our analysis these gaps can certainly not threaten the *unity* of the brain's internal model of the world and of the self- in-the world, for that unity is impressed upon this model by the factual unity of the world and of the body, since the brain's states of expectancy are continuously being checked against the world as it is factually. The world is not split by this brain lesion nor, therefore, is the brain's global model of the world, even though conditional expectancies requiring cross- hemispherical associations are now missing from the totality of those that constitute the model. Moreover, these gaps will not be mentally perceived as gaps – just as, for reasons explained earlier, the blind spot in the eye is never perceived as a hole in the visual field.

The conceptual confusions to which I have referred relate mainly to two questions, viz.,

1. to what extent is the unity of the self affected by the severance of the commissural fibres?

2. to what extent can consciousness be attributed to each of the two separated hemispheres?

The first question has already been answered. In connection with the second, it is not uncommon to meet the view that consciousness resides in the speaking and communicating hemisphere, while the other hemisphere merely acts as an unconscious automaton. But on our analysis the question itself is meaningless. Consciousness, as we have come to perceive it, does not reside in any specific locality or region of the brain. It lies in an operational mode of the brain *as a whole,* viz. one which amounts to the effective operation of a comprehensive and coherent model of the world and of the self-in-the-world. This includes those parts of the internal representation of the self which have been furnished by the faculty of second-order self- awareness. On this analysis, therefore, it is meaningless to attribute a separate mind to each hemisphere, as has been suggested by Pucetti (1973), for example. In other words, it is meaningless to describe the split-brain patients as 'split-mind' patients.

We must also be careful not to jump from the notion of a split-brain to that of a split personality. By split personality we mean a condition which relates primarily to a subject's motivation and attitudes and the goal's staked out in consequence. It is a condition in which, inexplicably, the subject seems to be under the control of two alternating and wholly distinct, yet internally self-consistent and hierarchically organized, set of motivational forces. It is not, therefore, primarily a cognitive disorder. And, although the frontal lobes are known to be involved in processes which affect our attitudes to things and the staking out of goals, their main role here would seem to be that of evaluating the brain's internal representation of current situations in the light of basic emotional attitudes determined lower down in the system.

***

9.4 - THE FREEDOM OF THE WILL

The supreme importance of second-order self-awareness is reflected in the notion of the *freedom of the will.* We touch here on an apparent antithesis which, more than any other, has tended to make people think of mind and matter as incompatible entities. In very general terms the problem may be described as follows. On the one hand, the brain is a physical entity composed of atoms, molecules etc., and from the standpoint of science it must be assumed that all processes and events that happen within the brain are rigidly governed by the laws of physics and biochemistry. Hence, whatever our behaviour may turn out to be, we must assume that it is the product of activities in the brain which have run their course in strict conformity to these laws.

On the other hand, in everyday life we also take our behaviour to be governed by our mind and we generally tend to think of this entity as a *free* agent. We feel that our intentional activities are the result of freely made choices, that we can act as we please.

Hence one may feel driven to ask: *how is it that our brains, which are made of molecules and atoms obeying the laws of nature, support thinking, which seems not to be governed by those laws?* Many people regard this question as central to the problems posed by the human mind.

No doubt this apparent antithesis between the notion of causal determinism and the notion of a free will has contributed significantly to the feeling that sensations, thoughts, phantasies, memories, volitions etc. are in some sense non-physical attributes of the people to whose mental life they belong.

However, in the light of the account of mental events developed in the preceding chapters, these two aspects of the matter are readily reconciled, and the contradiction is seen to be only an apparent one. For it follows from our definitions and hypotheses that voluntary activities will in fact be perceived introspectively by subjects as governed by a freedom of choice. Voluntary activities (as I have defined them) are activities which are determined by the brain's internal model of the world and of the self-in-the-world, their goals being set by our desires, i.e. by that part of the model which relates to the current needs of the self. More specifically, our action-decisions are determined by those of our desires which are being translated into action, i.e. by what we have defined as our *volitions*. Moreover, our volitions are a *sufficient* determinant of our voluntary actions, i.e. of the goals they are set to pursue. The link between them is an *unrestricted* one. Thus the realization of the goal set for an the action may be impeded by external circumstances, but not the link between the volition and the set goal. In that sense therefore we are *free to act as we will*. In the same sense we are *free* to *think* as we will. Moreover, we also *feel* free to act as we will, because we can become aware in introspection of the unrestricted nexus between volition and action.

I believe that this is all that needs to be said about the matter. For I agree with Quine (1960) that the 'freedom of the will' is just "the freedom to do as we will". In other words, the 'freedom of the will' denotes no more than the ability (as perceived introspectively) to perform actions whose goals are completely determined by the subject's current and conscious volitional state.or to have thoughts which are so determined. Seen in this light, therefore, the contradiction proves to be only an apparent one: there is nothing in this interpretation that would conflict with the assumption that the underlying physical brain processes operate in a deterministic fashion.

MacKay (1978) gives an illuminating counterexample to the claim that what he calls the 'I-story' and 'brain-story' respectively are mutually exclusive. "Suppose that a neurophysiologist sets up a computer to solve the Hodgkin-Huxley (H-H) equation under specific boundary conditions. Provided he does the job properly, he can truthfully claim that the behaviour of the computer is determined by the H-H equation and his boundary conditions. At the same time an electronic engineer analysing the chain-mesh of physical cause-and-effect in the machine can equally truthfully claim that *this* determines completely its behaviour."

Now the reader may well assent to the above account intellectually, but nevertheless feel uneasy. For it may seem to him that in any voluntary act the causal determinism of the underlying neural processes must make that act wholly predictable.

To dispel these doubts let us recall how much enters into the determination of a voluntary act: nothing less than the subject's global model of the world and of the self-in-the-world. According to our account, this consists of an operational mode of the brain which is the product of current sensory inputs acting on brain states which have been moulded and articulated by a life-time of antecedent experiences. Other internal representations enter as well, e.g. representations of the objects of one's desires. So much, in fact, enters, that there is simply no question of predictability even in theory, except in the trivial sense that an individual who had been endowed with the same genetic make-up as a given individual, who ever since conception had occupied an identical spatial position in the world and who had suffered identical experiences throughout, would have made identical decisions under identical circumstances.

***

9.5 - KNOWLEDGE AND BELIEF

In the course of our work we have arrived at precise and objective definitions of a variety of concepts relating to the consciousness in general and different categories of mental events in particular. If we could extend this conceptual precision to cover also the concepts of *knowledge* and *belief*, a number of questions often prompted by these concepts could be given a clearer formulation and, perhaps then also, a definitive answer.

In the treatment of *knowledge*, it is common to distinguish between knowing how (e.g. *how* to ride a bicycle) and knowing *that* (e.g. that the sun is shining). Since the former relates merely to acquired skills, only the latter is of primary interest in the present context. Wittgenstein took knowledge (in the second sense) to mean the possession of a true description of a state of affairs. We can replace this by defining a subject's knowledge as the totality of valid representations in his internal representation of the world and the self-in-the-world. And we can interpret the statement that A has knowledge *of* some state of affairs B as meaning that A's representation of B is part of this valid totality. Our criteria for what constitutes a *valid* internal representation were given in Chapter 4: the internal representation of an object, event, or situation is valid if the conditional expectancies of which it is composed conform to what is, or would be, the case - for example, if the expectancy that the object in my hand will break if dropped, conforms to what in fact would happen.

9.5 - Knowledge and belief

This definition of knowledge is more general than Wittgenstein's, because the word 'description' in his definition suggests that only representations cast into a propositional form (cf. Chapter 8) qualify as knowledge, whereas no such restriction is implied in our definition. And I am sure that is nearer to what we commonly mean by knowledge. On the common view, one can have knowledge of a locality without being able to put any of it into words.

On the common view, too, one can have knowledge of a fact without being conscious of that fact. Our definitions take care of this since we equated being conscious *of* something with the brain having singled out by a shift of *attention* some component of the brain's global internal representation of the world and the self-in-the-world.

In many cases the representation thus singled out would be present in a propositional form. But that is not a necessary condition.

*

Belief differs from *knowledge* primarily in the fact that beliefs can be mistaken, whereas knowledge cannot.

We can meet this requirement if we define *belief* as the possession of some internal representation of a state of affairs as part of the brain's global world-model, *regardless of the validity of that representation*.

Knowledge may thus also be described as the totality of our valid beliefs. Note that, according to this definition, beliefs, as much as knowledge, need not be present in the brain in a propositional form - though, of course, one cannot *discuss* a belief unless it is cast into such a form.

Why can 'belief' occur in the plural, whereas 'knowledge' cannot? The reason, as I see it, lies in the fact that it is usually just individual representations that are mistaken - not the totality of our internal representations. Hence, by contrast with the notion of knowledge, the common notion of belief has come to relate, not to the brain's global world-model as a whole, but only to individual parts thereof.

Can one have *unconscious* beliefs? According to our definitions representations which are part of the brain's global world-model would be unconscious if they are not covered by second-order self-awareness.

These observations also give a new slant to the meaning of *truth*. Here we must distinguish again between internal representation which have a propositional form (Chapter 8) and those which do not. In logic the universe of discourse of *truth* and *falsehood* is generally taken to be restricted to propositions. On the other hand, in everyday life we often speak of such things as having a 'true perception' of somebody's character. One could accommodate both these notions in a single concept by defining truth as the mark of any valid internal representation regardless of whether it is cast into a propositional form.

It is worth noting in this respect that the brain has no test for the validity of any of its internal representations except in a context in which they are applied as part of its global world-model and subjected to the reality-testing processes which I have described in detail in Chapters 5 and 6. The above suggestion, therefore, would combine elements of both the 'correspondence' and 'coherence' theories of truth.

To illustrate how definitions on the lines suggested above can place old and unresolved philosophical problems into a new perspective, I need only give one example, viz., philosophical questions of the type *how do I know* that I am in pain, that he is in pain, that other people have minds, that the image before my mind is the image of yesterday's picnic etc.? We have settled this question in objective, albeit very general, terms by having defined knowledge in terms of internal representations and having described in objective, non-mental, terms how internal representations of different kinds come to be formed and tested, including those which amount to the "perception of what passes in a man's own mind". We have also seen that the processes which participate in this formation need not themselves be conscious processes. And that is a fact worth recalling here, because it shows that it can be a mistake to follow the philosophical tradition of looking for an answer to the above question in purely *mental* terms.

***

9.6- ARTIFICIAL INTELLIGENCE

The achievements to date of AI in the production of artefacts whose overt achievements rival some of the overt achievements of the human brain, are not in dispute. In the application of sophisticated algorithms to the solution of complex formalized problems these artefacts may even greatly surpass the powers of the human brain.

What must be disputed, though, is the not uncommon belief that such artefacts are achieving a genuine modelling of mental events and mental processes, and that they can thus reveal to us how the brain achieves what it achieves. The common argument is that the brain *must* be functioning along the lines of these impressive computer programs, because it seems impossible to think of any other way in which the brain could achieve what it does achieve. The theory I have set out in this volume should at least prove that it *is* possible to think of other ways.

The simple truth is that to *emulate* an overt competence is not the same as to *simulate* the method through which that competence is achieved. Hydraulic jacks may be said to *emulate* the action of human muscle, but they can hardly be called *simulators* of muscle.activity. Anyhow, for the brain the formalization of a problem and the application of appropriate algorithms to its solution is a regimented exercize which is the exception rather than the rule in the normal flow of human thought.

Computers are manipulators of sets of binary digits which their programmers can cause to be representative of real-word variables if they so desire. And this can form the basis for formalizing real world problems to whose solution the computer's software can address itself. But this kind of representation is very different from the kind that I have defined for the internal representations of the brain. For the structure of the latter is an analog of the structure of the things they represent, whereas that is not so in the computer case. The computer's internal representations of external variables are merely of the nature of *symbols*, since we mean by a 'symbol' precisely a representation whose structure is not an analog of the structure of the thing symbolized. Hence it is misleading to talk of the mind, as some cognitive scientists do, as a manipulator of symbols. For Margaret Boden (1988), for example, this is the very starting point for a theoretical discussion. As she sees it, both minds and computers process symbols in accordance with a formal set of rules. And a truly scientific psychology would be a computational psychology. Obviously I find myself in total disagreement. Symbols and formal rules are used by the brain only in some of the highest brain functions, e.g. in verbal and mathematical thought. But the symbolic operations which a fully developed and trained brain may execute when it exercises its power of rational thought, must not be confused with the basic processes in the brain on which these competences rest, e.g., the processes that create the internal representations with which human thought operates. These are all part of what we call 'mind'. The former can be copied by computers, the latter very much less so in the present state of the art. Besides, writers who make a lot of play with the notion of symbol, generally tend to leave it undefined and invariably fail to give objective criteria for distinguishing a brain-event that operates as a symbol from one that does not.

Moreover, the brain is not the kind of system that solves the problem of producing an appropriate response to a given situation by breaking up the problem into a logical sequence of discrete steps, as the designers of computer hardware and software are wont to do, subsequently providing for each of these steps a dedicated module in the one case or an algorithmic sub-routine in the other.

Thus the brain does not establish the distance of a perceived object by explicitly using algorithms based on the laws of Euclidean geometry as a robot might do. Of course, if, on the basis of its learnt responses, the brain produces outputs which correctly reflect those distances, then those outputs will accord with the laws of Euclidean geometry. These laws will be implicit in the constraints to which the brain has become subject while learning to respond appropriately to the variety of distance cues available to it, such as binocular parallax, movement parallax, texture gradients or relative size. That learning process has taken the form of innumerable tiny but enduring changes in the brain's connectivities and degrees of neural responsiveness which has resulted in sets of conditional expectancies that faithfully reflect the distances concerned. On these expectancies the geometry of space has impressed itself *implicitly*. But there has been no *explicit* application of geometrical laws in the brain's construction of its internal model of the world, no *calculation* of individual parameters.

Hence, strictly speaking, it is also misleading to talk about the brain *computing* the distances of objects. For the word 'computing' has traditionally always been understood to denote a process in which a required answer is arrived at by way of a sequence of explicit logical steps.

It is a popular slogan in AI that whatever can be formalized can also be mechanized. Hence much of the search has concentrated on formalizations.

Computers can be programmed to model any logical system within the limits imposed by their memory capacity and processing speed. It is not too difficult to formalize the steps that are taken in a mathematical or logical argument, and even the basic rules of grammar can be attempted. Also, for example, one can formalize the possession of some given property by an object by flagging the object-symbol with a marker and programming the machine to respond appropriately when the object-symbol occurs coupled with the marker-symbol. Similarly one can formalize a relation between two objects by representing both as members of a class of ordered pairs of objects - the marker symbol now coupling the two object symbols. Such attributions can even be elegantly symbolized in diagrams like those used in Minsky's 'frames for representing knowledge' (Minsky 1975). Thus for any system whose operations concern elements that can be symbolized in this way and conform to rules, laws or algorithms which can in turn be formalized, AI has been very successful in exploiting these techniques. But the use of these techniques has always relied on a *serial processing* of the formalizations concerned.

As has been said, though, none of this amounts to the mind- modelling in a genuine sense. States of consciousness, including our knowledge of the world, amount to a 'set' or 'operational mode' of the brain as a whole in which certain functional relationships (mainly of a *representational* nature, as that term was defined in Chapter 4) are realized. And each such operational mode passes into the next through a multitude of delicate shifts of neural excitability in the numerous networks involved in the updating and evaluating of the brain's model of the world. The whole affair is a massive manifestation of *parallel* rather than *serial* processing. There is little here that can be formalized, hence mechanized in the sense here generally envisaged.

This integral or 'holistic' manner of the brain's operations, has clearly come out in our analysis. It also rules out as a possible model of brain function the kind of recursive processes that were made famous by the Turing machine.

The Turing machine was a brilliant thought experiment which demonstrated the extent to which in theory even extremely complex problems can be broken down into sequences of very simple and, in this case, *recursive* steps. It was devised by the

9.6 - Artificial intelligence

Cambridge mathematician Alan Turing (1936), one of the founding fathers of computer science. It was he who gave us the original idea of shifting onto the software the breakdown of the problems which the new automata were intended to solve - thus sealing the fate of computers as slaves of the human race, with the software acting as the medium through which the human master imposed his will onto the machine.

Since any problem-solving process is an adaptive process whose goal is to produce an output which satisfies certain specific conditions, and since the Turing machine can be regarded as proof that such an answer can in principle be found by a series of recursive steps, the suggestion has sometimes been mooted that perhaps the brain solves its own problems by means of comparatively simple recursive steps (Johnson-Laird 1983). It will be obvious from what has gone before that this suggestion is at variance with the bulk of our own conclusions.

[Briefly, the Turing machine is a logical machine, designed to compute any desired number. It is governed by a tape reader and a sectored data tape of infinite length, which the machine reads one sector at a time and which it can only move one sector at a time, either to the left or to the right. Each sector is either blank or carries one of a finite number of information-carrying symbols. The machine can read the symbols, erase them or replace them and add new ones on a blank sector. The internal structure of the machine, at any point of time, was conceived simply as one of a finite set of numbered 'states' contained on a suitably fashioned 'program tape'. This contained only four different kinds of instructions: (1) read the symbol on the currently scanned sector of the data tape; (2) replace the symbol by another, erase it, or print one if the sector is blank; (3) move the data tape either one step to the left or to the right; (4) switch the machine to the next 'state' (or set of instructions). The computed number can then be obtained from the symbols that remain on the tape when the program ends. Owing to the infinite length of its tape, the Turing machine has an infinite memory capacity.]

Although great progress has now been made in the field of parallel-processing computers, notably since the arrival of the transputer, and many of the notorious difficulties of programming such systems are being tackled with apparent signs of success, it is yet early days.

Noteworthy in this field are the efforts of the *connectivist* school of AI to simulate the holistic character of higher brain functions, as in shape discrimination. To this end they are busy investigating the network properties of systems consisting of a very large number of inter-connected units, each designed to simulate the basic properties of a neuron as a multiple-input, single output device with modifiable sensitivities at the receptor sites. The units of these networks are parallel computing elements which have the power to influence their neighbours to a degree that is computed as the product of an *activation* value assigned to the source unit and a *weight* assigned to the contact it makes with the neighbour concerned. Both quantities can be either positive or negative, so that the effect on the neighbour may be either excitatory or inhibitory. The networks are intended to learn to perform the discriminations desired by the designer by way of a gradual modification of the weights in accordance with tentatively supplied algorithms (Smolensky, 1988).

Positive results are reported in such fields as shape recognition, language processing, logical inference and motor control. However, although it is claimed that such results will eventually help us to understand cognition, our analysis must cast doubts on the ability of such 'neural computers' eventually to produce genuine 'mind-modelling' - at any rate so long as their designers still lack a clear conception of the nature of consciousness in general and of the different categories of mental events in particular.

A laudable feature of these attempts is that they have dropped the idea of the brain operating in a modular fashion except for that basic module, viz. the neuron. Only too often the tacit assumption is made that the brain consists of neat networks of modules above the neural level, each designed to do some specific job or take some specific logical step, as would be the case in the modules of a computer and in the subroutines of its software. Everything we know about the brain argues against this assumption. It is hard to reconcile, for example, with the recuperative power of brain functions after severe lesions despite the fact the lost neural tissue is never replaced. The brain is a complex adaptive organ which has evolved in the course of time, never quite knowing, so to speak, what would next be expected of it. It was never designed *ab initio* on a drawing board - a process that tends to lead to a modular design as the required functions are tackled one by one.

I am not denying that locally, in the cortex for example, neural networks may exist which are of a modular nature and designed to achieve some local effect. Indeed, in my LOGIC OF THE LIVING BRAIN I have drawn attention to the way in which local cortical networks based on certain kinds of backward- acting inhibitions may result in a local adaptive plasticity which can cause any specific output configuration of the network to become associated with any specific input configuration should this association prove to be rewarding in practice. Nor am I denying that in the primary processing of the stimuli reaching the peripheral receptors certain very basic functions, such as detecting shading edges or enhancing contrasts in the visual field, may be performed in a modular fashion at the input side. But in the production of the internal representations essential to consciousness, including the end-products of visual perception, the brain acts as an integral whole and there is only one relevant module: the brain in its entirety.

Another point to be mentioned is this. For each of us the most immediate reality is that of *subjective experience,* i.e. what we apprehend in second-order self-awareness. Yet, as Ullman (1980) has pointed out, the mediating processes in the computational/representational theories favoured in AI, do not operate on subjective experience. The nature of subjective experience remains for these theories a total mystery, whereas it has been fully accounted for in our analysis (cf. Chapter 7).

This brings me to a further, and related point. Anyone who has read samples of the 'dialogue' which Winograd (1972) has been able to conduct with his software robot SHRDLU about the robot's pre-programmed little world of cubes, pyramids and boxes will be struck by the life-like and intelligent impression given by the robot's replies to his 'friend's questions or instructions. A typical example:-

INSTRUCTION: *Find a block which is taller than the one you are holding and put it into the box.*
ROBOT'S RESPONSE: *By "it" I assume to mean the block which is taller than the one I am holding - OK.*

However, one also has to realize that while SHRUDU's words mean something to the programmer and experimenter, they mean nothing to the beast itself. Thus SHRDLU readily draws the inference that, since pyramids are pointed and (according to the rules with which it has been primed) single pointed objects cannot support other objects, a single pyramid cannot support a cube. But in the 'mind' of this robot pyramids and cubes remain just tokens or counters which it has been programmed to manipulate in certain ways, and which have no meaning bar the list of properties supplied by the programmer. And these properties again are just counters or tokens to be manipulated in the manner stipulated by the designer and programmer.

By contrast, in humans all objects and properties have meaning in the sense, for example, that they are represented in the brain's model of the world by a host of expectancies about what sensory and emotional experiences in the given circumstances would result from meeting them, attending to them, manipulating them, or attempting to change them - including expectancies of need satisfaction, for example. This context also includes the cultural and social contexts to which the subject seeks to relate himself imaginatively or emotionally. Until the programs of AI can simulate all this in real time the symbols with which the robot operates will remain for it no more than tokens devoid of meaning in the common sense of this word.

I have chosen SHRDLU as an example, because it was one of the first programs to produce a seemingly intelligent and meaningful dialogue, and it is still one of the best-known. Since those days many other and equally impressive programs have been devised, including story- or screenplay-writing programs, programs producing dialogue biased towards specific attitudes of mind or ideologies, diagnostic programs, etc. They are all open to the same criticism.

Winograd himself was only interested in questions of semantics, syntax and logic, and would not claim that his program simulated genuine mental activity. Indeed, in more recent times he has become very conscious of the limitations of current methods in AI when it comes to genuine mind-modelling and to simulating the higher brain functions (Winograd and Flores 1987).

The misconceptions I have discussed are often fortified by a deplorable tendency to describe AI programs in psychological terms, thus suggesting that a true simulation of the human psyche has already been achieved. One well-known book in AI is even entitled "The Psychology of Computer Vision"! Defenders of this practice will claim that they need a psycho-biological language because that is the only way in which those features of computer function can be identified that are analogous to the psychological aspects of people. They overlook that the analogies relate only to the problem-solving *accomplishments* of the human brain, and not to the play of internal representations and their transformations (partly conscious, partly unconscious) on which these human accomplishments rest – nor, therefore, to the underlying *psychology* of the accomplisher.

Finally, let me return to an earlier question: could robots eventually be produced that might be credited with consciousness? On the strength of the analysis I have given, one's immediate answer might be: yes, if robots could be produced that are capable of forming the categories of internal representations in terms of which I have defined the faculty of consciousness (including self-consciousness) in Section 4.6. But while this answer would then be technically correct, it would not follow that we would in consequence extend to such robots also the emotional and moral reactions that we extend to human beings and other living creatures credited with consciousness. Thus we would be unlikely to treat deliberate injury to that robot as cruelty, or its destruction even as murder. The reason being that deeply imbedded in our human consciousness of the world lies the culture into which we have been born and bred; also a sense of our own identity and that of the community to which we belong; a sense of how we are linked with our fellows and with the whole living world beyond; a sense of communion and of the empathy on which it rests; above all, a sense of all that has gone into bringing life to this level of richness, hence of the sacrilege of undoing any of this. Our robot would stand outside of all this. There would be no valid basis for feelings of empathy. We could love it only in the sense in which an artist might love his creations – a projection of our own being. It would not be loved in its own right.

***

9.7 - CONTRASTS WITH THE APPROACH OF ACADEMIC PHILOSOPHY

Not infrequently one meets scientists who fight shy of the problem of consciousness, and claim that it had best be left in the hands of the academic philosopher. They are mistaken on three counts. Firstly, because consciousness is a faculty of a living organism and that organism can never be fully understood by the natural sciences so long as that faculty remains a scientific mystery. Secondly, because, by contrast with the scientist, the academic philosopher feels free to look at a problem from any standpoint he pleases (subjective, objective, metaphysical), and discuss it in any terms he pleases, elegant metaphors being especially favoured. Hence the language of academic philosophy does not in general have the precision and unequivocal reference to observable events that would permit scientific inferences to be drawn from any generalization about mental events at which the philosopher may arrive. Thirdly, because the philosopher is not in the business of formulating hypotheses, whereas this is the scientist's way of arriving at his account of the natural world.

To say this, of course, is not to disparage the work of philosophers in their own proper field, though philosophers themselves are often divided as what that field should be. Speaking generally, I think one can say this: the philosopher grants himself a far greater degree of conceptual freedom than the scientist. For the latter is bound by a discipline which demands that all his concepts should be unequivocally related (or relatable) to the world of the observable. Hence the philosopher's natural province lies the realm of questions which we do not yet know how to formulate or resolve in that kind of disciplined language. Unless one admits the use of a less restrained language, these questions could not be discussed at all. With the advance of science, of course, the borderline between these two realms was bound to shift.

Traditionally the problem of the nature of mind and its relation to matter has been the province of metaphysics. Here the task was seen to examine the various logical possibilities that presented themselves in this respect. The defence of one or other of these logical possibilities has given us such diverse metaphysical schools of thoughts as the *materialist* theories of the mind, the *idealist* theories, the *dualist* theories and the *dual aspect* theories. Though more often than not the notion of mind itself was essentially left undefined.

One form of the materialist theory was the so-called *identity theory*, which asserts that mental events are strictly identical with brain events. This is also the upshot of the scientific analysis I have presented. And this conclusion follows simply from the way in which we have *prescriptively defined* consciousness and mental events in terms of objectively described internal representation formed by the brain.

Is it cheating to solve philosophical puzzles simply by way of prescriptive definitions? Certainly not in this case. By contrast with the philosopher, the scientist cannot allow his key concepts to go undefined (either explicitly or implicitly) in the kind of objective terms to which his methodology constrains him. and which has been the secret of his success. And he has to define his concepts *prescriptively* as an integral part of the development of his theories. Hence the only legitimate accusation that could be leveled against any such effort is that it fails to capture adequately what we commonly mean by terms like *mind* and *consciousness*. And in our case I would argue that this accusation has been fully countered in Chapter 4. The same cannot be said of the attempts made by some materialist philosophers (e.g. Armstrong, 1968) to define mental events objectively in purely behavioural terms.

In this century metaphysics has receded into the background in academic philosophy. Thus when I was a student the prevailing view in this country had become that the philosopher's prime duty was to analyze meanings. Gilbert Ryle's THE CONCEPT OF MIND (1949) was an outstanding example of this linguistic school of thought. Yet, while there is something of value to be found by everyone in this remarkable book, the analysis of meanings can be of lasting benefit only if it succeeds in isolating the different possible connotations of a vague concept, and clarifying each in terms of more precise concepts. And for the scientist this will be of real profit only if the latter concepts satisfy the standards of objectivity and precision demanded by science. From his point of view, it does not advance matters, for example, if mental concepts are merely explicated in terms of other mental concepts. Where Ryle attempted to go beyond this, he, too, tended to fall into the behaviourist trap.

Moreover, only too often such linguistic analyses have proved to be Sisyphean labours. Typically, the philosopher will pick on a concept about which he feels unhappy and seek to clarify it in terms of concepts about which he feels happier. Alas, as soon as these explicatory concepts thus move into the limelight and come to attract an attention they never received before, doubts begin to stir: hidden ambiguities suddenly come to light, rival interpretations raise their voice, and soon everybody begins to feel as unhappy about these explanatory concepts as they had felt originally about the concept it had been intended to clarify. And so the dance continued - a minuet without end. For example, the philosopher may feel that the difference between mental and physical events might be clarified by explaining the different way in which we get to know the one category of events compared with the other - only to find that 'getting to know' on close inspection proves to be as difficult a concept as was the concept of 'mental' to start with. Other representative examples of this genre, culled at random from the literature: "mental is what is introspectable", "mental is that of which we are directly aware", "mental is what is intelligible", "mental is everything that is purposive", "mind is the sum total of appearances at a place at which there is a brain". This failure to replace unclear concepts with genuinely clearer ones has left much of the outside world with the impression that philosophers merely 'spin webs of their own substance' around any concept on which their attention fastens (to paraphrase Francis Bacon) - and not without some justification. Or that philosophers merely talk to other philosophers.

Since the Second World War Linguistic Philosophy has been on the decline. It did not fulfil the hopes that had been pinned on it, and as Popper remarked: "Just to concentrate on language is like spending one's life just cleaning one's spectacles without putting them on and using them." And a new line of thought has asserted itself in some quarters: philosophy was not about the meaning of single words but about the structure of propositions and relations between propositions. This was all very well, and, of course, the value of one branch of philosophy, namely Propositional Logic, had never been in doubt. But once more, there was little to be gained by science: the realities of the natural world appeared in these deliberations only at one remove, namely as the truth condition of a certain class of propositions. Moreover, this preoccupation with propositions also had the detrimental effect of luring some cognitive psychologists into the belief that all the brain's internal representations have a propositional form - a belief which I regard as profoundly mistaken (cf. Section 8.5).

Uncertain meanings also tend to be fluid meanings and so fashions will come and fashions will go in academic philosophy. The nowadays again fashionable concept of 'intentionality' is a case in point. It was first introduced by Brentano over a hundred years ago in an attempt to describe the common mark of mental events.

Mental events were said to have *intentionality* in the sense that they always point beyond themselves, that they are always about something. Intentionality denotes this 'aboutness'. Thus thoughts are about holidays, about people, about quadratic equations; desires are about things expected to satisfy needs; volitions are about goals to be realized; memories about past experiences, etc. Yet it was never a very satisfactory concept. For example, if one looks at these assertions more closely one soon realizes that 'about' is never used in quite the same sense throughout.

However, in accordance with the propositional trends of philosophy I mentioned a moment ago, the concept of intentionality underwent a subtle transformation when it was revived by Chisholm (1957). It now became a feature of *language* rather than the feature of a human experience as Brentano had seen it, viz., a feature of certain classes of propositions expressing beliefs, hopes, fears, desires etc. Typically, it was asserted, such propositions contain a *that* clause: 'I believe *that* it rains'; 'he hopes *that* the train will arrive on time , but fears *that* it will not'. Typically, too, the truth of those propositions did not depend on the truth of the *that* clause: even though it may not rain, it can still be true that I believe it rains.

None of this was of benefit to scientists struggling to discover what happens in the brain when people perceive, think, imagine, remember, etc. Moreover, some categories of mental events do not fit into intentionality pattern, viz., *sensations*. One does not have the sensation *that*........ In our account, on the other hand, sensations have been comfortably absorbed as a special class of projections into the subject's internal representation of the current state of the self.

The main fact which the vague notion of 'intentionality' hints at, but also obscures, is that mental events are essentially *representational* events: states of the brain which operate as representations (in the sense we have defined) of actual or possible objects, events or situations.

From a scientific point of view, therefore, the best strategy must be the one we have followed, viz., to concentrate on the notion of internal representation and to see what can be made of it in precise and objective terms.

Dennett (1983) has proposed an 'intentional systems theory'. But this is not systems theory in the sense in which the term is generally understood in science. His 'system' is not the living organism, not the brain/body system. It is classificatory system, viz., a hierarchy of beliefs and desires, structured according to what those beliefs and desires are about.

In philosophical circles a number of disputes have been raised by the identity theory. Differences of opinion have centred on such questions as whether the identity between mental states and brain states is to be regarded as a contingent identity or as a logically necessary one; whether mental states can be said to be quite literally identical with brain states; whether it is permissible to say that statements about mental events should in principle be translatable into statements about brain events, whether the relation is a reversible one, etc. The reader will find rich illustrations of these and other philosophical disputes connected with the materialist theories of the mind in a collection of essays edited by Borst (1970), also in an illuminating review of current issues by Dennett (1978). All the questions I have cited above have been given an unequivocal answer by the account of mental events we have developed in the preceding chapters and the *prescriptive* definitions it has introduced. For the analysis has made the identity a literal identity and it has postulated a translatability (Section 9.2) which is in principle a reversible one.

By contrast with the materialist view, *dualism* asserts that mind is a separate entity, which exists outside matter and yet somehow manages to interfere with matter in a way that produces actions controlled by conscious desires and thoughts, the processes of thought themselves proceeding in time free from any constraints arising out of the laws of physics and chemistry. To most scientists this view is quite unacceptable, because it forces the conclusion that all over the brain there exist points at which lawful events can be disrupted by some ill- defined entity operating outside the laws of the natural world. Nor can one accept attempts to render the intervention of the mind compatible with physical theory by falling back on the indeterminacy which the quantum theory assumes for the atomic level of physical systems. The suggestion being that the mind renders determinate those physical variables which the Heisenberg Indeterminacy Principles declares to be indeterminate. One has to reject this on several grounds. Firstly. because quantum theory does not pronounce any particular variable to be indeterminable *per se*. It merely postulates that the precise determination of some variables rules out any simultaneous precise determination of certain other variables. Secondly, because, in general, this *microscopic* indeterminacy is statistically lost in the *macroscopic* or *molar* behaviour of large assemblies of elementary particles. And the behaviour which the mind produces is such a *molar* phenomenon.

On balance, it is probably fair to say that the dualists have mainly failed to convince the materialists on account of their failure to show *how* this separate entity called 'mind' could control the neural events in the brain that govern our muscular contractions and hence our observable actions, while the materialists have in the past failed to convince the dualists through their failure to show even in outline *how* properties of the mental kind can emerge in a physical system like the brain - which, of course, has been one of our major concerns in this book.

Equally unacceptable to science is the school of philosophical *idealism* which asserts that the whole universe is within the mind: that it exists only as a construct of the mind and, hence, as a determination of the mind. From the standpoint of our account of the mind, this school in effect confuses the brain's global model of the world with the reality it models. This by-passes all the questions that have concerned us in this book. It can have nothing to say, for example, about the nature of the brain's internal representations and the mechanisms that control them, or about the concrete question of how John's thoughts come to determine John's physical actions.

One school of philosophical thought which has gained much support in recent times, goes under the name of *functionalism*. According to this school, mental events are brain states distinguished by their functional role. This, of course, is a view with which we concur, since we ourselves have defined the brain's internal representations in functional terms. However, as we have also seen, if this approach is not to remain barren, we require two things. Firstly, the concepts used must be defined with the semantic precision demanded by the natural sciences (which in this case means that the notion of *function* must be defined, as we have done, in terms of the underlying spatio-temporal and causal relationships). Secondly, if it is to lead to real insight about what happens in the brain, it must be followed up by fertile hypotheses about the *structure* of the brain-states discharging the asserted functions, because any given function can, in general, be discharged in a variety of different ways and by a variety of different kinds of hardware.

A final view worth mentioning is the *dual aspect* view. On this view mind and matter are just different aspects of the same thing. Just as a suspension of water droplets in the sky appears as a *cloud* when perceived from the outside but as *mist* when perceived from the inside. We have established the sense in which this is true in our discussion of the nature of self- awareness.

A modern variation of this is the suggestion that the relation between mind and brain is to be compared with the relation between the software and the hardware of a computer, i.e. between the programs and the machine on which they are run - or, if you like, between the 'logic states' and the physical states of the machine. However, the analogy does not really bear close scrutiny. The software of a computer consists of a hierarchy of instructions designed to break down the logical steps required to solve a particular problem into an appropriate sequence of the kind of very primitive logical operations that a microprocessor can perform with phenomenal speed (basically just fetching 0s and 1s from particular locations, pushing them around a number of different registers, combining them according to rules of Boolean algebra, and finally dumping the result into some new location – all in sequences dictated by the operating system in conjunction with the entered program instructions. Now in the brain every problem-solving activity, i.e. activity that has to produce the right responses in a given situation, is produced by means of simple units with limited capabilities, viz. the neurons. Yet, on our account of the nature of mental events, it would patently be stretching analogy beyond acceptable limits if we were to describe these events as a system of logical steps designed to break down the problems the brain has to solve into the simple problems individual neurons can handle.

Of course, analogies *can* be powerful instruments where other means of expression fail. Take, for example, Satre's definition of his brand of Existentialism: "We mean that man first of all exists, encounters himself, surges up in the world - and defines himself afterwards." Though such a statement defies translation into precise and objective terms, its three metaphors nevertheless convey a great deal.

In the apt and poignant use of metaphors lies the strength of a great deal of poetry and of virtually all good literature. Already Aristotle praised the metaphor as an effective way of illuminating the unknown in terms of the known, of the unfamiliar in terms of the familiar. Thus in difficult areas of investigation metaphors can point the mind in the right direction, even if they cannot map the path to be followed. However, metaphors offer no reliable basis for cogent deductions, and the extent to which any system of thought with scientific aspirations has to rely on metaphors, is generally a measure of the distance that it still has to travel. And metaphors can also mislead. Especially when one scientific discipline borrows notions from another - as when cognitive psychology borrows from computer science - it is apt to absorb unwittingly a host of tacit assumptions which are valid in the one field but not in the other. Of course, some philosophers have been fully aware of these weaknesses. John Locke, the father of British empiricism, for example, condemned metaphors outright. "Figurative speech", he wrote, "serves but to insinuate wrong ideas, move the passions, and thereby mislead the judgment".

***

PART III

CULTURE

Chapter 10

DEEPER LEVELS OF CONSCIOUSNESS (1)

INTRODUCTION TO PART III AND SUMMARY

The failure to understand the nature of human consciousness is not the only major gap in the scientist's account of human nature. Another gap is the failure of science to understand those human sensibilities and responses which may broadly be described as the spiritual dimensions of human life. I use the word here in a purely secular sense, viz., as a convenient bracket for two main categories of human sensibilities and responses:

> **1.** those *moral* sensibilities and responses which cause the norms of what we regard as a civilized life to diverge from the laws of the jungle, i.e. from the kind of human conduct that would have resulted if the evolution of human feelings and behaviour had remained solely governed by the yardstick that governs the evolution of all animal behaviour, viz. their contribution to the genotype's fitness to survive;

> **2.** those *aesthetic* human sensibilities and responses that, similarly, cannot be explained in terms of their genetic cost-effectiveness: foremost the sense of beauty and the enjoyment of works of art. I use 'beauty' here in the broad sense in which it relates to the attraction that may be caused by the perceived form and perceived compositon of an object, and is caused below the level of our conceptual knowledge and evaluation of the object.

In so far as these traits go beyond the genetically cost-effective, they amount to uniquely human traits which could come into being only. because man's *cultural* evolution is not governed by the same forces as his genetic evolution. The main reason being that cultural goods are not transmitted from one generation to the next via the genes, hence their evolution is not governed by the Darwinian laws of natural selection, for which the survival of the genotype is the ultimate criterion of merit. These human traits mark a radical departure in the evolution of life on this planet, for they have thus caused civilized man to become more than a mere survival machine. Yet they are not mere sports, and in this third part of the work I shall try to delve into their roots, look for their rationale, and try to see them in their proper biological perspective. For I believe that some of the concepts we have formed in the earlier chapters can help us to see all this in a clearer light. Nevertheless, we are dealing here with *affective* aspects of consciousness, not the *cognitive* ones that have occupied us so far, and this increases the difficulty of accounting for them in strictly objective and biologically plausible terms. Hence this part of the work has been set apart from the rest.

It is an important topic. We live in an age which is dominated by science, and the failure of science to understand this dimension of human life has tended to blunt our very sense of what it means to be human. Not surprisingly, this failure has also aroused considerable antagonism towards science itself. It has tended to leave people with the feeling that the advance of science has profoundly disturbed our spiritual ecology: that it has increased our knowledge of the external, natural world only at the expense of our knowledge of the inner world of the human psyche – and especially of those levels of consciousness at which we experience the tonic effect of beauty, compassion and other manifestations of the human spirit.

*

According to the prevailing theory of evolution, in essence still the Darwinian theory, the different traits of animals have in the main developed because in the given ecological circumstances they were *genetically cost-effective*. That is to say, they evolved because they enhanced what may be called the *biological fitness* of the genotype, i.e. its ability to survive and to leave offspring which in turn survived.

In this sense, then, animals may be described as just sophisticated survival machines. In the social species, like the ants and the bees, these traits came to include various types of social and seemingly altruistic behaviours, including even self-sacrifice by individual insects in the defence of the colony. That sacrifice is genetically cost-effective because it protects the survival of the insect's genetic cousins – hence of the shared genes, i.e. the genotype.

The evolution of this kind of 'altruistic' behaviour through the 'Principle of Kin Selection' was first publicized by W.D. Hamilton (1963) and later became one of the explanatory principles of the sociobiology of which E.O. Wilson (1975) has probably remained the best-known exponent. However, it is a fallacy to assume, as some followers of this school have only too rashly assumed, that this line of explanation can be satisfactorily extended to the human case

Although man, too, has social instincts whose origin could be explained along parallel lines, e.g. devotion to the care of his offspring, and respect for the blood-bond and tribal loyalties. Yet the sovereign and autonomous sensibilities and responses I am here considering transcend the requirements of biological fitness. Neither the tonic effect on the human psyche of beauty nor the enjoyment of the arts can be explained in terms of their phylogenetically adaptive value. Nor are cost-benefit calculations of genetic advantage the hidden yardsticks by which moral excellence is measured in our society. Neither the major religions nor the humanist movements that have shaped our culture have spread because they enhanced the believers' genetic contribution to the next generation. Nor are they at heart tribal or racist, as would have been the case had it been otherwise. They claim *universal* validity for their beliefs, and they justify their values in terms of *universal* human cravings. Again, the ethos they have fostered in our society demands respect for all men as human beings, and indeed respect for all living creatures. It seeks to alleviate all suffering, animal as well as human.

Even though Judaism has a strong ethnic component, it nevertheless teaches that all men, not just the Israelites, are engaged in a covenant relationship with God, the unique role of the Israelites being presented merely as that of chosen witnesses. It is not a racist creed: members of other races, too, have traditionally been admitted into the Jewish community.

Nor can these religions and humanist movements be regarded as mere organs of social utility. They have related their claims, not primarily to our corporate life, but to the fulfilment of the individual, to the soul's quest for peace and, in religious terms, union with God. Though it is here understood that the individual can realize his full human potential only in harmonious relationships with his fellow men, and that the community, too, is a spiritual entity.

In this sense, therefore, civilized man seems to have become an altogether different creature: a creature looking beyond the material necessities of life and one that seeks (or seeks out) more subtle, abstract, and perhaps inwardlooking forms of harmony, order and communion than those required for the efficient functioning of the body and of the collaborative associations on which all social species depend for their survival and propagation.

Of course, at the physical level we are all survival machines in the sense that the primary functions of the body are devoted to sustaining its life. And without life there can be no quality of life. In that sense survival comes first in most human beings. The spiritual side of human nature can flourish only when the basic material necessities of life have been satisfied, but even in a crisis here, people tend to feel the pressure of moral norms.

In religious circles, this spiritual dimension of human life is regarded as that part in man which is most akin to God. But humanists, too, often speak of 'spiritual values', although the phrase is here understood in a secular sense (as we have understood it above), and humanists have tended to confine themselves to listing these values rather than search for a generic definition. To avoid misconceptions arising out of this uncertain connotation of the word 'spiritual,' I have chosen a neutral label for the sensibilities and responses concerned, and borrowed from Maslow (1973) the terms *meta-sensibilities* and *meta-responses* respectively – the prefix here serving merely to remind us that they go beyond the human traits whose development can be explained in terms of their genetic cost-effectiveness. From a strictly Darwinian point of view, therefore, they are an extravagance. And, for the reasons I have stated, they lie beyond the proper scope of Sociobiology, the discipline that seeks to explain social responses in the animal world in terms of their phylogenetically adaptive value.

These meta-sensibilities and meta-responses raise four important questions which I shall attempt to answer:– How did such a 'Darwin-defying' development in the evolution of the human species come to be possible in the first place? What are the psychological roots of this development? How did the associated value systems, or 'meta-values', come to impress themselves on the advance of civilization as in some sense superior to the pragmatic values that relate to the material necessities of life? Finally, how is this whole development to be seen in the general biological context of the evolution of life on this planet?

In Section 10.1 I shall begin by taking a closer look at these 'meta-traits' and seek an answer to the first of these questions. I shall argue that these moral and aesthetic sensibilities and responses could only develop because they are essentially the product of cultural and not phylogenetic developments. And since cultural goods are not transmitted from one generation to the next via the genes, the Darwinian criteria of selective advantage do not apply to the course taken by their evolution. The rate at which they propagate does not essentially depend on the effect they have on the life-expectancy or fertility of their carriers. And they can freely cross racial or tribal frontiers.

In Section 10.2 we shall then turn to the question whether a common root can be found for the whole spectrum of these meta-sensibilities and meta-responses. Here I shall suggest that we can find valuable pointers in a number of basic psychological forces to which different schools of dynamic psychology and personality theory have, each in their own style, drawn our attention. According to the theory to which these pointers have led me, this common root lies in a thrust to minimize the level of anxiety, conflict, frustration and cognitive dissonance which the individual experiences. *This thrust makes biological sense:* other species, too show a strong aversion to these 'biological negatives' as we may call them, and the biological function of such reactions is readily understood. But to account for the uniquely positive expression these aversions have found in the evolution of human civilization we have to invoke three additional and uniquely human factors :–

1. Through the faculty of the imagination man's aversion to these 'negative' conditions comes to be supplemented by visions of, and a longing for, the corresponding 'positives', e.g. a longing for qualities of life in which security, self-fulfilment, freedom and harmony have replaced anxiety, frustration and conflict.

2. Through one particular power of the imagination, viz. the power of *empathy*, man is able to enter into kinds of relationship to the surrounding world which are not open to creatures lacking that power. I shall suggest that this plays an important part in our aesthetic sensibilities as well as in our moral ones, though in a different form. And here we can fall back on the objective account of the faculty of the imagination and empathy given in earlier parts of this work.

3. Since, for reasons already indicated, man's cultural evolution is not subject to the restraints of genetic cost-effectiveness, these basic longings have been able to impress themselves on that evolution to the unique extent to which this has manifestly been the case – though the forces that have brought this about need to be examined.

In the final section of the present chapter I shall add some comments on three of the key concepts which have thus emerged: the concepts of frustration, conflict and anxiety – the nature of the imagination and of cognitive dissonance having already been examined in objective terms earlier in this work. I shall stress that each of these three concepts, too, can be defined in objective terms.

In Chapter 11 I shall then look at our aesthetic and moral sensibilities separately and suggest how either category may be interpreted as a cultural expression of the basic aversions I have mentioned in a form, and to a degree of penetration, which is the product of the three factors I have just listed.

In view of the role which religious belief has played in the evolution of the ethos of the Western world, and in view of the fact that even in this scientific age people still feel drawn to religion, one has to ask whether there exist basic human needs which religion might claim to meet more comprehensively than any purely secular account of the world. The issue concerns us in the context of what has gone before, because if these needs can be identified then we shall have to examine whether they, too, can be explained along the same lines. In section 11.4 I shall discuss what I believe those religious needs to be and how they can be accounted for in terms of the theory I have outlined above.

***

10.1 - A CRUCIAL DEPARTURE FROM THE BIOLOGICAL NORM

Human culture is not a uniform phenomenon, and what are regarded as the laws of a civilized life in one part of the modern world may differ significantly from those accepted in another. This does not necessarily mean a disagreement on fundamentals: often it reflects no more than a difference in the relative weights attached to shared moral precepts. To give an elementary example, familiar to most travellers: in situations in which speaking the truth may disappoint someone's expectations, the members of some societies may tend to be more anxious than you or I might be to please rather than convey the truth. Indeed, the most common differences in moral judgments spring from the different weights people attach to the rival claims made by conflicting moral principles, the principles themselves not being in dispute. Even so, I have to declare the culture I am talking about and, to restrict the field, I shall confine myself in the basic norms of our own society, and thus to the value systems which have here evolved under the pressure of its Judeo-Christian and humanist movements of thought and feeling.

As concrete examples of what I have called our 'meta-sensibilities' and 'meta-responses', I may cite the sense of beauty, the sense of the sublime, and the enjoyment of art; the ability to feel compassion, not just for one's fellow men but for all living creatures; a basic respect for the individual and his rights; the care for the aged, the needy and the infirm; the sense of justice; the disinterested pursuit of truth for the sake of truth; the search for self-expression in creative works, the love of perfection; and visions (however dim) of a world in which harmony and peace reign supreme. I would not claim that this list is complete, nor, of course, that these meta-sensibilities and meta-responses are in any way uniformly shared traits. But they typify what we regard as our highest cultural values. In its purest form the resulting ethos transcends any mere set of 'sociobiological' attitudes and responses. Indeed, in the highest traditions of this ethos, human compassion, friendship, love of beauty and art are only seen as pure, or 'true', to the extent to which they remain uncontaminated by considerations of practical utility and expediency. Thus the altruism that springs from human compassion bears no relation to the 'altruism' of the worker bee which sacrifices its life in the defence of the hive, and which has genetically been programmed to do so only so that its genetic cousins may live and spread their shared genes. Human altruism transcends the tribe- or race-oriented ethos that would be an analog of the bee's 'altruism'. Nor has our sense of beauty anything in common with the bee's predilection for brightly coloured flowers. We are dealing here with qualities of experience that appear to be able to act as a tonic to the psyche, and which strike those who are sensitive to them as qualities of an intrinsic and enduring worth – even as qualities from a devotion to which one can derive a unique sense of stability and continuity of purpose in one's life.

10.1 - A crucial departure from the biological norm

In the history of Western thought these seemingly sovereign and autonomous values have frequently been conceived under the classical triad of *beauty, truth and righteousness,* and the ethos which is enshrined here in the concept of righteousness is certainly more than just an organ for the preservation of the genotype. If it were only the latter, one's moral obligations to others would be strictly proportional to the degree of their genetic kinship, and our general rules of conduct would be no better than a codified form of jungle law. The laws of a civilized life protect the weak; whereas the laws of the jungle decree that the weak shall perish so that their deleterious genes shall be removed from the gene pool. There exists no altruism in the jungle except as a sophisticated form of genetic selfishness. The ideal of the good society that has evolved from the Judeo-Christian and humanist tradition is that of a caring a loving society, not that of a ruthlessly efficient and Spartan one in which each individual is drilled from birth solely for the power game. Although both the institutionalization of a religion, and the use of a religion to underscore a sense of ethnic identity, tends to drag that religion into the power game (and to corrupt it accordingly) the pure Judeo-Christian as well as the humanist ethos transcends the power game altogether and seeks to curb it in the name of nobler principles.

*

The crux of the matter, then, is that these meta-sensibilities and meta-responses amount to human traits whose evolution has been governed by criteria of selection very different from those that govern the phylogenetic evolution of animal species. Before we ask what these criteria are, we must ask how this remarkable departure from the biological norm came to be possible in the first place.

The answer lies in the fact that we are dealing here with *cultural* and not phylogenetic developments, and with developments which neither originate in the genes nor are transmitted through the genes. The birth and propagation of the religious and humanist movements which have placed their stamp on our civilization was not the triumph of a genetic mutation. Because, through the development of speech, art and teachable skills or knowledge, a second and *exosomatic* mode of transmission from one generation to the next came to pass in the human species. Human knowledge, experience, skills and values are transmitted from one generation to the next by word of mouth, literature, art and by way of example. All of these are ways which by-pass transmission through the genes and thus escape the criteria of selection that govern the rate of propagation through the population of genetically transmitted traits – the sole area to which Darwin's theory applies.

This development has had two consequences. Firstly, whereas the genetic complexion of a population can change only slowly, over a long succession of generations and at a rate dependent on the mutation pressure (even though man himself can affect the selective factors), the cultural evolution of mankind can proceed at a much faster pace and often with new points of departure. Secondly, the propagation of cultural goods does not respect tribal or racial barriers. Since they are not transmitted through the genes, cultural goods can freely cross racial frontiers. An idea first formulated by a Jew can readily be accepted by a Gentile, and vice versa.

In short, man's moral and aesthetic sensibilities and responses have been free to evolve in 'Darwin-defying' directions only because they evolved as a *cultural* and not phylogenetic changes. This does not, of course, explain why and how they came to evolve in the particular directions in which they did evolve: that is a separate question which I shall take up in Section 11.3.

*

Since these considerations show that in the discussion of the evolution of human spirituality, Darwinism is an irrelevancy, it seems remarkable that still so many scientists seek to explain or justify our spiritual values in Darwinian terms. I mean not only the erroneous belief that sociobiology can provide a complete explanation of human behaviour - as expressed, for example, by Wilson and Ruse (1985). I mean also those well-meant, but simplistic attempts of some sociobiologists and ethologists – also leading humanists – to make the world safe for the human spirit by seeking to establish along Darwinian lines some kind of biological rationale for those spiritual values. Such theories tend to be counterproductive in the long run. They diminish our spiritual nature by seeking to interpret it as part of our animal nature.

The following statements made by the well known neurophysiologist Colin Blakemore (1976, p.743) typify this diminished view:–

> "Virtually the only things that can be defended as uniquely human traits are his continuous sexual appetite, his formal taboo on incest, and his language."
>
> "Many of the most precious elements in human social behaviour, such as formalized sexual bonding and family structure, rich communal ceremony and *the embracing of ethical principles within religious codes,* can be viewed as deliberate cultural exaggerations of inherited components of behaviour that favour kin selection" (my italics).

In his CIVILIZED MAN'S EIGHT DEADLY SINS Konrad Lorenz (1974), one of the best known ethologists of this century, even went so far as to interpret the very notion of sin in Darwinian terms. In his eyes, sin is whatever conflicts with the criteria for fitness that operate in phylogenesis; man does not qualitatively differ much from the animals; all ethical values are identical with biological values. A typical example is his condemnation of competition within the species (the third deadly sin). It is hardly an account that squares with the kind of ethos here under discussion. Indeed some critics have described it as the perfect philosophy for an aspiring Master Race.

Gross misconceptions of this kind I have illustrated have brought not only Darwinism into disrepute, but also science itself. Jointly they have left the world with the impression that according to science, as one critic has put it, "there is more to be learnt about man from watching a tribe of baboons than from the works of Shakespeare".

***

10.2 – THE ROOTS IN THE HUMAN PSYCHE

What then are the psychological roots of our meta-sensibilities and meta-responses? What is it that people crave for when they look for something more deeply and enduringly satisfying than the mere satisfaction of their physical needs and appetites? One might expect to find an answer to that question in the findings of dynamic psychology and personality theory. Unfortunately, we are confronted here by such a diversity of different schools of thought that no coherent picture emerges about the basic motivational forces that might be here at issue. Nevertheless, the theory which I shall state below has evolved from pointers found in this field.

The diversity of different schools of thought here is indeed bewildering. For instance, merely the question of the batteries of tests that should be used to discover basic human traits is met by a bewildering variety of opinions, notably opinions about the criteria of acceptability which are to be applied to the 'evidence' different investigators offer in support of their theories. Is one to accept verbal self-reports as evidence, for example? Or Rorschach's well-known tests with inkblots?

The investigators also differ significantly in their intuitive choice of categories. For example, the first study which claimed to give a complete list of man's basic needs was that of Murray (1938). And in Murray's classification man's sense of beauty is bracketed with taste, smell, and thumb-sucking as 'sentience'. This in turn, is bracketed with hunger, thirst, sex, urination and defecation as a 'viscerogenic need'! Again, Cattell (1965) lists 'appeal' (attraction) and 'disgust' as basic motivational forces alongside such forces as hunger, thirst, curiosity, sex, gregariousness, protectiveness, longing for security, self-submission and self-assertion, whereas most investigators would regard disgust, appeal and anger as the consequence of human traits rather than traits in their own right.

Murray's list of basic needs had 40 entries. Cattell's list, on the other hand contained only 16 basic drives (or 'ergs' as he called them). It was based on an attempt to tackle the problem with the latest mathematical techniques of factor analysis. To the best of my knowledge, no one in the field has come up with a list longer than Murray's, but some have settled for a list shorter than Cattell's. An extreme case has been the work of Eysenck, one of Britain's leading behaviourists. The best known of his personality classifications have been based on just two variables or 'axes': the 'introversion - extraversion' axis and the 'neuroticism' axis.

Is there then no common ground to be found in this diversity? Following a survey of the leading schools of his day, Maddi (1968) has suggested that the main body of personality theories can be divided into

> a) those in which the *avoidance of conflict* in one form or another is assumed to be the overriding factor in the determination of our emotional life (Freud, Murray, Sullivan, Rank, Bakan);
>
> b) those in which the *urge for freedom and self-fulfilment* is the ultimate motivational determinant (Rogers, Maslow, Fromm, Jung, Adler, White), and,
>
> c) those who see as major human determinant a *longing for consistency* in one's emotional or cognitive life. (Kelly, McClelland, Fiske and Maddi).

And one common finding appears to run through all of these theories, viz.,

> d) the existence in man of a *fundamental aversion to anxiety*.

For example, Adler's theory of perfection-seeking as a basic human motive may be regarded as just a special version of a fulfilment-theory.

However, one major division among the conflict theorists has to be noted: in Freud, Murray and Sullivan the relevant conflicts are psychosocial ones, whereas in Rank, Angyal and Bakan they are intrapsychic.

When I first encountered this study, it struck me as a singularly revealing one, because from the very broad spectrum of different theories he had examined Maddi had abstracted *four basic motivational factors which made biological sense*. For, if we regard the longing for freedom and self-fulfilment as a derivative of an aversion to frustration, the ultimate driving forces suggested by his analysis may then be described as an aversion to four *biological negatives*, i.e. four conditions which obstruct the full functioning of the organism, or which are symptoms of its failure. They are: *conflict, frustration, anxiety* and *cognitive dissonance*.

10.2 - The roots of the human psyche

This makes biological sense because the first three of those aversions are well established traits in all of the higher orders of the animal kingdom, with an obvious selective value. And as to the fourth, the aversion to cognitive dissonance, we can interpret this as an aversion to the experience of disconfirmed expectancies. That aversion, too, makes biological sense, at any rate according to the account of cognition which I have given earlier in this volume.

The very basic theory of morals and aesthetics which I shall outline in this and the next chapter assumes that these four motivational forces indeed lie at the root of our moral and aesthetic sensibilities and responses, and it sets out to explain the nature of this connection.

Here the theory will rely on two arguments. The first is an appeal to the fact that while in the animal world any species-specific expression of these basic aversions will always be constrained by the demands of genetic cost-effectiveness, no such constraint hinders the expression of these aversions in the evolution of human culture (for reasons I have explained above). Secondly, it postulates that in man the basic aversions I have listed result in much more than just an avoidance of these negatives. They result also in a longing for the corresponding *biological positives,* i.e. a longing for situations, conditions, degrees of freedom and a way of life, in which the above negatives are absent or minimal. Thus the aversion to frustration becomes a longing for freedom and self-fulfilment, for these are the very opposite of frustration. A crucial point to be made here is that this transformation of an aversion to negatives to a striving after the corresponding positives requires the faculty of the *imagination* – a faculty which may be present in the infra-human species only in a rudimentary form, if at all. For the imagination enables man to conceive (however dimly or abstractly) of states of affairs in which the abhorred negatives are absent or minimal.

The basic theory to which we are thus led may be summarized as follows. Man has an innate aversion to such biological negatives as conflict, frustration and anxiety. It is part of his animal endowment and as such intelligible in biological terms. From the nature of his cognitive functions he also derives an aversion to cognitive dissonance, i.e. the experience of disconfirmed expectancies. Through the power of the imagination his aversion to these four negatives produces visions of, and a longing for, the corresponding positives, i.e. visions of, and a longing for, conditions and qualities of life in which order and harmony prevails and the abhorred elements are absent or minimal.

In particular, I would list the following as three major forms in which this search for positives finds expression in man:–

1. From the aversion to frustration springs a longing for self-fulfilment, hence also a longing for a fuller and richer inner life, i.e. a life in which one's potentialities are fully realized and one's faculties fully engaged.

2. From the aversion to conflict springs a longing for a life free from conflict, and from this a longing for emotional and motivational stability, i.e. a craving to find a strong and stable direction in one's life. This assertion is not far removed from the common assertion that a strong and stable purpose in life is one of the powerful integrating forces that can hold body and soul together.

3. From the above aversions also springs a longing for social security, and, in more general terms, for a harmonious communion with the surrounding world. I mean by this the kind of assurance that comes with a sense of affinity with the surrounding world and with the feeling of being at one with that world.

In these positive expressions, I suggest, lies the ultimate root and rationale of the bulk of our meta-sensibilities and meta-responses. It is relatively easy to see how these particular moral sensibilities fit into this overall picture. Because it is not unreasonable to suggest that they ultimately derive from on visions of a better world: a world of peace and harmony in which conflict is minimized through a mutual concern for one another's well being – a brotherhood of man. It is more difficult to see how our aesthetic sensibilities, e.g. the sense of beauty, relate to all this. I shall deal with both these issues in Chapter 11. The key-concept in both cases will that of *empathy*, a concept which I shall consider in detail later.

Of course, the different forces mentioned above may compete with each other. Thus owing to the full engagement of one's faculties which adventure offers and the strong sense of direction (hence inner harmony) that can come from the acceptance of a major challenge, some people may prefer an adventurous life to a sheltered and secure one. Such choices will be settled by each individual according to his or her temperament. For, coupled with the above conclusions also goes the suggestion that differences between individual personalities in these dimensions arise mainly from differences in the relative influence which different components of these core tendencies have in the individual's emotional life.

***

10.3 – FRUSTRATION, CONFLICT AND ANXIETY

Before I try to probe deeper into the psychological roots of the cultural phenomena with which we are concerned, in this chapter, I must add a few observations on the key concepts which have now emerged: the concepts of *frustration, conflict* and *anxiety*. I shall make the point that on the strength of our earlier analyses all three terms can be understood in an objective sense.

FRUSTRATION

Broadly speaking, we can look upon the typical behaviour pattern of any organism as a sequence of actions in which each is triggered by the current situation in conjunction with stimuli produced by results of the preceding action. However, not in all cases are the cues for the next action in a given sequence always instantly available. Often they first require a brief search, i.e. orienting reaction. A bird, in the process of building a nest, for example, has to search for the right kind of moss or twigs for its nest before it encounters the stimuli that will trigger the next stage in its nest-building program. A frustrating situation, therefore may be defined as one in which there is an unexpected absence of positive clues for further action, notably one in which the scanning and search that are part of the resulting orienting reactions fail to produce such cues. This state of affairs is detectable by the persistence of the orienting reactions. An *impasse* has been reached: from one situation eliciting orienting reactions the organism merely passes into another. The fly buzzing against the window is an obvious example. In higher organisms, capable of learning, the occurrence of frustration in this sense is a punishing experience. This can be as effective in influencing subsequent behaviour as can be the more obvious forms of failure and punishment. It differs from the gentler kind of failure we have in those cases in which an error committed by one action in a sequence can be corrected by the action that follows, as when a false move in riding a bicycle is corrected by the next move. Indeed, apart from acute discomfort and pain, the state of affairs I have described is probably the most common and powerful indication of failure, and from a biological point of view it is easy to see how organisms come to develop aversive reactions to situations which prove to be frustrating in the sense I have explained.

In the human case, too, frustration may be identified with a situation of the kind outlined above, i.e. a situation in which the obstruction consist in the lack of cues for further action: the imagination fails to come up with any possible courses of action that promises to bring one closer to one's goals – as in the case, for example, when one goes on trying to teach a pupil who fails to make any noticeable progress.

In a wider sense one can speak of frustration also in every case in which the execution of some plan of action meets unforeseen, repetitive or unexpectedly fatiguing obstructions: it is frustrating to try to run with a leg in plaster.

Since we have been able to deal with the nature of the imagination in objective terms, viz. in terms of an objectively defined categories of internal representations there is nothing in the above concepts that bars us from accepting frustration, too, as an objective concept on either of the above interpretations.

CONFLICT

Having fixed their sights on observable behaviours or behaviour-tendencies,, experimental psychologists tend to define a state of conflict as the occurrence of opposed response tendencies of equal strength; whereas in everyday life we tend to think of conflict in psychological sense as a mental state in which the individual harbours two or more desires of which he realizes that they call for mutually incompatible steps. Hence he has to make a choice between the alternatives, each carrying a penalty. Often the penalty is trivial: the majority of the conflicts we meet every day are minor ones which cause merely imperceptible hesitations, because we can resolve them by dealing with the one desire first and the other afterwards. I cannot eat and drink simultaneously, so I eat first and drink afterwards, or vice versa. These *internal* or *intrapsychic* conflicts become critical only when the subject realizes that the steps called for by the one desire may prevent for ever, or for a very long time, the realization of the other. I may be tempted to rob the bank, but also realize that any such action is likely to prevent for a very long time a life free from fear, or even a free life altogether – two things I value greatly.

Since we have here conceived of internal conflicts in terms of categories of internal representations each of which has been covered in objective terms earlier in this work, our notion of *internal* (or *intrapsychic*) conflict, too, can be accepted as an objective one, and thus one which we are entitled to use in our hypotheses about the roots and rationale of the meta-sensibilities and meta-responses we have discussed.

According to the conclusions we arrived at in Section 10.2 one of the roots our meta-sensibilities and meta-responses was an aversion to conflict and this conflict was conceived as an *intrapsychic* one, i.e. a conflict between mutually incompatible desires. However, since moral norms are importantly concerned with the minimization of *psychosocial* conflict, i.e. with the minimization of actions whose goals conflict with the goals pursued by other members of the community, our theory has to show how the one can lead to the other. It will be recalled in this connection that an aversion to intersocial conflict figured as a human core-tendency in the work of Freud, Murray and Sullivan.

This is no great obstacle. The key factor here lies again in the power of the human imagination and in the resulting powers of prevision. Because owing to these powers every psychosocial conflict tends to generate an intrapsychic conflict. If I have a desire which conflicts with the desire of another member of the community in the sense that my actions would obstruct his actions or vice versa, then I am automatically confronted with two alternatives, each of which carries a penalty: I can either try and have my own way, with the prospect of incurring that person's hostility and opposition, or I can abandon my goal and suffer the resulting feelings of frustration. This is an intrapsychic conflict. It follows that any aversion to intrapsychic conflict will naturally produce also an aversion to intersocial conflict.

The power of selfless love to reduce psychosocial conflict is obvious: if I love people well enough I shall be identifying with their interests and make their concerns my own. Hence there will be a minimal tendency for desires to arise in my mind that conflict with their desires. It is not difficult to see, therefore, how an ethos that rests (*inter alia*) on an aversion to intrapsychic conflict would become one amounting to an aversion to intersocial conflict, and ultimate one in which the power of love to bring order out of chaos occupies a central role – as it does both in the religious and humanist movements that have shaped our culture.

One final point has to be made. It is important to distinguish between a social *conflict* which arises from a clash of interests or desires, and a voluntarily entered social *contest,* which is ultimately based on social agreement. The aims of the players in a chess match are deliberately chosen to conflict, but the players enter the conflict-situation by mutual agreement. Hence there exists in this case no intrapsychic conflict. This is an obvious point, but it is nevertheless one worth making because the objection is sometimes leveled against the conflict hypothesis in personality theory, that individuals engaged in commerce and trade frequently *enjoy* the conflicts inherent in the pursuit of competitive enterprises. The answer, of course, is that what they here enjoy is a voluntarily entered *contest* which operates within a broad framework of mutually accepted rules, rather than a truly social *conflict.*

ANXIETY

Freud is said to have been the first to have introduced the notion of anxiety into psychology. Anxiety, as he saw it, is a internal signal for danger. And Freud spoke of fear if the danger came from the outside world, and of anxiety if the danger came from within – from repressed sexual urges, for example. The modern idiom tends to follow the common linguistic practice which recognizes 'fear of...' but not 'anxiety of...'. In other words, fear has an object, anxiety has not. In general, fear will have anxiety as a concomitant. On the other hand, anxiety may also be caused by the experience of the impasse that exists in cases of extreme frustration and the resulting feeling of helplessness. Sometimes the cause of an anxiety is wholly obscure and we then speak of an *anxiety neurosis*.

In the genesis of both fear and the resulting anxiety, one can again see the far-reaching effects of internalization through the powers of our imagination and prevision. We fear things because of their anticipated consequences. Indeed, this internalization has condemned man to a mental restlessness which is for ever seeking to anticipate and explore contingencies, even when the belly is full and all is quiet around him; and it takes more than a plate full of gruel to lay this mental watchdog to rest. In fact, through the development of the faculty of the imagination man the *warrior* has become man the *worrier*. This is a powerful driving force behind the longing for social security and for a state of happy communion with the surrounding world.

The state of arousal which characterizes a state of anxiety manifests itself both at a physiological level (increased heart rate, respiratory rate, adrenalin secretion, etc.) and at a behavioural one (alertness, search, etc.). Clearly, in terms of such syndromes *anxiety*, too, can be construed as an objective concept.

These, then, are the main aspects of conflict, frustration and anxiety that concern us in the present context. And as I have pointed out, although my description of these states of the individual were couched mainly in mental terms, they were all of a kind that can be conceived in objective terms on the basis of our earlier analysis of the different kinds of internal representations that form the basis of our mental life.

***

Chapter 11

DEEPER LEVELS OF CONSCIOUSNESS (2)

11.1 - THE SENSE OF BEAUTY

My object in this section is not to develop a fully-fledged theory of aesthetics but merely to suggest that it is possible to make biological sense of man's aesthetic sensibilities in terms of the basic motivational forces we have isolated in Section 10.2. Even so, the field of aesthetics is such a broad one that I have to be selective. I shall therefore confine myself to just one manifestation of our aesthetic sensibilities, but also one which, I believe, touches the very heart of the matter. I mean *the sense of beauty*, the pleasure we can derive from the mere form and perceptible composition of an object prior to any conceptual analysis or evaluation of the object concerned, and regardless of the object's significance as a help or hindrance in our ongoing activities.

Primarily I shall have the beauty of natural objects in mind. In this respect I am departing from the mainstream of aesthetic theory, which tends to concentrate on our responses to works of art and the nature of the creative act itself. But the special factors that enter here largely presuppose the factors I shall be considering, and their complexity is such that I could not do them justice in a limited space, even if I felt qualified to attempt this. Towards the end of this section I shall briefly indicate how what I have to say in the present context also reflects on our perception of the nature of art.

What is it about beauty that lifts the spirit, and about ugliness that depresses it? Why may an abundance of ugliness even injure our health?

For science this is a tricky subject because of the very subjectivity of the response. "Beauty", they say, "lies in the eyes of the beholder". Even so, one can try and isolate in objective terms the kind of features in the appearance of an object that trigger a pleasurable response prior to any rational conceptualization and evaluation of what is perceived. For that is part of the essence of a purely aesthetic response. This fact also puts the matter beyond the reach of those cognitive scientists (and philosophers) who are preoccupied with conceptual knowledge. As Baumgarten had already perceived in 1735, the exclusive preoccupation with conceptual knowledge, of which Descartes was the main exponent at the time, tends to divert attention from the sensory and perceptual levels at which our aesthetic responses originate.

Later theorists have made the same point. Thus Croce has argued that pure aesthetic intuition must be completely devoid of concepts. Aesthetic pleasure springs from the impact made by a sensory perception before that perception has been interpreted by the intellect. Indeed, conceptual analysis of a perceived shape or form tends to undermine the intuited quality of that shape or form.

Some schools of thought deny this and maintain that there is no such thing as an innocent eye. All perception, they say, is theory-laden, for we see what we expect to see. But, of course, that aspect of perception has already been covered by the Expectancy Theory of perception which I have set out in Chapter 5, and the relevant point which I made then is that the expectancies which are effective here may operate below the level of consciousness and conceptual knowledge.

Croce also stressed, as have other theorists, notably the Gestalt theorists, that beauty is a quality of the whole object, not a mere summation of the qualities perceived in its components. Although the texture of surfaces may also enter the equation, that contribution, too, depends on its relationship to the whole.

I shall develop my analysis and hypotheses in *three steps,* in each of which I isolate one of the three factors which I believe to be mainly responsible for the primary attraction exercised by objects of natural beauty.

1. In the first of these steps I shall consider the nature of the *order* and *unity* we may perceive in the appearance of an object. This order and unity is of obvious importance in connection with the *significance* we come to attach to the object concerned. But I shall argue that it is also a source of aesthetic attraction in its own right.
Here I shall draw on the twin concepts of *information* and *redundancy,* such as they have become established in modern information theory. The reader who is not familiar with these two concepts will find a brief account in Appendix E. I shall argue that the perceived order and unity of a figure or shape resides in the redundancy of the information the visual inputs convey, and that this redundancy is perceived in terms of confirmed expectancies as the eyes scan the figure or shape. This leads to the suggestion that one factor contributing to the attraction of beautiful objects lies in a feeling of satisfaction which springs from the confirmation of current expectancies, hence from the experience of what may be called *cognitive consonance.*

2. In the second step I shall suggest that we have an intuitive propensity to see such order and unity as the manifestation of, and a window into, some *creative* and *purposeful* activity. This is an anthropomorphic projection which has much in common with the faculty of *empathy.* From the extent to which the object invites such an anthropomorphic

11.1 - The sense of beauty 261

projection, in turn, we derive a sense of *affinity* with the object which adds to its attraction.

3. Finally, in the third step I shall suggest one additional, but rather more contingent factor, namely the impression a beautiful object can make of being a *communicator* or "*speaking presence*", to use Roszak's phrase, on account of the qualities I have outlined above.

STEP 1.

I suggest that the sensory attraction of a beautiful object springs in the first instance from the perception of ORDER and UNITY in the appearance of the object. And, as I shall explain, this is not only because only the perceived unities in the world of sense impressions enable us to see the wold as a world of separate shapes and objects.

To articulate this suggestion I must fall back on the basic concepts of information theory. One of these is that the amount of *information* conveyed by a signal or message may be equated with the amount of *uncertainty* it removes. Thus the perception of any the outcome of throwing a dice conveys information because it removes uncertainties And since the degree of uncertainty can be expressed in terms of the number of alternative outcomes that might have occurred (suitably weighted according to their probability), information theory is able to attach a quantitative measure to the amount of information conveyed by a signal or message (cf. Appendix E).

In the cases I shall discuss the introduction of probabilities is not an expedient one, since the range of possibilities is not sufficiently well defined. Nevertheless, we can treat the concept of uncertainty itself as an objective concept because we can define it as the negative of an expectancy held with confidence - two concepts which we have been able to define in objective terms earlier in this work.

From its concept of *information*, in turn, information theory derives the all-important concept of the *redundancy* in a signal or message. This denotes the excess of elements used in the signal or message to convey the given amount of information, compared with the minimum number of elements that could have conveyed the same amount of information had the signal or message been coded in the most economical way. Thus in a string of signals with redundancy the uncertainty which some members of the string remove will in part already have been removed by some. earlier member .

When applied to the perception of a figure or shape, the notion of redundancy is a fruitful one, because it can help us to understand the nature of the *order* and *unity* which one may perceive in the appearance of the object. To see this, consider a simple figure or shape with a high degree of order, like a circle or a regularly shaped fern or flower. Such regular figures or shapes have a high redundancy Because as the eye runs from one part of the circle or flower to the next it meets no surprises: what it sees looking at one part has to some extent already been suggested by what it saw looking at the previous part. Hence the perception of the present part does not add as much information (remove as much uncertainty) as would have been the case in a totally irregular shape.

As I have explained in Chapter 5, objects are perceived as a familiar *unit* through the linked set of confirmed expectancies which their perception elicits. Hence, at the basic visual level, the redundancy in a shape or figure which the scanning eye meets holds the key to the perception of that shape or figure as a *unit*. It is the redundancy in the stimuli the eyes receive when they track a line, for example, and the set of expectancies fulfilled as they move from one segment to the next (each exciting 'bar-sensitive' elements in the visual cortex), that makes the brain respond to the line as a unity. And the line as a whole derives from these features an attention-holding quality which also potentiates it as a cue for further action.

The importance of this unity, and therefore redundancy, in perceived figures or shapes and their elements is obvious. And it is plausible to suggest that this is a source of attraction for the eye. However, a second source of pleasure is also the pleasure of *fulfilled expectancies* as the eye scans a figure or shape with high redundancy. We may describe this as the pleasure of *cognitive consonance* – a pleasure which could be explained in terms of the *aversion to cognitive dissonance* which I discussed in Chapter 10. Of interest in this connection is a theory which Birkhoff (1932) suggested a couple of generations ago. He argued that the aesthetic measure (M) of an object could be expressed as a function of its order over its complexity. In informational terms we could render this as redundancy over information, thus:

$$M = \frac{\text{order}}{\text{compexity}} = \frac{\text{redundancy}}{\text{information}}$$

In a broad sense one could say that this formula expresses the concentration per unit figural space of the unifying relationships perceived in a figure or shape. However, I believe that all of this is only part of the story of aesthetic attraction. The perception of order and unity is only one element in the perception of beauty, albeit an important one. And this brings me to our second step.

STEP 2.

I believe that the next important element in the perception of beauty is a tendency to see this order and unity as the product of a *creative act*, as the manifestation of some *purposive*, even *mind-driven* set of activities or processes. Looking at a fern or symmetrical flower, for example, we tend to feel intuitively that some *end* or *goal* must be served by so seemingly planned a shape, e.g. that it has some *functional significance* - even though we can form no concept of the nature of that goal or function.

Perhaps, Immanuel Kant had the same aspect of beauty in mind when he said that beautiful objects give the effect of purposiveness, albeit without being the satisfaction of purposes. The poet Schiller went further when he said that beautiful things have an order which only a mind could produce - thus making the point that the purposiveness we see in beautiful objects tends to give the impression that there has been some kind of mind at work.

It must be stressed again in this connection that this must be seen as the way beautiful objects strike us at a *deep intuitive level*: it has nothing to do with our rational understanding of the object and of its true origins or properties. Indeed it may be quite contrary to what our reason suggests about the nature of the object we see before our eyes. It is a purely intuitive and subconscious response which occurs at a level of our reactions at which objective knowledge and rational beliefs have not yet come into play.

I suggest that this intuitive tendency to see conspicuous cases of order and symmetry as the product of a creative act and purposeful activity, also adds to the attraction exercised by the object because it is a gateway to an empathic relationship to the object When the symmetry of a flower or fern (or any other conspicuous type of orderliness in a perceived form or shape) strikes the us intuitively as the product of some purposive and mind-driven activity, we are in fact projecting into the object, or the forces that have created it, an aspect of our own being. And from the feeling that we are here witnessing the work of some kindred entity we derive, in turn, a sense of *affinity*, indeed *communion* with the object. In consequence the object becomes for us a vital presence, a kind of companion - even a *'promesse de bonheur'*, as Marcuse has put it. Perhaps this is the secret of the consolation which the experience of beauty can afford to the lonely and distressed.

The process I have described is of the nature of an anthropomorphic projection, and as such a species of empathy. Thus it could be cited as an example of imaginative activities operating below the level of consciousness.

Seen in this light then, it would seem that some of the attraction which objects of transparent unity, orderliness and harmony can exercise on man, springs from one of the three psychological forces I have isolated in section 10.2, viz., a longing for a state of communion with the surrounding world and for the comfort and assurance that a sense of communion may bring. It might be objected that in this feeling of affinity and communion the perceiver is merely extracting from the object something he himself has put into it. But this is not so: the perceived object, too, makes an essential contribution, viz. by virtue of those qualities in its shape or form which we are able to perceive as order and unity. For only those qualities have made the whole process possible.

Although I have taken here but a few concrete examples, it seems to me that all of this applies quite generally to the perception of objects in whose shape or form we perceive order, harmony and some unifying principle, though naturally in unequal measure. In all cases the objects seem to point beyond themselves in the sense I have described.

The intuitive perception of an object as the product of a purposeful and unified activity tends to be especially strong in the case of what we are wont to describe as *functional beauty*, such as the beauty of an elegant teapot, a bird in flight, or the graceful leaps of the antelope. For here the perceived shape, form and, where this applies, movement, seems to express most explicitly a unifying function or purpose.

Here, too, the aesthetic pleasure seems to be inversely related to the complexity of the phenomenon (as suggested under STEP 1 above), for one of the contributing factors to the pleasure of the perception is the *economy* with which the function of the object, or of its parts and their movements, seems to find expression in what we see before our eyes. Nothing is wasted, everything appears to be tailored to the function the object or its parts fulfil. Indeed, among the Greeks *suitability* and *aptness* often came to be regarded as synonymous with beauty itself.

Empathy elicited by inanimate objects, such as the kind I have suggested above, differs in two ways from person- to-person empathy, i.e. from the empathy we can feel for another human. Firstly, because it is solely invited by the appearance of the object (whereas this may be at most an auxiliary stimulus in the human case), and, secondly, because it amounts to an attempted identification with the flow of things of which the beautiful shape or form is the creation, whereas in the human case it is a sharing of joys and sorrows amounting (as the word itself suggests) to an identification with another's emotions and feelings.

The two species of empathy also have one important element in common: provided no other factors intrude, both can contribute to feelings of love, and in both cases the relationship between the empathy and the love can be a reciprocal relationship: in both cases loving through knowing and knowing through loving can go hand in hand. I have had to add the above proviso, because a good empathizer is not necessarily a good altruist.

In both cases, if love does follow, it will tend to be more than a mere feeling of attachment: in the human case it culminates in a frame of mind in which the concerns of the loved one become the concerns of the lover, and desires are stifled in the lover that would conflict with the desires of the loved one. In the love evoked by beautiful things, too, one can detect an element which goes beyond mere feelings of attachment and which has a kind of selfless quality: for it tends to produce a wish for the preservation of the objects in their own right, independently of the pleasure they can give us personally. It amounts to a 'love of being' with almost religious overtones, for it makes us see the destruction of beautiful objects as something approaching sacrilege - feelings not aroused by the destruction of objects to which we have become attached merely through familiarity, for example, or through their proven usefulness.

Anthropomorphic reactions of the kind I have described can also go further than the instances I have illustrated, and this, too, is of aesthetic significance. For example, when a mountain range strikes us as *majestic* and *imperturbable*, or a landscape as *serene* and *tranquil*, we encounter again reactions which are essentially based on anthropomorphic projections, and it is not unreasonable to suppose that they add to the aesthetic attraction of the phenomena concerned on much the same grounds as the anthropomorphic projections I have analysed above.

STEP 3.

There is a third factor which, I feel, calls for consideration. Although intuitively we tend to see the order and unity in a figure or shape as the product of some purposive activity, we have, as has been said, no cognizance of the nature of this purpose. However, I believe that at times we do intuitively and subconsciously read one particular purpose or intention into those activities, viz. an *intention to communicate*. In these cases the informational aspect supplements the redundancy aspect. In WHERE THE WASTELAND ENDS Roszak (1973) used a telling metaphor when he described a beautiful object as a "speaking presence" and as a "creation that speaks to us in a kind of silent language". As a concrete example, take the beauty of the starry sky. What is the magic of those twinkling lights? Primarily, I suggest, that intuitively we respond to a light in the darkness (especially a twinkling light) as a *signal*, albeit a signal we cannot decipher intellectually. And any stimulus accepted as a signal strikes us intuitively as the manifestation of an intention to communicate. Thus out of the depth of infinite space the Universe seems to speaks to us, and in doing so establishes for us a kinship to something (in this case) infinitely greater than ourselves, to something majestic in its grandeur and imperturbability. Perhaps this is the secret of the kind of aesthetic experience which we call the experience of the *sublime*.

What I have said above about the stars applies, in a transferred sense, also to the role which *colour* can play in the perception of natural beauty. When in the midst of a mass of green foliage we perceive a bright flower, be it red, yellow, blue, or pink, one immediate impression, I suggest, is again that of a "speaking presence". And once again the emotional impact of this is a feeling of *company*, a feeling of *not being alone*. Colour may have aesthetic significance for at least two further reasons. Firstly, because it can add unity to the elements or parts of a perceived figure or shape. Secondly, because colours can have a variety of emotional effects based on deep-seated associations of one kind or another. They can suggest light, danger, sombreness. And all of this goes in the final mix that decides the aesthetic impact made by the objects we perceive. In the pictorial arts, of course, these effects tend to be exploited deliberately.

One brief word must here be added about *art*, because in art some of the elements I have just mentioned are particularly obvious. Here the intention to communicate is accepted already as part of the very essence of a work of art. Equally obvious is the element of empathy, because successful art enables us to see the world *as through the artist's eyes* and to feel our way into his or her emotions. In a sense, of course, a work of art also communicates with its own creator, because part of its value to the artist lies in the power of a work of art to articulate or accentuate for its creator things only dimly felt and relationships only dimly perceived at the onset of the creative act. However, I would be the first to admit that this is not the full story of the impact made, or intended to be made, by a work of art.

Finally, I must return to the fact with which I began: the *subjectivity* of our aesthetic responses. On account of this very individual character of our aesthetic sensibilities and responses some people claim the roots and rationale of aesthetic responses can never be analysed in objective terms - indeed that it is idle to look for any kind of rationale here.

It will be clear from the preceding pages that I do not share this view. For I have done nothing less than offer an objective account of the roots and rationale of certain characteristic kinds of aesthetic sensibilities and responses - here those which we identify with the sense of natural beauty. It seems to me that the subjectivity of aesthetic experience can be accounted for simply in terms of the multiplicity of factors that contribute to that experience, and the different degrees to which people will be sensitive to them.

***

11.2 - THE MORAL SENSE

For a biologist it is tempting to regard the moral norms of a society as mere instruments for the efficient functioning of the social order in a species that has come to depend for its survival on collaborative associations between its members. Believers in this type of functional theory tend to cite certain crosscultural studies of anthropologists which suggest that what different societies regard as right and wrong has to be understood in terms of their functional requisites. On this view, then, ethical values are *relative* and morality is a mere organ of social utility which has evolved in accordance with its genetic cost-effectiveness - not unlike the 'altruism' of the worker bee. On this view then, no society can legitimately pass moral judgments on the norms of another society, especially one occupying a different ecological niche.

However, whereas up to a point this interpretation may apply to some of the primitive, and occasionally rather exotic, societies anthropologists have studied, it clearly does not hold for the major cultures of both the Oriental and Western world. For in the evolution of the cultures that have given us the writings of Lao-tzu, the Judeo-Christian Scriptures, the Koran of Mohammed, the teachings of Buddha, the complex beliefs of Hinduism, and the secular beliefs of Western Humanism, a line of development has gained the upper hand, which has been marked by a growing *internalization* and, at the same time, *universalization* of moral values.

By *internalization* I mean a tendency to see the ultimate root and rationale of moral values, not in the cooperative needs of society, but in the needs of the individual human psyche, and especially in the individual's craving for that kind of harmony of desire and experience which by some is described simply as happiness, self-fulfilment, inner peace, or bliss, and by others as a union with God. This is seen as a craving shared by all humans, hence as the expression of a need resident in the "common human frame", as Hume put it. Moral principles thus came to be seen as principles of *universal* validity, as precepts about social relationships which flowed from aspirations fundamental to the human psyche.

In Chapter 10 I have already suggested that these spiritual needs, and the visions of a better world that flow from them, have their ultimate roots in four basic aversions, viz. an aversion to conflict, anxiety, frustration and cognitive dissonance. In the present context only the first three are relevant, since the aversion to cognitive dissonance operates only at the cognitive level, e.g. in the manner I have described in the last section. Thus I am suggesting that the intimations of a state of happiness, inner peace, self- fulfilment or bliss, mentioned above, are no more than intimations of a state of the psyche in which conflict, anxiety, frustration are minimal, culminating in visions of a world of social harmony in which mutual concern and the power of love to bring order out of chaos has removed the very causes of these three biological negatives.

The full realization of this ideal is no doubt unattainable on Earth. For the humanists this is the bottom line, whereas the religious believer pins his faith on the full realization of the ideal in the Kingdom of Heaven. Why is it unattainable on Earth? In our terms the short answer is that in the physical world it is impossible for everything to be directively correlated to everything else. In this basic fact lies the ultimate root of the necessity of evil or, rather, of some inextinguishable residue of evil.

The type of conflict to which the above statements refer is an intrapsychic one, not a psychosocial one. However, in Section 10.3 I have explained how the aversion to intrapsychic conflict leads naturally also to an aversion to psychosocial conflict. Because, as I said at the time, every *external* conflict, i.e. conflict with another human being, creates an *internal* conflict, since it confronts the individual with the choice of either insisting on having his own way at the risk of incurring the other's hostility, or yielding to the other's desires and having to accept the resulting frustration of his personal aims.

In consequence the inward-looking craving for personal bliss tends to result in outward-looking ideals of social harmony. The importance of self-less love, a cornerstone of both the Judeo-Christian and humanist outlook, enters at this point. For selfless love has a strong component of identification-love, i.e. a disposition to identify with the interests of the loved ones. Hence selfless love tends to extinguish the desire to do things that conflict with the interests of others. Once again, this draws our attention to the importance of the human faculty of the imagination, this time in the form of empathy. One can also see how a morality which flows from empathy via love can reach out beyond one's neighbourhood to the whole of mankind and even to the whole of the living world.

Another important fact follows from this interpretation of the roots of our moral values, viz. the need for *collectively accepted* moral principles, i.e. moral *norms*. Because the minimization of psychosocial conflict demands shared principles, permitting mutual understanding and mutual regard. (Immanuel Kant turned this round when he suggested that moral rules of conduct are rules of conduct of which you could wish that they were observed by everybody.)

Moral principles are communal concerns also in another sense. Man's spiritual needs are not very articulate needs, nor, in general, will the individual have a clear perception of the path that needs to be followed to achieve their gratification. The perception of that path is a slowly growing cultural good and communal possession. Nor has that growth proved to be a steady and unilinear process. It has proved to be a fragile plant, apt to wither under the impact of tribal or racial forces and hostile political climates. Historically it has been intimately linked with the evolution of religious belief, punctuated by the insights contributed by exceptional men. Often these were articulators rather than innovators: their messages would have fallen on deaf ears, had they not expressed in an articulate and vivid way something that their contemporaries had already dimly felt themselves.

Before I move on, let me give a summary of the theory as it now stands:-

1. The universalized and internalized ethos of the religious and humanist movements which have shaped our society is the product of a cultural development in which certain fundamental needs and aspirations of the human psyche have asserted themselves. They were able to do so to a unique degree because this development was a *cultural* phenomenon, hence a development that could proceed free from the constraint of the cost-benefit calculations of genetic advantage that govern the propagation of the *innate* traits of any living species. (The forces that actually caused them to assert themselves to that unique degree, will be discussed in Section 11.3.)

2. These needs and aspirations have their ultimate roots in an aversion to anxiety, conflict and frustration which is common to other living species as well, but which in the human case has developed in new directions. Through the faculty of the imagination the aversion to these biological *negatives* produced in man also intimations of, and a longing for, the corresponding biological *positives* - for states of affairs in which frustration, conflict and anxiety are absent or minimal. As particular expressions of these longings I have cited

> a) a longing for freedom and self-fulfilment, for a life in which one's potentialities are fully realized and one's faculties fully engaged - the antithesis of a state of frustration;
>
> b) a longing for emotional and motivational stability, i.e the craving for a stable direction in one's life;
>
> c) a longing for the emotional security (absence of conflict and anxiety) that comes from harmonious relationships with the surrounding world. I have explained how this longing for conditions of minimal psychosocial conflict flows naturally from the aversion to intrapsychic conflict.

3. These longings, in turn, have tended to generate visions of a better world, a world of harmony and minimal social conflict - visions of a brotherhood of man.

4. The result has been an ethos which rests on shared visions of this kind, and which reflects the community's accumulated insights into the path that has to be followed in pursuit of such ideals. Among this wisdom is the knowledge that, though we may each wish for ourselves a rich and harmonious inner life, we are unlikely to find this unless we also wish it for others.

Of necessity the visions of social harmony to which this ethos relates will include those relationships that are needed for the efficient functioning of the social order. Thus the ethos I have discussed is bound to include a considerable element which is of social utility in the pragmatic and common biological sense. But it obviously goes well beyond this. Indeed, a morality which is no more than an organ for the efficient functioning of the social order, amounts to no more than a codified form of the law of the jungle.

*

Alongside the historic trend towards internalization and universalization of moral values there has also run a trend towards *rationalization*, i.e. towards the construction of coherent and unifying systems of beliefs from which specific moral precepts could be deduced, and which could thus serve as rational guidelines for those willing to obey the moral law.

Religious dogma was the sole purveyor of such rationalizations until the philosophers turned their mind to the problem and began an independent search for unifying formulae. Thus ethical theory was born. But although in the history of Western thought *ethical theory* began with the Greek philosophers, notably Plato (427-347 B.C.), it took another thousand years before the point came to be forcefully made (Hume, 1711-1776) that ethical theory should solely concern itself with the *description* of existing moral systems and analysis of their implications. It should not engage in the *prescription* of moral systems. Hume also made the point that in the matter of ultimate values, no *prescriptive* statements can be deduced from any kind of *descriptive* statements: no statements about what *ought* or *ought not* to be can be deduced from statements about what *is* or what *is not*. Bertrand Russell (1935) took up the same theme in his discussion of science and value: science is about facts and as such it can describe or take note of what ends people pursue. Above all, it can point out what actions can serve what ends. But it can never establish the value of the ends themselves: Given the *intrinsic* value of the ends, science can point out the *instrumental* value of particular courses of action. Given that we want to remain in good health, science can point out how to set about it. But it cannot establish the intrinsic value of good health or any other attribute or object, because no statements about facts can do so. On this view, then, one cannot look to science for a justification of our moral values.

By contrast, I would suggest that an account of moral values which succeeds in making scientific sense of those values (as my account would claim to do), does amount to a kind of justification. Rational beings need systems of beliefs and guidelines that are intellectually plausible. And in a scientific climate (such as ours) that means beliefs which make scientific sense. A mere appeal to authority can be no substitute for those who want to understand the rationale of the moral precepts they are expected to follow. Moreover, the doctrine that God ordains what is good or evil was already shown by Kant to be untenable from a detached intellectual point of view: " Even the Holy One of the Gospels", he wrote in his FOUNDATIONS OF THE METAPHYSICS OF MORALS, "must first be compared with our ideals of moral perfection before we can recognize him as such."

Alfred Ayer (1973) put it more strongly:

"Moral standards can never be justified by an appeal to authority, whether the authority is human or divine. There has to be the additional premise that the person whose dictates we follow is good, or that what he commands is right".

11.2 - The moral sense

In other words, we cannot grant God moral authority on the strength of his goodness and righteousness, unless we begin with some prior notion and appreciation of those qualities.

*

Conceived as an academic discipline, moral philosophy needs to remain purely descriptive.But by 'a philosophy' we also often mean just one's general outlook on life and despite Hume's injunctions few moral philosophers have resisted the temptation also to articulate what they have felt that outlook *should* be. Where the norms they advocated departed from those prevailing at the time, they became moral *reformers*. We have a notable example, and one that is relevant in the present context, in the doctrine of *Utilitarianism* which Bentham encapsulated in the formula that the greatest good lies in the greatest happiness of the greatest number. It is a deceptively simple formula, but also the one to which popular ethics has to this day proved to be most receptive - partly, no doubt, because of its very simplicity and direct appeal to our common perception of human nature.

Textbooks tend to divide Moral Philosophers into *subjectivists* and *objectivists*. According to the narrowest form of *subjectivism*, propounded already by Hobbes (1588-1670), moral judgments are mere expressions of personal approval or disapproval: a purely emotive response. This is also known as the Boo-Hurrah theory, a theory still favoured by Logical Positivists. Hume rejected it on the grounds that when we make a moral judgment we step outside the personal situation and choose a point of view that we take to be shared by others: we appeal to shared principles, principles which Hume himself believed to be universal and rooted in the common human frame. In this last respect the theory of moral values I have set out above agrees with Hume. But we have gone further than he did, for I have attempted to show in some detail how the moral perceptions we have discussed relate to that common human frame - how they derive from it.

According to the pure *objectivists*, on the other hand, a moral judgment is not an emotive response, but essentially a *perception* of qualities, relations, or laws that exist independently of the perceiver. For Plato it was the perception of the Form of the Good. Plato's *Forms* were conceived as abstract structures which encapsulate what is permanent in a world which the senses can only perceive as constantly changing, e.g. mathematical laws and standards of perfection. The Form of the Good was one of these.God presided over the realm of forces. He was conceived by Plato as the Form of Forms.

For Aristotle it was the perception of what would constitute one's total self-fulfilment, whereas for Thomas Aquinas it was the perception of God's providence and his eternal laws. And in our century G.E. Moore propounded the view that the good is a simple unanalysable quality which people can perceive with greater or less clarity, depending on the eye they have for it. Thus perceiving the difference between good and evil was rather like perceiving the difference between red and green.

The theory I have outlined in this chapter is clearly opposed to *objectivism* in this special sense. Rather, it bridges the gap between subjectivism and objectivism. Because on the interpretation I have given, moral sentiments may be described as an emotive response which reflects what the individual has absorbed of the community's perception of the path that needs to be followed if certain basic needs and aspirations of the human psyche are to be met - modified, as the case may be, by the individuals' own intuitive grasp of what morality is all about. On this view, then, moral sentiments rest on a mixture of objective and subjective elements.

The present theory is also opposed to the school of moral *relativism* mentioned before. This maintains that moral norms are entirely a matter of contingent social needs to which no universal criteria can be applied. It is a view often favoured by theorists searching for moral theory suitable for a pluralistic society. They are mistaken in their premisses, because, as I have stressed, one can believe in a set of moral principles which one regards as universally valid, and yet allow for considerable variety, including ethnic variety, in the relative weights people may attach to the individual members of that set.

*

To conclude this section, let us just check up on the key concepts I have introduced, both in this section and in our treatment of the sense of beauty. For it may seem to the reader that in both these sections I have observed less scrupulously my declared intention to conduct our analysis solely with the aid of notions which can be clearly defined in strictly objective terms. But this is not so. The main explanatory concepts I have used were those of *frustration*, *conflict* and *anxiety*, and in Section 10.3 I have explained that each of these concepts can be construed in an object sense. Then I brought in the faculty of the *imagination*, and that was defined in objective terms in Chapter 7. I also appealed to a particular application of the imagination, viz., the faculty of *empathy*. But the different kinds of internal representations which this faculty encompasses are all of a kind that has been covered in our discussion of the faculties of the imagination and self-awareness. Thus to have sympathy with another human being, is to imagine as part of the current state of the self particular feelings one believes to be present in the other. And in a strong case one can react emotionally to these imagined feelings as if they really were one's own - the secret of truly shared joys or sorrows.

Finally, in various contexts I have talked about *harmonious* relationships. Here I have to add that my concept of harmony has throughout been the same as that which I introduced in Chapter 2, when I described the processes which sustain the life of an organism as "working in perfect harmony". I then meant the high degree or *còordination* and *integration* that these processes manifest. Both these concepts are teleological concepts which can be given an objective non-teleological definition with the aid of the concept of *directive correlation* first introduced in Chapter 3 (cf. Appendix B). Hence, in the sense in which I have consistently used these words, 'harmony' denotes a certain objective set of relationships, while 'harmonious' signifies an attribution of such relationships.

***

11.3 - A DISTINCTIVE STEP IN THE EVOLUTION OF LIFE

How, from a detached biological standpoint, are we to see the historic development of our moral and aesthetic sensibilities and responses as a distinctive step in the evolution of life on this planet?

Let me first clarify one point. The phrase 'evolution of life' can mean two things. It can mean simply the changes in life-forms that go on all the time and at all levels of the scale of life. As recent events have brought home to everybody, even viruses are evolving all the time. But often we also mean by the 'evolution of life' the progression over the millenia from simple forms of life, like the protozoa, to much more complex and sophisticated ones, a development that has climaxed in the emergence of the species *homo sapiens*. In the present context I mean evolution in this second sense.

From the standpoint of general biological theory, the problem of why and how organisms come to develop aversive reactions to states of anxiety, conflict and frustration, poses no major problems. The biological rationale of such reactions is readily understood. It is all part and parcel of the machinery through which the organism monitors the efficacies of its activities and mobilizes its responses to failure. The phylogenetic evolution of these mechanisms can be readily explained along Darwinian lines, and equally readily can it be explained how some such aversions may also come to be acquired ontogenetically, i.e. as learnt internal reactions to situations in which failure prevailed over success.

Hence it is not difficult to understand in orthodox biological terms how humans, too, come to have a deep aversion to anxiety, frustration and conflict. I have also explained how a uniquely human element came to be added through the human faculty of the imagination. For this enabled man to form internal representations, however ill-defined, of states of consciousness and ways of life in which these negatives were absent or minimal, and which would thus come to impress themselves on the mind as desirable ends. As derivatives of this I have cited such human traits as a longing for self-fulfilment, for a stable direction in one's life, and for a sense of communion with the surrounding world. I have also explained in these terms the emergence of moral beliefs which are based on visions of a brotherhood of man and of world in which peace and harmony have displaced chaos and conflict. Finally, I have explained that this ethos was able to develop and to assert itself to the extent to which it has in the evolution of our civilization, because it was a cultural phenomenon, and thus one that could develop free from the constraint of cost-benefit calculations of genetic advantage that governs the propagation of the innate traits of a species.

This leaves one important question unanswered: given that moral norms were thus *able* to develop in this direction, what actually *caused* them to develop so? What factors gave this universalized and internalized ethos a selective advantage in the cultural realm over the tribal or racial ethos that would have triumphed had genetic cost-effectiveness been the decisive factor?

I suggest that this selective advantage derived from two facts. Firstly, these moral beliefs had their roots in *universal* human cravings. Secondly, their precepts claimed to be generally valid regardless of circumstance. They were *categorical* rather than merely *hypothetical* imperatives. Specifically, they were of the form: under all circumstances the way towards a life free from conflict, anxiety and frustration is the way of love and mutual concern. These facts are decisive, for it seems to me that *in the long run, and in a large and mixed society, the propagation of those precepts will be most favoured which relate both to the most widely shared needs and to the most generally valid means of meeting these needs.* Thus the need to feed and clothe oneself is also a universal need, but the precepts for meeting that need are hypothetical rather than categorical: they differ from case to case and circumstance to circumstance. They take on one form for the farmer and another for the industrial worker, yet another for the pensioner. They differ in Europe from Africa, etc.

However, there is one important proviso. These selective advantages in the evolution of human civilization can become effective only in conditions in which the moral beliefs in question can freely compete with rival beliefs, i.e. if the social climate is one that permits freedom of thought and speech, freedom of information and ready access by all members of the community to the system of beliefs concerned. In the history of our civilization that, of course, has been a major impediment. In particular the propagation of moral systems tied to religious beliefs has often been by the sword rather than free choice. It follows as a mere corollary that in the long run an ideology which ignores, or conflicts with, the basic human cravings I have discussed, can maintain itself in power only by denying the freedoms I have mentioned.

*

11.3 - A distinctive step in the evolution of life

To return now to our original question: how, from a detached biological standpoint, are we to see the historic development of our moral and aesthetic meta-sensibilities as a distinctive step in the evolution of life on this planet?

I have already given a partial answer to this question: the development amounts to a major discontinuity in the evolution of life. For it has made man more than just a sophisticated survival machine, and it has made his rules of conduct more than just a codified form of the laws of the jungle.

But have these human traits made man a *higher* form of life in the sense in which the term is used when we describe the mammals as a higher form of life than the insects, and these in turn as a higher form of life than the protozoa?

In Chapters 2 and 3 I have outlined the objective sense in which 'higher' may here be understood. I made it clear at the time that the progression from the protozoa right up to (say) the mammals is not marked just by an increase in the complexity of the living organism, i.e. the number and variety of its constituent parts. This would count for nothing were it not for the concomitant increase in the scope and degree of the mutual adaptation, coordination and integration of the activities of the parts - thus giving us not just more parts, but more parts *working in perfect harmony towards common goals*, notably the goal of survival.

All of these are relationships which the analysis which given in Chapters 2 and 3 has enabled us to define in objective terms, mainly with the aid of the concept of directive correlation. They are the essence of what we mean by *organic order* (see also Appendix A.)

Now if we think of this organic order merely as part of the machinery of survival, then man undoubtedly qualifies as a higher form of life than (say) the apes on account of the scope of the coordination, adaptation and integration of activity which his cognitive, rational and linguistic faculties, his technical skills and powers of communication have made possible, but *not* on account of the aesthetic and moral meta-sensibilities I have discussed. As I have said elsewhere, particularly in times of war we realize how much this ethos may actually conflict with the demands of survival.

On the other hand, if we abstract from the goal of survival, and think purely in terms of the progressive increase, as we pass from the lower to the higher forms of life, in the scope and degree of organic order *as such*, i.e. in the scope and perceptible degrees of harmony in both internal and external relationships (regardless of the ultimate goal this may serve), then we can see in man's sense of beauty and moral sense the further extension of a continuous line of development in the evolution of life on this planet. For, looking at the matter from this standpoint, we can see in the harmony and communion to which our spiritual nature aspires, a further expansion of the kind of organic order that I have discussed in Chapter 2 and have there described as the most distinctive feature of all living systems (see also Appendix A). From this (still strictly biological) standpoint, therefore, the evolution of the spiritual side of human nature can be seen as the continuation of a trend (albeit in new dimensions) which has characterized the evolution of life since its first beginnings. And, seen in this light, it represents a singularly significant step in what may plausibly be called *biological progress*.

***

11.4 - THE RELIGIOUS DIMENSION

One factor not yet considered in my remarks about the evolution of cultural norms is the religious dimension: the close ties of these norms with the evolution of religious belief. Yet religion has always been more than a pathfinding exercise in the pursuit of ultimate bliss. It has offered a cosmology whose roots reach right back to primitive man's attempts to interpret the world around him and understand the forces that moved it. This element has followed its own historical course, but with feedback to the ethical dimension.

From the start these interpretations were subjective, anthropomorphic, animistic, teleological. For man's natural tendency was to interpret all the activities of the natural world in terms of the motive forces which he subjectively experienced as the causes of his own behaviour. Since his own activities were governed by conscious purposes, he assumed that all activities around him were of this kind: that all things were driven by purposes and that objects moved because they were striving to achieve some goal. Hence it was natural for him to see the objects around him as the possessors of souls, indeed the whole natural world as populated by spirits and replete with symbols of the divine. At the same time he remained conscious of his own relative impotence *vis à vis* the powers of nature. Hence the spirits in the world around him had to be placated and their goodwill assured by appropriate rituals and sacrifices.

Although there appears to be some evidence for early monotheisms, the general pattern seems to have been one in which such an early multiplicity of spirits sustained itself for some time before a process of coalescence began that terminated in one single spirituality, a single deity held to pervade and govern the cosmos. At first this deity was conceived as a high deity lording it over a multitude of lesser deities: the monotheism was inclusive and sometimes still pluriform. But eventually only the high god remained, now conceived as creator and unmoved mover of the world, the supreme legislator of the Universe. And so it has remained in the three great monotheistic religions of our time: Judaism, Christianity and Islam.

In the course of time, however, ever clearer perceptions developed of natural causation and of the causal principles that govern the activities of not only inanimate objects, but of the living organism as well. By the 17th century the human body, too, had come to be perceived as a kind of machine (albeit a machine controlled by what seemed to be a wholly different entity: the mind). Thus the seeds were sown for a scientific explanation of the world, whose exponentially mounting successes have since proved conclusively the superiority of its methods and conceptual framework over all other forms of explanation of the natural world. A new cosmology was born.

Yet religion has survived despite the success of science, and not only in societies in which it is imposed from above or societies ignorant of modern science. It has survived even in societies in which it was actively opposed from above. Evidently, for a decisive number of people, religion continues to meet felt needs which other systems of thought cannot meet or cannot meet so well in their eyes. The following is an outline of what I believe these particular needs to be, and how they appear to relate to the motive forces which I have depicted in the preceding sections as the roots of the moral or aesthetic sensibilities and responses that are characteristic of our culture. I suggest the following six:-

1. The felt need for a philosophy which can serve the individual as the basis of moral certainties, and which can offer the individual a set of stable guidelines in the conduct of his life, given the major theme of that life. For the main integrating power in any individual human life springs from the stability and intensity of the ultimate purposes that direct it. Moreover, the special asset of religion in this respect is that it encompasses all aspects of human life, not only our moral nature, but also, for instance, the significance of truth, the inspirational power of beauty, and the power of love (as expressed, for example in the concept of a loving and merciful deity). Thus in the eyes of many people religion answers more fully the kind of questions they find themselves asking.

It could well be that this need for a stable direction in one's life and the related need for a basis of moral certainties, have both been potent influences favouring the evolution of *monotheistic* religions: a multiplicity of deities can hardly give a single direction to one's life or furnish a single basis for moral certainties.

2. A felt need to come to terms with one's solitude, for, up to a point, every individual is an island. Yet it is not in the nature of man to be wholly self-sufficient, and the longing to belong - the longing to be part of a larger whole while yet maintaining one's identity - is perhaps one of the most consistent of human impulses. I have tried to explain its origin earlier in this chapter. Of course, the way it expresses itself tends to differ greatly from case to case.

Many people appear to be content to feel at one with their immediate social environment, or with a community of people sharing the same aims. But in others, and especially in those who are at odds with their social environment, the search may be for a sense of communion with something that runs deeper, something more general and consequently also more abstract: humanity, nature, or God, as the case may be.

3. The lasting attraction of a philosophy which gives a *teleological* account of the world and of the forces that govern it, i.e. an account which interprets all phenomena as the product or manifestation of some purposive activity. For this enables people to read some kind of significance, meaning or purpose (a) into their own existence and (b) into the objects and events of the surrounding world.

People may respond emotionally to the first because from the purpose which religion enables them to attribute to their own existence, they can derive that stable sense of direction to which I have alluded under (1) above. They may respond to the second because it helps them to orient themselves imaginatively towards the surrounding world in the kind of empathic fashion that fosters the affinity with that world which I have described in Section 11.1. Thus the belief that all things happening in this world are part of a divine and purposeful order, and that this order touches one personally in the experience of beauty, love and goodness, amounts to an imaginative re-orientation which can lend a new significance to even the commonest objects. It amounts to a different way of seeing the world - a difference which has been compared with the difference between seeing someone with love and seeing the same person without love.

To see the whole world in this teleological way, and all its creatures as both purposed and purposive, was the native treasure of our primitive ancestors. It was severely bruised by the growth of objective knowledge. But on the above grounds it is not difficult to understand that man has continued to feel a need for something like it.

4. The need to come to terms with man's insufficiency in respect of many of the immaterial qualities of life he desires most - a key note in all major religions.

By man's insufficiency I here mean the fact that many of the qualities of life we ultimately desire most, like health, peace of mind, harmony of desires, love, vitality, an interest in life, a sense of direction, are all of a kind which lies beyond our direct control in the sense that we cannot fabricate them at will – no more than a gardener can fabricate a flower at will. Unlike the machines that we have the power to plan, design and fabricate to our own specifications in a logical sequence of steps, the qualities of life I have mentioned cannot be planned and assembled with nuts and bolts according to some rational and preconceived design and logical procedure: like the flowers in our gardens, they can only be *cultivated*, but never *fabricated*. To take this analogy further: we can create the right soil and climate for their growth, but for the rest we have to trust that 'nature' will do its stuff.

That goes for our social life as well. We can no more fabricate a friendship, or, indeed, a free and democratic society, than we can fabricate a flower, or with piecemeal reconstitute one that we have trodden under foot. No democratic constitution will produce a truly free and democratic society unless the moral climate is one in which people accept the responsibilities that freedom imposes. And, again, that moral climate cannot be fabricated, it can only be cultivated.

On the other hand, it is also true to say that basically the facts are friendly in all these respects - just as basically the facts are friendly to the gardener who knows how to provide the right soil and climate for his seedlings. Love and friendship will grow by their own internal mechanisms, so to speak, provided we throw no spanner in the works - the ultimate notion of sin. In religion, of course, this friendliness finds expression in the notion of a *benevolent* God.

5. The power of hope to act as an antidote to the debilitating effect of despair in a crisis. In moments of great stress, frustration or helplessness, prayer and trust in God can undoubtedly do much for the believer to prevent the disintegration of the will that might otherwise have resulted from the ensuing despair. Hope, of course, is one of the three pillars of the Christian attitude of life, sharing this place with faith and charity.

6. The closely related need to come to terms with death. By this I mean the attraction of beliefs, such as the belief in an eternal life, which prevent the certainty of death from ridiculing the absoluteness of any moral commitment; from condemning as futile the pursuit of any goals whose fruits one will not be there to taste; from condemning as irrational any loving and caring for others until the very end of one's life; and, indeed, from making nonsense of the entire human enterprise. Man is the only creature which knows about the inevitability of its death. But, as Thomas Mann has said: "For the sake of love and goodness let no man's thought be ruled by death".

If we look at these six needs individually, we can readily see how each links up with the needs and aspirations of the human psyche which I have discussed earlier in this chapter and whose biological roots I have sought to trace.

Naturally, the needs I have listed will not be felt by every individual in equal measure. There are those, for example, who do not feel any need for a philosophy that can give a sense of direction to their life, because they are already driven by some inner force which fixes that direction, or by a passion to achieve something whose value seems to them wholly self-authenticating such as the pursuit of art or scientific truth, or just a happy family life.

*

In attributing the survival of religion to the fact that it satisfies certain basic needs, I have accentuated the *regulative function* of religious beliefs, i.e. their effects on the hidden chemistry of one's emotional life. These effects will only occur if there is a full personal commitment to the adopted creed. For the acceptance of religious belief which is here at issue is more than an intellectual assent to a number of propositions. It is a surrender to the implications of those propositions in the actual conduct of one's life. In fact, for some people the intellectual assent may even be the most difficult part about it - a point to which I shall return presently. Hence the commitment is something of a leap, the "leap of faith" as Kierkegaard put it. The acceptance that God exists and, for the Christians that Jesus is alive, is part of that leap. It lies in the nature of this leap that these propositions are accepted as axiomatic - a fact that can make it very difficult to find common ground in discussions between the believer and the sceptic. All fruitful discussions need some common premises to which the participants can appeal.

The regulative function of religion has been singled out by a variety of thinkers. Thus Galileo, for whom the difference between the regulative and the cognitive function of religion was a matter of great personal concern, made the point with an elegant aphorism when he wrote to the Grand Duchess Christina: "the intention of the Holy Ghost is to teach us how one goes to Heaven, not how heaven goes." Similarly, Kant described religious dogma as a "regulative principle of practical reason".

We encounter the same idea in more recent times. Thus the philosopher/mathematician A.N. Whitehead (1925) has described religion as "part of the art and theory of our inner life"; while Lee (1946) described it as a "route-finding exercise" - the route in question being the route to the kind of inner harmony and order and sense of purpose in their life which the route-seekers desire. Julian Huxley (1961), too, must have had something like this in mind when he called religion a kind of "spiritual ecology". In again in different terms, and stressing the act of faith, Masterson (1971) has described religion as an "existential option".

For a scientist trying to understand the motive forces of human behaviour the regulative role of religion is no doubt the crucial one. However, in modern times a new factor has entered that needs to be taken into account in assessing the appeal of religious beliefs. I mean the undeniable conflict between the cosmology of religion and the cosmology of science - at any rate if the former is accepted as a literal account not only of the origin of life (Creationists vs.Darwinists!) but also of the ultimate powers that govern the general flow of things. In a society such as ours, in which thinking rationally has increasingly come to be identified with thinking scientifically, it is hard to maintain that to-morrow's weather may not be solely determined by the conditions listed by the weatherman, but also by the number of farmers praying for rain. And it is certainly not a qualification of his forecasts the weatherman himself would feel obliged to add for the sake of a truly realistic prognosis.

The problem is that no critical and reflective mind can live happily with a set of mutually conflicting axioms. Moreover, the fact that a belief works, is not in logic a sufficient ground for claiming it to be true in the same sense in which a scientific proposition may be claimed to be true. This is demonstrated by the well-known placebo effect: the mere belief that a particular pill will prevent seasickness can make even a neutral pill have that effect.

One may regard religion as no more than an image-driven way of relating oneself effectively to the conditions of human existence and to some of the fundamental demands of the human psyche which I have outlined above. In other words, as a set of images that can be effectively used to give a direction to one's life capable of meeting some of the basic needs and aspirations of the human psyche. But the fact remains that these images will achieve their regulative efficacy only if the believer accepts them as more than apt metaphors.

There is no shortage of literature attempting to reconcile the two cosmologies. But in most of it the quality of the arguments has tended to fall short of the quality of scientific thought, either at the logical or the semantic level. And to the critical mind a 'reconciliation' of religion with science which is achieved only at the cost of sacrificing the stringent logical and semantic canons of scientific thought does not qualify as a reconciliation at all.

I have mentioned this problem not only because it is an important factor that has to be taken into account in assessing the influence of religion on the cultural milieu of our time. It is also of special interest in connection with any work that deals with the relation between mind and matter. What bearing, then, do the conclusions of our analysis of mental events have on the issues that are here at stake? The main one, I suggest, is that the Conditional Expectancy Hypothesis enables us to resolve the conflict between the two cosmologies if we start with the premise that two sets of axioms (here those of science and those of religion) can conflict only if they are applied to overlapping universes of discourse. It follows that the two cosmologies would *not* conflict if

1. the axioms of science are taken to apply exclusively to expectancies relating to the outside world, including expectancies relating to the behaviour and fate of people in a context in which these people are regarded as part of the furniture of that outside world (the common stance of the psychological and behavioural sciences), while

2. the axioms of religion are taken to apply exclusively to expectancies relating to one's inner world of subjective experience and to the consequences of one's actions in this sphere, i.e. in a strictly existential context and so as part of the art of our inner life, to use Whitehead's phrase.

***

11.5 - CONCLUDING REMARKS

I said at the beginning that to arrive at a clear scientific perception of the nature of consciousness and of the neural correlates of mental events, one has to adopt a top-down approach and follow a methodical analysis which from the start is conducted in terms that can be clearly defined in the language of the natural sciences - the language of physical events and their relations in time and space. To prepare for this, we first had to turn to general systems theory in order to clarify some of the general characteristics of the organization of living systems. Introducing the mathematical concept of *directive correlation* (Section 3.2) we were able to define some of these characteristics with the degree of precision a stringent systems analysis must demand. This gave us a basic set of precise physical concepts which were relevant to our enquiry and could be used later to give equally precise definitions to any special concepts the theory needed to introduce, such as the concept of internal representation.

Of the specific hypotheses I introduced, two occupied key positions. They were

1. the assumption that the brain can form the three kinds of internal representations described in Section 4.5; and
2. the Conditional Expectancy Hypothesis (Section 5.2).

The second of these hypotheses may perhaps be regarded as the most daring. Yet, it followed naturally from the sheer logic of the problems the brain has to solve if it is to equip itself in the best possible way for matching its outputs to the requirements dictated by an constantly changing environment. Nevertheless, we did not ignore the main condition any hypothesis must satisfy, viz. that it should explain what it was designed to explain and not be categorically falsified by any known observation. It is also clear from the nature of our approach that alternative hypotheses would qualify as rivals only if they were formulated to the same semantic standards.

I hope I have been able to show that proceeding along these lines it is possible to cover the phenomenon of consciousness in all its major aspects, including the subjectivity and privacy of conscious experience, and the reflexivity of self-awareness. Indeed, the theory has given us in physical terms a synoptic scientific account of the nature of human consciousness and mental events generally. The bridge between the 'mental' and the 'physical' which this has established was itemized in Section 9.2.

At the same time, though, the theory has left us with an equally synoptic view of the vast number of specific problems that remain to be solved, especially at the neurological end, and loose ends that need to be tied up before the ultimate goal of arriving at a full understanding of the biology and physiology of our mental life would come within reach. However, since our analysis has also pointed the directions in which future research should look for an answer, that exposure, too, may perhaps count as a gain.

Some of the fundamental questions that people tend to ask about the nature of 'Mind' have been answered explicitly by our analysis, others implicitly. Some the analysis would expose as meaningless - mostly on account of their committing a *logical category error* through the use predicates outside their proper universe of discourse, outside the realm of entities to which they can properly be applied (as in *happy soups*) Again others would be rejected for including terms too vague to permit a definite answer.

The following questions will serve as illustration of this variety. All are raised in Gregory's MIND IN SCIENCE, for example (Gregory, 1981).

What is the relation between mind and body? I have answered this, the most divisive question in the history of philosophy, explicitly in terms of the three categories of internal representations which were defined (and explained in physical terms) in Chapters 4 and 5. I also exhibited the link in the steps listed in Section 9.2 for translating statements about mental events into statements about physical events.

What is the survival-value of consciousness? This has been answered implicitly by showing at various points the great advantages the organism derives from operating with an internal model of the world-and-the self-in-the-world.

Does consciousness actually *do* anything? Or does its privacy make it too odd for inclusion as causal? Consciousness defined in terms of internal representations of category 1 (model of the world-and-the-self-in-the-world) and category 3 (self-consciousness, second-order self-awareness) is certainly actively involved in the determination of our actions.(Of all *voluntary* actions, in fact.-m cf. Section 9.4) The inaccessibility to the outside observer of anyone's internal representations of category 3 (except through the subject's introspective self-reports) makes no difference in this respect.

Can computers have consciousness? This was answered explicitly in Section 9.6.when I outlined the conditions a computer would have to satisfy.

Is knowledge of ourselves a construction? Yes: both first- and second-order self-awareness are a construct consisting of linked sets of acquired states of conditional readiness or expectancy of particular kinds.Is it based on inferences? Yes, in the sense that expectancies amount to a form of inference.from past experience.

Where is Mind? If Mind stands for the stream of consciousness, hence for totality of our internal representations, then the question commits a logical category error. For the representations themselves are a mode of operation of the brain/body system (Section 4.4). And although modes of operation may *relate* to spatial entities, they are not themselves spatial entities. They are not entities which can legitimately be described as being to the left or to the right of one another, to be nearer or more distant.

Is it odd to say that experience is located in the brain? Yes, quite illegitimate - and for the same reasons as those stated in the last paragraph.

How can brain states and mind states be identical if one can occupy space and the other does not? The answer is the same as above: they can be identical if the mind states are conceived as operational modes of the brain and the meaning of 'brain-states' is taken to cover also the current operational mode or 'set' of the brain

If we think that it is difficult to conceive that spaceless mind can affect matter, is this just prejudice? No, it .is just woolly thinking. If, as I have argued, the brain's internal representations are modes of operation of the brain/body system, then there is no difficulty in conceiving that they influence the organism's physical behaviour.

If A's brain is put into B's body, would A's Self move into A's body? The obviously woolly term here is 'the Self' As a minimum we must assume that the Self here includes the faculties of (first- and second-order) self-awareness, plus the image A has of his own identity, a representation in which memories of his past life obviously play a large part. Now, since all the important functions of the brain we have considered in this volume are based on internal representations which consist in the main of act-outcome expectancies acquired during the life-time of the individual, it follows that no transplant would leave a properly functioning brain unless everything was transplanted that would be needed to leave the acquired act-outcome expectancies intact *and valid*. And since these act-outcome expectancies relate to the whole brain-body system, there is very little in that system that would not have to be included in the transplant! It would be easier therefore to think of the brain-transplant from A to B as a reverse transplantation of this uncritical part of the brain/body system from B to A – in which case A's Self,would obviously not be affected.The example illustrates the kind of enticing question that makes very little sense once you get down to particulars.

Do we act because we are conscious of a situation, or are we conscious of a situation because we act.? *Both*, according to the analysis I have given. Our voluntary actions depend on the expectancies incorporated in our model of the world, and our expectancies depend on our past act-outcome experiences.

The questions addressed in Part III of this work were of a different kind. Here we were concerned to see the 'spiritual' dimension of our culture in a biological perspective, and to search here for its roots and rationale.Attempts to find this rationale in Darwinian terms, i.e. in terms of genetic cost-effectiveness,had to be rejected on several grounds. We had to find our answer at a deeper level.

This,too, is an important issue. For in an age dominated by the scientific world-view a positive answer to this question could be the foundation of an enlightened rationalism and thus speed the ultimate triumph of reason over the irrational forces that continue to keep this planet in a state of turmoil.

***

PART IV

APPENDICES

Appendix A

FURTHER APPLICATIONS OF THE CONCEPT OF DIRECTIVE CORRELATION

The following are additional examples of teleological concepts which can be given a precise definition in non- teleological terms with the aid of the concept of *directive correlation*. Some have been chosen because they may be needed in the analysis of concrete behaviour patterns along the lines suggested in this volume, the others because they relate to the general nature of organic order and harmony.

1. Coordination

When a number of activities are coordinated - and it does not matter here whether we think of the physical movements of a single individual or of the co-ordination of the activities of separate individuals, as in a football team - we can generally say this:

  i) the activities have a common goal;
  ii) in pursuing this goal each activity takes account of some or all of the others;
  iii) these relationships continue over a finite period of time.

Clearly, what is meant here can be expressed as a *directive correlation in which the activities in question figure as the correlated variables, and which persists over a finite period of time*. The part of the overall concept that has to be covered in a separate clause, though, is the notion that each activity in this correlation "takes account of some or all of the others". Now in effect this means that the *joint causal determinants of this directive correlation (cf. Section 3.2) are internal to the set of activities concerned*.

2. Regulation and Control

A thermostat is a typical example of a regulator. It 'regulates' the amount of heat supplied to a room or appliance so as to ensure the constancy of its temperature. In the same sense, we look upon the way in which the body preserves the constancy of its environment as the manifestation of complex 'regulatory' functions.

The concept of 'regulation' in this case is clearly, again, a teleological concept. And it is not difficult to capture its most common connotation. For in the sense in which the term is here used, 'regulation' means really no more than a *directive correlation serving the constancy of some particular physiological variable or set of variables*.

The concept of *control* is a closely related one, although it tends to have a wider connotation. Controlled variables are not necessarily kept at a constant value. But they are kept within limits which satisfy some specific requirements - although these requirements may vary with time. And that is really all we need say about it.

3. Trying and Searching

Whenever we think of an organism as 'trying' to do something the implication is that there is a measure of uncertainty in its responses: there is a measure of trial-and-error in what it does. And, as has been said, the 'trial' part of these responses contains an element of randomness. But there is also a directive element in the selection of trials that come to be reinforced or come to act as trigger for an activity certain of its goal. And this selection will rest on criteria which are themselves directively correlated to the circumstances in which these events take place. However, the activity of 'trying' will typically also manifest a directive correlation in a different sense. For the very initiation at the particular juncture in question of a trial-and-error routine is likely to be of the nature of an *appropriate* response in the prevailing circumstances (see the definition of this term in Section 3.4).

Searching is also a teleological concept. For searching is basically a kind of trying. But it has a specific goal, namely the goal of bringing some particular object into the orbit of the organism's perceptions.

4. Organic integration

Consider a complex action, as when one picks up a lighter and proceeds to light a cigarette. It is a typical case of what one would describe as an 'integrated' sequence of individual actions. How can we define this type of integration in terms of directive correlations?

Now, we can look upon such a complex action in two equally legitimate ways. On the one hand, we may regard it as a single action with a single goal - lighting the cigarette. On the other hand, we may equally well regard it as a set of part-activities, each having its specific goal - moving the arm to bring the lighter within reach of the hand, grasping the lighter, lifting the lighter to the cigarette, and inhaling to draw the light to the cigarette. And we would say that this was an *integrated* sequence of movements. The concept of an *integrated* sequence of activities, therefore, stands for a relation between these activities which enables us to attribute an individual goal to each, and at the same time an ultimate goal to the whole sequence. And the characteristic of each sub-goal is that it either facilitates the next action in the sequence, or creates a situation satisfying some necessary condition for the success of that action.

From this integral relationship between directive correlations, therefore, may be derived the concept of *hierarchies* of directive correlations unified by a single apical goal. This is the characteristic manner of integration of organic activities. We can see therefore in what precise sense one can say that in the living organism innumerable adaptive, regulative and coordinative processes combine to unite its component parts and their activities into an actively self-maintaining and self-regulating whole. Formally, the picture that has emerged is that of a physical system whose parts and part-activities are connected by complex hierarchies of directive correlations which have the preservation of the whole as ultimate goal at the apex of the hierarchy. It is these abstract relationships that weld the system into the kind of a dynamic biological unit that we typically mean by 'organism'.

A characteristic concomitant of *aging* is the gradual shrinkage of the *degrees* of these directive correlations that precedes the final, irreversible collapse, i.e. death. What knocks a young organism over knocks an old one out.

5. Organic Order and Harmony

In the above passages, we have captured the essence of that higher kind of order which I described in Chapter 2 as marking the distinction between living- and non-living systems. And we have done so in terms which have since been defined in precise and objective terms. However, in the above passages I have focussed exclusively on the relation between the part- activities of single organisms, whereas organic integration can also, of course, be a feature of the activities of groups of individuals operating in collaborative or symbiotic associations of one kind or another. But what was said then also applies here: in an exactly similar sense such a social or symbiotic group can become united into a self-maintaining and self-regulating unit through hierarchies of directive correlations linking the activities of their members. The ant colony is an outstanding example. It is from such hierachies of directive correlations that the colony derives its similarity to an intelligent beast.

What then is the abstract difference between the single organism and the 'social organism' when both are considered as entities that derive their organic unity from hierarchies of directive correlations? Mainly one: in the former the correlations unite the activities of entities which are mechanically connected, while in the latter case the entities concerned are not so connected.

In short, in terms of directive correlations we can capture the essence of that distinctive kind of order which is typical of living organisms and which we commonly describe as *organic order* or *organic harmony*, for the harmony that is here meant is precisely the orderliness that has been outlined above. We have also here touched upon the essence of the distinction between *higher* and *lower* forms of life. For this distinction is not primarily one of complexity, as is often asserted by those who have failed to grasp the nature of the directiveness of organic activities. It is, in fact, primarily a distinction in the *degrees* and *scope* or *range* of the directive correlations that sustain their life - especially of two categories of directive correlations: those that mark the ontogenetic adaptability of the organisms concerned, and those that mark the choice and execution of their outward directed activities. For the higher organisms are mainly distinguished by their greater ability to learn from experience and by the greater number of currently relevant factors they can take into account in the choice and execution of their outward directed activities.

6. Teleological explanations

There are two kinds of explanations in common use: causal explanations and teleological explanations. "The dog ran to the kitchen because it was hungry" is a causal explanation. "The dog ran to the kitchen in order to find food" is a teleological one. The first type needs no comment, but the status of the second has often been a subject of dispute. Our explications can clear up this matter simply because they permit the latter type of explanation to be readily translated into one of the former type. In the present example the translation would yield: "The dog ran into the kitchen because it was in a state conducive to a particular kind of goal-directed activity" (here: the search for food, where *'search'* is to be understood in the objective sense defined above).

***

Appendix B

SOME BASIC CONCEPTS OF SET THEORY

I. SETS AND CLASSES

The object of this appendix is to clarify some of the basic logical concepts we have used in our analysis, such as the concept of non-numerical variables, sets, classes, and mappings - also to show that what has been said about an organism's responses to the *properties* of objects can also be taken to cover responses to *relations* between objects.

My overriding aim in this book has been to observe the strict semantic standards of the natural sciences. These sciences, of course, owe a great deal of their success to the fact that they deal with *measurables* upon which the computational methods of mathematics can be performed. On the other hand, it would unnecessarily restrict the validity of some of our concepts, e.g. the concept of *directive correlation* introduced in Chapter 3, if the variables mentioned in our definitions had to be assumed to be exclusively numerical variables. It is important that our definitions should apply to objective attributes of living organisms and their behaviours regardless of whether those attributes can or cannot be readily defined in numerical terms. Thus in some contexts I have introduced variables taken to represent different patterns of behaviour. Provided objective criteria are available for distinguishing these patterns, it is perfectly legitimate to introduce a discrete variable B, each of whose values is taken to denote one of those patterns.

Fortunately, mathematics itself has given us the necessary conceptual tools to deal with such cases, viz. of the concepts of *set theory*. These concepts have a degree of generality and inherent clarity so great that they were able to serve as logical basis for virtually the whole of mathematics. In this work we need them mainly at the points at which we have to work with non-numerical variables as in the example given above. But there are also other cases in which these concepts are helpful. For example, in cases when it helps to treat *properties* as classes, and, *relations* as classes of ordered pairs (triples, quadruples etc.), thus treating both on an equal footing (see below).

Few ideas are more basic and more readily understood than is the general idea of *class-membership,* i.e. the idea of some particular object being a member of some particular class of objects. Although in some mathematical treatises the terms 'class' and 'set' are treated as synonymous, the more general practice is to us 'class" for the abstract entity, i.e. for the abstract condition that decides membership, while the term 'set' is reserved for any concrete collection or aggregates of objects conceived as a unity. A set may be described, for example, simply by listing its members, as when I list the people I have invited to a party. In other words, the terms 'class' and 'set' tend to be used in a connotative and denotative sense respectively. Everything that is said in the remainder of this appendix about relations between *sets* applies equally to relations between *classes*. In this work I often use 'category' as synonymous with 'class'.

*

The primitive concept of set theory is the concept of *membership* i.e. the relation between an entity and the set of which it is a member.

The members of a set are generally called its *elements*. These may be concrete (e.g chairs) or abstract (e.g. triangles). They may also themselves be sets. A line, for example is a set of points, and a set of lines, therefore, is a set of sets (of points).

If all members of a set A are also members of a set B, set A is said to be *included* in set B, or to be a *subset* of B. Thus my grandparents are a set which is included in the set formed by my ancestors.

The set of all elements which are members of both a set C and a set D is known as the *intersection* of C and D. Thus the set formed by my two grandfathers is the intersection of the set formed by my grandparents and the set formed by my male ancestors.

The set of all elements which belong either to E or F is called the *union* of E and F. Thus the set formed by my grandparents is the union of the two sets formed by the parents of my mother and the parents of my father.

The set of all possible combinations that can be formed by pairing a member of a set G with a member of a set H is called the direct or *Cartesian product* of G and H Thus the Cartesian product of a set of four males and five females will have 20 members.

A set which has no members is called an *empty* set or *null* set. Thus the intersection between the set formed by my grandparents and that formed by my uncles is an empty set.

The totality of elements to which a particular classification can legitimately apply, or from which a particular set of sets can legitimately draw its members, is called the *universal* set or *universe of discourse,* generally designated by the letter U. When applied to classes this becomes a very important concept for anyone who wants to avoid talking nonsense. For instance, to talk about happy soups is nonsense because soups do not belong to the universe of discourse of the attribute *happy*. The universe of discourse of the attribute *happy* is *people*.

Appendix B 293

Finally, the set of all members of the universal set which are not members of a given set A is called the *negation* of A. It is also known as the *complement* of A in U.

*

The main relations between sets (or classes) can conveniently be symbolized with the aid of *Venn diagrams*, as illustrated in Fig. A.1.

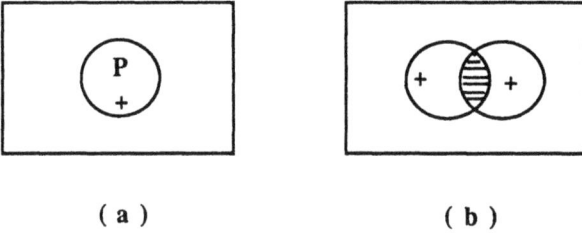

(a) (b)

Figure A.1 - *Venn diagrams*

In (a) the circle denotes a set P of elements while the rectangle denotes the universal set. Hence the space between the rectangle and the circle represents the complement of P. When it is intended to indicate explicitly whether a particular set is empty or not empty, the convention in Venn diagrams is to use shading in the first case and a + sign in the second. Thus the Venn diagram of Fig. A.1 (b) indicates that the intersection between the two sets shown is empty, though neither set is empty - in other words, that these sets have no common members.

II. PROPERTIES AND RELATIONS

In all studies of the natural world, I have said, we have to deal with properties (or features) and relations, and both can be thought of as *classes*. Thus we may think of 'brittle' as the name of the class of all objects which tend to break on impact.

Similarly, a *relation* between two objects may be viewed as a *class of ordered pairs* of objects. Thus the father-son relation may be viewed as a class of ordered pairs of males - namely pairs in which the male mentioned as the first member has sired the male mentioned as the second.

Each ordered pair (a;b) of a relation R is again called an *element* of R. The set of all first members is called the *domain* of R, and the set of second members is called the *range*. Domain and range may overlap, as indeed they do in the above example: the same male may be somebody's father as well as being somebody elses son.

The father-son relation is a two-term or *dyadic* relation. Some relations link more than two entities. The relation between, (as in 'John stands between Peter and Dick'), for example, is a *triadic* relation. Such relations, then, may be viewed as classes of ordered triples.

Thus, one of the main advantages of this class-centered approach is that it enables us to treat properties (or features) and relations under a single heading and on the same footing. Hence, in this work we need not draw a radical distinction between the ability of an organism to discern particular properties in its environment and its capacity to discern particular relations. For example, the capacity of an organism to respond in some particular way to some particular dyadic relationship among objects of its environment, may be viewed simply as a capacity to respond in this way to a particular class of ordered pairs of objects. This can be a fruitful way of looking at the matter when the organism has a physical procedure available which will take it from the first member of any given element of this class of pairs to the second member (or *vice versa)*, since such a procedure will automatically establish the respective pair of objects as an ordered pair in the organism's perceptions. For instance, in such relations as *to the left of*, *to the right of*, *above*, *below*, higher organisms have such a procedure available in the eye movement which takes them from a visual fixation on the one object to a visual fixation on the other.

There is nothing forced or artificial in adopting such a class-centered approach to properties and relations whenever it proves helpful in the course of our analysis of, for example, the brain's internal representations of the world. Classes are very much part of the common furniture of the scientist's account of the world which routinely distinguishes different classes of elementary particles, chemical compounds, stars or galaxies, and different species of plants or animals.

Numbers, too, may be viewed as classes. Thus the number 2 may be viewed as a name for the class of all pairs, the number 3 as a name for the class of all triples, etc. Moreover, mathematicians have found ways of defining these classes without resorting to the number concept itself.

The basic rules of the practical logic which people use when they reason or argue a point, may also be thought of in terms of classes, for these rules derive ultimately from certain self-evident implications of certain fundamental relationships between classes - relationships like inclusion, intersection or union. I mean such implications as the fact that if a class A is included in class B ('all men are mortal')and x is a member of class A ('Socrates is a man') , then x will also be a member of class B (ergo: 'Socrates is mortal) - (see Appendix C).

III. MAPPINGS

Of particular importance for anyone dealing wit the subject of representations are the different kinds of correspondence or mappings that may exist between two sets of elements (e.g. between external objects and the brain's internal representation of those objects).

Any rule which assigns to every given element of a set A one or more elements of a set B is said to be a *mapping* or *map* of A to B. The set A is called the *domain* of the mapping, while the set B is called the *range*. The element of B which corresponds to a given element of A is said to be the *image* in B of that element. If every member of B is an image of some member of A then the mapping is said to be *onto* B. If not, we say *into* B. Or, if we don't want to commit ourselves in this respect we can simply say (as we did above) *to* B.

Typical mappings can be *one-many*, *many-one* or *one-one*. The terms speak for themselves. For example, if A represents a group of people and B is a list of their birth certificates, then there exists an one-one mapping between them because to each birth certificate there corresponds one and only one person. But if B is a list of surnames, then there may be a many-one mapping from A to B because two or more members of that group may have the same surname.

A one-one mapping of A onto B is commonly known as a *one-one correspondence* between A and B. If every member of B is an image of some member of A, then it is also sometimes said that B maps A. This enables us to say, for example, that a roadmap 'maps' the road network or that a photograph of a painting 'maps' the painting.

Sometimes one wishes to compare the elements of a complex whole, i.e. of a set of interrelated parts, with those of another complex whole - for instance, a geographical map may be compared with the landscape it purports to represent. It may then be expedient to distinguish between an *isomorphic* and an *homomorphic* mapping. We say that a complex whole C maps isomorphically on a complex whole D if

i) the elements of C map one-one onto the elements of D, and

ii) for every relation R between the elements of C there exists a relation R' between the corresponding elements of D.

The mapping is *homomorphic* if the mapping from C to D is a many-one mapping which preserves relationships in the above sense. Thus a scaled down version of a piece of sculpture is an isomorphic mapping of that sculpture, whereas a photograph or schematic drawing of it would be an homomorphic mapping only.

An object in real space may be mapped either isomorphically or homomorphically into any other space of adequate dimensionality. Thus, according to the theory developed in this work, the brain's internal representations of external objects may be described as homomorphic mappings of those objects into the space of conditional expectancies.

IV. SINGLE-VALUED FUNCTIONS AND DETERMINISTIC SYSTEMS.

Many-one and one-one mappings correspond to what traditionally in mathematics is called a single-valued function, for which the customary symbolism would be

$$B = f(A).$$

Given the speed of a car, for example, its distance from starting point at any given point of time will be a single- valued function of the time traveled:

$$d = f(t) .$$

The notion of single-valued function is crucial to understanding what is meant by a *deterministic* system in the natural sciences. For, it here means a dynamic and closed system whose state at any given point of time is a single- valued function of its initial state. This need not be a one- one mapping, of course. It can be many-one: several possible initial states may result in the same end-state.

V. VARIABLES.

A set of mutually exclusive possibilities of which one and only one is realized at any one point of time, is known as a *variable*; and the member that is realized at any given point of time is known as the *value* of that variable at that point of time. Thus the time and distance that has elapsed since the beginning of a journey are typical examples of physical variables. So is the temperature of the air and the hardness of the road surface. Variables may be *continuous* or *discrete*. Although time is a continuous variable, on your digital watch it is read out as a discrete one.

Each of the possible values of a given variable may itself consist of a set of more than one simultaneously realized numerical quantities. An example is the *vector* variable which, in terms of three numbers, defines the location of a point in three-dimensional space .

However, in the context of this work it is important to note that at the set-theoretical level of thought we can talk about *physical variables* without having to limit this expression to numerical and measurable variables in the usual sense of these terms. This should be bourn in mind especially in connection wit the definition of *directive correlation* given in Chapter 3. It applies especially to *discrete* variables. For instance, if an organism is known to be able to respond to a given stimulus with a discrete number of alternative behaviour patterns this can be regarded as a behavioural variable in the strict sense of 'variable' defined above.

***

Appendix C

SOME BASIC CONCEPTS OF FORMAL LOGIC

In our analysis of language functions (Chapter 8) I made the point that to pass from the ability to *think* (in the sense of forming internal representations of possibilities as well as actualities) to the ability to *reason*, i.e. to *think logically*, the brain has to assimilate the properties of classes and their formal relationships. The following is a brief outline of the logical basis on which this assumption rests.

Probably the most common type of logical argument we employ in our daily life is of the form

> *All drunken drivers are dangerous*
> *This man is a drunken driver*
> *Therefore this man is dangerous*

The logical theory that deals with this kind of deduction, and with the implications of such general statements as *All A is B* and *No C is D*, is known as *Predicate Logic*. Its study dates from the days of Aristotle.

The above deduction contains two premises and one conclusion. This form of deduction is known as a *syllogism*. The general theory of syllogistic reasoning allows for *four* basic forms of premises, all expressed in general terms, namely:

> *All A are B*
> *Some A are B*
> *No A are B*
> *Some A are not B*

The terms *all* and *some* are called the *quantifiers*, while the no may be viewed as a sign of negation acting as quantifier, since *No A are B* can also be regarded as the negation of *Some A are B*.

All is called the *universal* quantifier, and since modern logicians prefer to express matters in terms of variables, this quantifier would be rendered as *for all x, x is ....* Hence *All A is B* would be rendered as *for all x, x is A and x is B*.

Some is called the *existential* quantifier, formally rendered as *there exists an x such that x is.....*

*

In all syllogisms we can distinguish *subject terms* (S), *predicate terms* (P) and the *middle terms* (M). The first two derive their name from their place in the conclusion. The middle term is the one that is shared by the premises, but never occurs in the conclusion. Thus the *'figure'* of the syllogism

>All A are B
>Some A are C
>Therefore some B are C

may symbolically be represented as

>M – S
>M – P
>-----
>S – P

It follows that the syllogism has four possible figures, namely:

| M – P | P – M | M – P | P – M |
S – M	S – M	M – S	M – S
S – P	S – P	S – P	S – P

Since each premise can have any one of the four forms listed above, and since there are two premises in each figure and four possible figures, it follows that there are only 64 possible pairs of syllogistic premises.

Although the theory of syllogistic reasoning tends to refer only to general terms, it does allow for the insertion into the syllogism of singular terms like *This man* (as in my opening example) and names, as in the classical syllogism

>All men are mortal
>Socrates is a man
>Therefore Socrates is mortal

*

Venn diagrams (see Appendix B) can help us visualize the different figures of syllogistic reasoning. Their power to symbolize classes which may or may not be empty, a universal class, the complement of a class, and the intersection and union of classes, enables them to symbolize the allowable pairs of syllogistic premises and demonstrate the conclusion.

In these Venn diagrams the set of individuals to which the subject-, predicate- and middle-terms respectively apply are represented by overlapping circles, thus allowing for every possible combination of common membership. Following standard practice empty sets or subsets are shaded while + signs indicate that a set (subset) has at least one member. Thus Fig. A.2 symbolizes the premisses

All B are A

Some B are C

for all of circle B is shaded except its intersection with circle A, and the + sign indicates that the intersection of B with C has at least one member. The conclusion *Some C are A* is immediately obvious from the diagram.

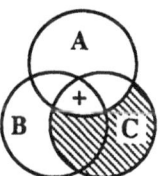

Figure A.2.

These illustrations make an important point: they show that the rules of logic, in so far as they cover the predicate logic of the most common deductions we tend to make in our daily life, really spring from the properties of classes and of the relationships between classes, such as those explained in Appendix B. Thus the formalization of predicate logic may be viewed as a special application of a *calculus of classes*.

In both its phylogenetic and ontogenetic development the human brain has had to adapt itself to the way the world is, and that includes the way the world logically is. And as we have come to see: the way the world logically is springs from the properties of classes and thus from the intrinsic nature of the act of classification. Hence it springs from the fact that it is in our nature (as in the nature of all other animals - though at a less sophisticated level) to respond to the world in a way that divides its objects into categories of one kind or another. Many of our common human classifications are based on the application of criteria of which we are conscious and for which we may have verbal formulae. Many others are not. Thus one does not have to possess any linguistic faculties in order to react to the world in a way which conforms to the syllogism: *All bulls are dangerous, this animal is a bull, therefore this animal is dangerous.*

However, predicate logic is not the only logic used in reasoned arguments, though it is the most basically human logic. In addition we have the logical rules which follow from the way our language permits us to link individual propositions together by means of such *logical operators* as *and, or, not, neither...nor* etc. A typical example of such a rule is the one which tells us that if two proposition p and q are both true, than the proposition *p and q* will also be true. These rules are known as *propositional logic* or the *calculus of propositions*. This calculus differs from predicate logic most notably in that it treats all propositions as units, and as units which have only two 'values': true or false. The truths it establishes are independent of the contents of the propositions.

This propositional calculus is a rather mechanical calculus because it defines its logical operators simply in terms of their stipulated effects on the truth of the propositions they generate, given the truth value of the constituent propositions. Thus, the logical operators *and, or, not* and *implies* may be defined as in the *truth table* illustrated in Fig. A.3, where T and F stand for 'true' and 'false' respectively.

p	q	p and q	p or q	not p	p implies q
T	T	T	F	F	T
F	T	F	T	T	T
T	F	F	T	F	F
F	F	F	F	T	T

Figure A.3 -*Truth table definitions of* **and, or,** *not and* **implies**

One important point has to be made in this connection. The business of the logician is to *systematize* and *normalize* the rules of logic with the object of presenting us with a competent, tidy and internally consistent body of rules, and preferably one that can be formalized in terms of a minimal set of axioms and definitions. His business is *not* to *describe* how people reason. In fact, some of his theoretical concepts must appear as extremely artificial to the layman. For example, his formal definition of 'implies', shown in the last column of Fig. A.3, differs greatly from what the layman would tend to understand by an implication. According to that formal definition any false proposition can be said to imply any other false proposition. For example, *Paris is in Russia* can be said to imply *Napoleon is still alive*.

Appendix C 303

   This underscores the point, already made earlier in the text, that it would be quite wrong for anyone studying the nature of rational thought to assume that when people reason, and reason validly, they do so by applying in their minds the concepts that fill the textbooks of logic. Their competence may, of course, be greatly enhanced by reading such books, particularly in intricate cases calling for logical inferences. But the roots of that competence lie elsewhere. They lie in the normal processes through which man learns to adapt to the way the world is and, therefore, also to the way the world is logically, viz., to the logical properties of classes we considered in Chapter 8.)

***

Appendix D

GENERATIVE AND TRANSFORMATIONAL GRAMMARS

Any systematic study of the relation between the grammar of a language and the meaning of the resulting sentences needs some way of representing that grammar itself in a systematic form. One would not get very far if one were simply to list all the rules that might be gleaned from, say, a grammar of English compiled for the use of foreign students. Most of the theoretical work done by modern linguists in this field relates to the problem of how to construct an adequate *generative* grammar. By this is meant a finite set of rules which are able to generate all (and only) the sequences of words that are permitted in the language under consideration, i.e. regarded as well-formed sentences by the speech community whose language is being examined. To do so satisfactorily the generative grammar must be *descriptive* of what is *found* to be permitted rather than *prescriptive* of what *shall be* permitted. For the linguist, therefore, the process of discovering such a generative grammar is a question of trying out different hypotheses in the hope of finding a set that fits observed linguistic usage. It is not his aim to influence that usage. Following the work of Chomsky, many linguists hope that, by pursuing this process at a sufficient depth and level of abstraction, they may even reach a basic framework and set of rules which is universal to all natural languages.

To achieve their object the set of rules must be of a kind that enables us to generate a sentence systematically, beginning with the fact that its basic structure will be that of a subject followed by a predicate. A typical case would be a noun phrase followed by a verb phrase. Hence, if we symbolize the sentence as a unit by S, a noun phrase by NP and a verb phrase by VP, rule number 1 of our generative grammar could state that S may be replaced by (or *re-written* as) NP + VP. Symbolically:

$$S \rightarrow NP + VP$$

Additional rules might then state what structures could be substituted for NP and VP respectively, and so one can proceed in a hierarchical fashion from the top downwards.

For example, the following set of *re-write rules* enable us to generate the sentences

the wild bull will charge fiercely,
the bull has killed the spectator

in the manner of the tree structures shown in Fig. A.4 (a) and (b) - as well as generating many other sentences of a similar kind.

$$S \to NP + VP$$
$$NP \to (Art) + (Adj)N \text{ (brackets denote optional items)}$$
$$VP \to V + Adv$$
$$VP \to V + NP$$
$$VP \to Aux + V$$

LEGEND:
S   = Sentence
NP   = Noun phrase
VP   = Verb phrase
(Adj)N   = Adjective (optional) + Noun
Art   = Article (determinate or indeterminate)
V   = Verb
Aux   = Auxiliary verb

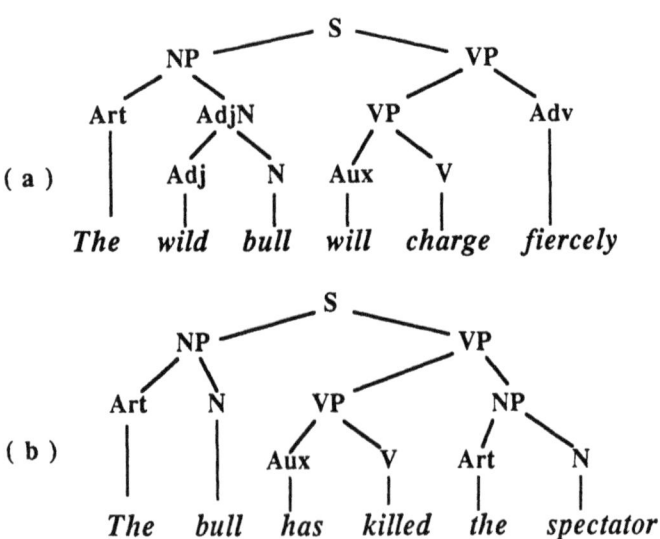

Figure A.4 - *Two examples of ('generative') phrase-markers.*

The tree-diagrams here illustrated are called *phrase-markers*. They can obviously be used to symbolize the analysis of a given sentence structure as well as the synthesis of a new one. (In Chapter 8 I distinguish them as *generative* phrase-markers from the *logical* phrase-markers I defined in Section 8.4)

A generative grammar which seeks to express in terms such as these the rules governing permissible word combinations, is called a *simple phrase-structure grammar*. However, following Naom Chomsky, many modern linguists go a step further. They believe that in order to arrive at a theory which is both adequate and economical, one has to assume that beneath the *surface-structure* reflected in the phrase-marker - there lies a *deep structure* which has its own re-write rules and which is related to the surface structure of the sentence according to specific *rules of transformation* or *T-rules*. These rules spell out the ways in which the surface structure of a sentence may be derived from the assumed deep structure. This transformation could be effected, for example, by copying some elements of the deep structure while deleting others, or by adding one or more new ones. Or they could specify a reordering of the components. Theoretical grammars which make that assumption are known as *transformational generative grammars*.

To give a concrete example, consider the sentences

The player is beat ing his opponent.
The player has beat en his opponent.

Note that -*ing* is the proper ending when a tense of *be* is used, whereas -*en* is the correct ending when a tense of *have* is used. Now, one of Chomsky's suggestions was that the best- fitting and most economical theoretical grammar results if in the deep structure of a sentence both *be* + -*ing* and *have* + -*en* are each taken to occur as a single unit. The whole matter could than be handled by a single rewrite rule for the deep structure

VB -> TNS (*have* + *en*) .(*be* + *ing*) V

(where TNS, stands for 'tense endings'), plus a reordering transformational rule which would dispatch the different affixes to the positions they would occupy in an acceptable surface structure. This might be symbolized as

TNS (*have en*) (*be ing*) *beat*.

This would then yield, according to requirements,

> *have* + TNS *beat* + *en*
> *be* + TNS *beat* + *ing* , or
> *have* + TNS *be* + *en beat* + *en,*

which, in turn, would yield

> *has beaten*
> *is beating*
> *has been beaten.*

No re-write rules and phrase-structures, of course, can be considered in isolation. They must be considered in conjunction with the so-called *lexicon* of the language in question - that is to say, a compendium and classification of the words occurring in the vocabulary of that language. For example, the lexicon has to specify which words are available as nouns, adjectives or verbs respectively. And, for instance, it has to say which verbs can be coupled with an object noun phrase (as in *he beat his opponent*) and which cannot (verbs like *fainting, sleeping, dying*, for instance.) Hence the re-write rules of the deep structure plus the lexicon are jointly called the *base* of the transformational grammar.

However, serious problems arise in connection with the notion of deep structure as here conceived. For, it would seem natural that the *meaning* of a sentence should be determined by its deep structure. Or, to put it another way, we need a term for the structure of the brain's internal representation and their resolutions from which the construction of the sentence proceeds. And what better name for this than 'deep structure'? That, indeed, is the sense in which I define the term for my own use in Chapter 8.

By contrast, when 'deep structure' is understood in the sense of the transformationalists, the postulated transformations from the deep structure to the surface structure may change the meaning of the sentence. Well known cases in point are the transformations from an active to a passive form of a sentence when the sentence contains a quantifier – such as *some, all, a few, many*. This change of meaning is clearly seen in such active and passive pairs as

> *Many pilots do not fly gliders.*
> *Gliders are not flown by many pilots.*

Appendix D

Linguists have responded to this problem in two contrasting ways. While some have simply acquiesced in the belief that both the surface- and deep-structure of a sentence contribute to its meaning (hence that both structures feed into the semantic component of the grammar), others have felt that this assumption offends the very notion of a deep structure and that the situation is to be met by assuming that the suggested notions of deep structures were at fault. The former theory is known as the *Extended Standard Theory*, while the latter is known as *Generative Semantics*. The analysis I have given in Chapter 8, and the notion of deep structure I have there introduced, place us firmly into the second camp.

***

Appendix E

SOME BASIC CONCEPTS OF INFORMATON THEORY

A message is informative to the extent to which it tells us something we did not already know. Information theory therefore, seeks to quantify the information content in terms of the uncertainties it removes, relying on uncertainties being quantifiable in terms of probabilities.

Consider an event for which there are n possible alternatives and assume for a start that these are all equally likely. The probability of each is then 1/n. Thus when tossing a coin there are two possibilities each with a probability of 1/2 and when throwing a dice there are 6, each with a probability of 1/6.

Information theory now defines the uncertainty which is removed when I am told of the outcome of these two actions as

$$H(\text{single coin}) = \log_2 2 \text{ bits},$$

and

$$H(\text{single dice}) = \log_2 6 \text{ bits}$$

respectively, where 'bit' is the chosen unit and performs as a contraction of 'binary digit'.

The choice of a logarithmic function here conforms with the ordinary sense in which we think of information. For example if we are being informed of the outcome of throwing *two* dice we would intuitively feel that we were given *twice* as much information as when we are only told the outcome of one of these throws. And this is precisely what the logarithmic formula yields. For, since with two dice there are 36 possible alternatives, the uncertainty removed by telling us the outcome of both dice is

$$H(\text{two dice}) = \log_2 36,$$

which is twice H(single dice).

In general terms, for any event x in a set of equally probable alternatives, each occurring with the probability p,

$$H(x) = \log_2 1/p$$

If the probabilities of the different alternatives are not equal, the weighted average is taken, and the uncertainty is defined as

$$H(x) = \sum p_i \log_2 (p_i)^{-1}$$

where p_i denotes the probability of the i-th event in a set of n possible alternatives.

If, by way of estimate, in a series of n trials the probability of each event is equated with the relative frequency of its occurrence in the series, so that

$$p_i = n_i/n,$$

the formula for the average uncertainty removed by the occurrence of the event (or information stating the occurrence) becomes

$$H(x) = \log_2 n - (\sum n_i \log_2 n_i)/n$$

It is easy to show that $H(x)$ is maximal if all the probabilities are equal, and that the quantity

$$R = 1 - \frac{H(x)}{H_{max}(x)}$$

will be zero if $H(x) = H_{max}(x)$.

This quantity is known as the *redundancy* of a message. It is important in information theory because it can serve as an inverse indicator of the degree to which the information contained in a message was conveyed in the most economical way. However, a low redundancy is not always desirable. For the greater the redundancy the less will be the effect of any incidental error in the message on the information it conveys Thus information which has to be relayed across a noisy channel is often cast by communication engineers into a form which has been calculated to have a redundancy high enough to neutralize the effect of the interference suffered in that channel. Written English, too, has a considerable degree of redundancy, because different letters of the alphabet occur with different relative frequency, and often in preferred combinations, so that the occurrence of the first member of the combination already removes some uncertainty about the nature of the next. For example, owing to the frequency of words like 'the','that','there', etc., the occurrence of the letter 'h' after a 't' removes relatively little uncertainty. Indeed, taking observed frequencies into account, Shannon has

Appendix E 313

estimated that written English has a redundancy of about 75%. And one should realize that it is only due to this redundancy that one can detect printer's errors, spelling mistakes and the like. For in a message without redundancy the slightest change would alter the information conveyed and there would be no way of telling whether the resulting information was, or was not, the intended one. Only through the redundancy in our language can we spot instantly the printer's error in *Tke tree fell*.

Of particular relevance to Section 11.1 is the fact that the redundancy contained in any signal, be this a verbal message or a perceived visual form, is *a measure of the degree of patterning or organization of that message or visual form*. Of course, in the realm of aesthetic experience there are few cases in which one could actually estimate in numerical terms the degree of redundancy of the object that appeals to us, be this a landscape, a painting, a sculpture a poem or a musical composition. Nevertheless, we have in redundancy a concept which can be useful even in such cases owing to the precision of its derivation.

***

Appendix F

PHYSICAL SYSTEMS AND CAUSAL RELATIONSHIPS

Throughout this volume I have treated the brain as a physical system and discussed a variety of causal relationships. Both concepts call for some general remarks, as does the concept of causal determinism which cropped up in our discussion of the freedom of the will (Section 9.4).

PHYSICAL SYSTEMS

Any set of interrelated or interconnected physical entities may be called a physical *system*. The elements may be either concrete, as in the case of the Solar System when we think of the planets as concrete entities, or abstract, as when a pendulum is conceived as just a straight line linking a fixed point at one end with a point-weight at the other. Physical theory generally relates to abstract systems, the physicist having abstracted from the total situation those elements that seem to him most relevant in the given investigative context.

In general, physicists will think of any physical system as having both constant and variable elements. The system itself will be defined or identified by its constant elements. Thus the physical system consisting of a ball falling in a vacuum, is defined by the three elements: ball, vacuum, gravity. But the scientist will find the greatest interest in the transient and changing elements, since it is *their* behaviour he will seek to understand and, if possible, predict. These elements are the *variables* that figure in his equations. The current values of the variables of the system jointly constitute what is known as the *current state* of the system. Note that this notion differs from the common practice in psychology, where 'state of mind' is generally taken to denote a more or less enduring set of mental attitudes or emotions, whereas here we are talking about a transient condition of the system. Since this is the sense also generally used in systems theory, I must therefore make it clear that throughout this book I am using 'state' in this transient sense.

The variables of a physical system are subject to *constraints* . These may derive both from the defining constants of the system and from the theoretical assumptions made by the scientist.

Abstract systems may be conceived either as *open* or as *closed* systems. A closed system is assumed to be free from outside influences whereas an open system is not. This distinction is of great importance in thermodynamics. Systems theory, on the other hand, tends to deal with hybrid systems, viz. systems which are closed except for a set of specifiable inputs. It is obvious that in general living systems will have to be treated as open systems owing to their ceaseless exchanges of energy with their environments. But in brain theory it is often permissible to think of the brain as a hybrid system of the kind I have described.

CAUSAL RELATIONSHIPS

Since the notion of causation tends to occur throughout the natural sciences as well as in our daily life, it may be helpful to clarify the scientific sense in which it is applied at various points of our analysis.

Looking back into history we find Aristotle distinguishing four kinds of causes, which he called material, formal, efficient and final causes respectively. As an example, consider a man making a statue. Then, as Aristotle saw it, the material cause of the statue is the marble, the formal cause is the essence of the statue to be produced, the efficient cause is the effect of the chisel on the marble and the final cause is the end which the sculptor has in mind.

To-day only Aristotle's notions of 'efficient' and 'formal' causes are of common currency. When it is asserted that a road accident was *caused* by fog, we have an example of the first, while an example of the second would be the assertion that centrifugal forces are *caused* by the laws of inertia. The modern notion which comes closest to Aristotle's 'material causes' would be that of the 'properties of matter'. But Aristotle's notion of 'final causes' has vanished altogether in consequence of our changed perceptions of the nature of teleological processes.

The sense in which the expressions 'cause' and 'causation' are generally to be understood in our analysis is that of *efficient* cause and causation. It is the sense in which one would say, for instance, that a fire was caused by an electrical fault, that a landslide was caused by erosion, or that a death was due to the administration of poison.

Although this general sense of the term is familiar enough it is not so easy to define. In general the event or state of affairs cited as the cause of some other event or state of affairs characteristically satisfies two conditions:

a) it is a *contingent* event or state of affairs, and

b) it *completes a set of sufficient conditions* for the occurrence of the event or state of affairs cited as the effect.

The most common form in which causation finds expression in the abstract models of the physical scientist is by way of formulae which enable him to express the state of the system at some particular time t as a single-valued function of its initial state. Examples were given in Section 3.2. It should also be noted in this connection that *causal necessity* is not a factual relation but merely an affirmation of the fact that the scientist's models are conceived as systems abiding certain laws, e.g. the laws of physics or chemistry such as they happened to be conceived at the time.

CAUSAL DETERMINISM

The distinction between deterministic and indeterministic systems often leads to confusion. And yet it is of more than technical interest. One does not have to be a scientist to be curious to know, for example, whether the universe as a whole is deterministic in a sense which would imply that it runs like a clockwork and everything that happens is somehow inescapable, predetermined and wholly predictable in theory. One might also wish to know whether this applies to all human activities, i.e. whether all our physical actions are predetermined by physical events. And one may wonder whether this would imply that we cannot really be held responsible for our actions. On the other hand, if the Universe is not wholly deterministic, would this offer scope for some deity to bring its influence to bear on the course of events? And if the body is not wholly deterministic, would this give room for some immaterial entity called 'the mind' to intervene in the physical course run by brain-events and thus to control our actions as an independent agency?

Some of these uncertainties can be removed if one agrees to restrict the meaning of 'determinism' to a strictly technical sense in which it can be applied only to the theoretical models the scientist uses to describe the world. Such a theoretical system will then be called deterministic if, according to the theory, the state of the system at any given time t can be represented as a single-valued function of the initial state of the system, as explained above. It follows that in general only closed systems will be deterministic in the above sense and hybrid systems when conceived over time slices in which the inputs are either constant or rule-following or assumed to occur only at the point of time taken as the initial point of time. Thus the brain could in principle be conceived as a deterministic system but only in these limiting contexts.

Now, it is well-known that this kind of determinacy has become inoperable in atomic and molecular physics. The quantum theory which has replaced classical physics in this field, includes Heisenberg's *Principle of Indeterminacy*. According to this principle there exist certain pairs of variables in such systems (e.g. the position and momentum of any small particle) which are such that a precise determination of the value of either variable precludes a precise determination of the value of the other. Thus the more accurate is our determination of the one, the less accurate must be that of the other. It follows that to the extent to which the current and initial states of the system in question are taken to be defined by the values of such variables, neither state can ever be determined with accuracy and the conditions for causal determinacy as defined above cannot be satisfied. In this sense, therefore, the system is indeterministic. However, it has also proved to be the case that if the current and initial states of the system are defined *stochastically* - that is to say, not in terms of discrete values of the variables concerned, but in terms of the set of *probabilities* of any given variable having any one of its possible values at the time in question, then we have a new set of parameters. And in terms of these *probability distributions* it is then generally possible to express the current state of the system as a single-valued function of its initial state. In stochastic terms, therefore, the system may once again be called a deterministic one, and since these probability distributions move in a lawful way, such systems offer no scope for a "God of the gaps" to interfere with the movements of the system - nor, in the case of the human brain, for the intervention of some ghostly entity called 'the mind' to interfere with the course run by the neural impulses.

***

Bibliography

Anderson, J.R. Arguments concerning representations for mental imagery. *Psychologial Review*, 85, 251-277, (1978).
Armstrong, D.M. *A Materialist Theory of Mind.*, London: Routledge and Kegan Paul, 1968.
Armstrong, D.M. & Malcolm, N. *Consciousness and Causality,* Oxford: Basil Blackwell, 1984.
Augros, R. and Stanciu, G. *The New Biology.* New Science Library, Boston: Shambala Publications, 1987.
Ayer, A.J. *Central Questions of Philosophy.* London: Weidenfels & Nicholson, 1973.
Austin, W.H. *The Relevance of Natural Science to Theology* London: Macmillan, 1976.

Baddeley, A.D. *The Psychology of Memory.* New York: Harper and Row, 1977.
Barsalou, L.W. and Bower, G.H. Discrimination nets as psychological models. *Cognitive Science,* 8,1-26 (1984).
Bartlett, F.C. *Remembering.* London: Cambridge U.P., 1932.
Battista, J.R. The Science of Consciousness, in Pope, K.S. and Singer, J.L. (eds.) *The Stream of Consciousness*, New York: Wiley, 1978.
Bertalanffy, L. von, *Modern Theories of Development,* (transl. Woodger, J.H.), London: Oxford U.P., 1933.
Birkhoff, G.D. *Aesthetic Measure,* Cambridge (Mass.): Harvard U.P.,1932.
Blakemore, C. The BBC Reith Lecture, 1976; *The Listener,* 9.Dec. 1976.
Boden, M. *Computer Models of Mind. Computational Approaches in Theoretical Psychology.* London: Cambridge U.P., 1988.
Borst, C.V. (ed) *The Mind-Brain Identity* Theory, London: Macmillan, 1970.
Bower, T.G.R, Broughton, J.M. and Moore M.K.: The coordination of visual and tactile input in infants. *Perception and Psychophysics,* 8, 51-53 (1970).
Brooks, L. The suppression of visualization by reading. Quarterly *Journal of Experimental Psychology,* 9, 280-99 (1967).
Bruner, J.S. Going beyond the information given. In *Contemporary Approaches to Cognition.* Cambridge (Mass.): Harvard U.P.,1957.
Bullock, D. and Grossberg, L. Neural dynamics of planned arm movements: emergent invariants and speed-accuracy properties during trajectory formation. *Psych. Rev.,* 95, 1, 49-90 (1988).

Buser, P.A. and Rougeul-Buser, A. (eds.) *Cerebral Correlates of Conscious Experience. Amsterdam:* North-Holland Publishing Company, 1978.

Cannon, W.B. *The Wisdom of the Body* (2nd ed), New York: Norton 1939.
Cattell, R.B. *The Scientific Analysis of Personality,* London: Penguin, 1965.
Corballis, M.C. Recognition of disoriented shapes. *Psych. Rev.,* 95, 1, 115-123 (1988).
Craik, K. *The Nature of Explanation,* London: Cambridge U.P.,1943.
Crook, J.H. *The Evolution of Human Consciousness,* Oxford: Clarendon Press, 1980.
Crowne, D.P. The frontal eye-field and attention. *Psychological Bulletin,* 93, 232-260 (1983)

Dawkins, R. *The Selfish Gene,* London: Oxford U.P.,1976.
Delgado, J.M., Roberts, W.W. and Miller, N.E. Learning motivated by electrical stimulation of the brain. *Amer. J. Physiol.,* 179, 587-93, 1954.
Dennett, D.C. *Content and Consciousness.* London: Routledge & Kegan Paul, 1969.
Dennett, D.C. Current issues in the philosophy of mind. *American Philosophical Quarterly,* 15,4, 249-261 (1978).
Dennett, D.C. Intentional systems in cognitive psychology: the 'Panglossian Paradigm' defended. *The Behavioural and Brain Sciences,* 6, 343-391 (1983).
DeRenci, E. *Disorders of Space Exploration and Cognition.* New York: John Wiley, 1982.
Deutsch, J.A. The cholinergic synapse and the site of memory. In Deutsch, J.A. (ed) *The physiological basis of memory,* New York: Academic Press, 1983.
Donchin, E. and Coles, M.G.H. Is the P300 component a manifestation of context updating? *Behavioural and Brain Sciences,* 11, 357-374 (1988).

Eccles, J.C. (ed) *The Brain and Conscious Experience.* New York: Springer, 1966.
Eccles, J.C. and Popper, K.R *The Self and its Brain,* Springer International, 1977.
Eccles, J.C. An Instruction-selection hypothesis of cerebral learning. In Buser, P.A. and Rougeul-Buser, A. (1978).
Evart, E.V., Shinoda, S.P. and Wise, S.P. *Neurophysiological clues to Higher Brain Functions.* New York: Wiley, 1984.

Farah, M.J. The neurological basis of mental images. A componential analysis. *Cognition,* 18, 245-272 (1984).
Farah, M.J. Is visual imagery really visual? Overlooked evidence from neurophysiology. *Psychol. Rev.,*95,3, 307-317 (1988).

Finke, R.A.and Kosslyn, S.M. Mental imagery acuity in the peripheral visual field. *J. of Experimental Psychology: Human Perception and Performance*, 6, 244-264 (1980).
Fodor, J.A. *The Language of Thought*, Hassocks, Sussex: Harvester Press, 1975.
Fodor, J.A. *The Modularity of Mind,* Cambridge (Mass): MIT Press, 1983.

Gaffan, D. Recognition memory in animals. In Brown, J. (ed) *Recall and Recognition.* New York: Wiley, 1976.
Gallistel, C.R. The Organization of Action, Hillsdale N.J.:Lawrence Erlbaum Associates, 1980.
Garner H. *The Mind's New Science.* New York: Basic Books, 1985.
Georgopoulos, A.P., Lurito, J.T., Petrides, M., Schwartz, A.B. and Massey, J.T. Mental rotation of the neuronal population vector. *Science,* 243, 141 - 272, 1989.
Gevins, A.S. et al, Human neuroelectric patterns predict performance accuracy. *Science*,235, 580-585 (1987).
Gibson, J.J. *An Ecological Approach to Visual Perception.* Cornell U.P. 1976.
Gibson, J.J. *An Ecologist's Approach to Perception,* Boston: Houghton Mifflin, 1979.
Goodman, N. *Languages of Art.* Indiapolis:Hackett, 1976. Stoughton, 1977.
Gregory, R.L. Perception as hypotheses, *Phil. Trans. Royal Soc.(B)*, 280, 181-197 (1980).
Gregory, R.L. *Mind in Science.* London: Weidenfeld & Nicolson, 1981.
Gregory, R.L. (ed.) *The Oxford Companion to the Mind.* London: OUP,1987.
Grossberg, S. How does the brain build a cognitive code? *Psychological Review,* 87,1-42 (1980).

Hacker, P. Languages, Minds and Brains. In Blakemore, C. & Greenfield, S. (eds.) *Mindwaves.* Oxford: Blackwell (1987).
Hamilton, W.D. The genetical theory of social behaviour. *J. of Theor. Biology*, 7(1), 1-52 (1963).
Hardy, A. *The Biology of God.* London: Jonathan Cape, 1975.
Hawkins R.D. Cellular neurophysiological studies of learning. In Deutsch, J.A. *The Physiological Basis of Learning.* New York: Academic Press, 1983.
Hebb, D.O. *The Organization of Behaviour.* New York: Wiley, 1949.
Held, R. and Hein, A. Movement-produced stimulation in the development of visually guided behaviour. *J.Comp. and Phys.Psych.,* 56, 872-6 (1963).
Helmholtz, H. von. *Handbuch der Physiologischen Optik.* Leipzig: Voss, 1866.
Herrnstein, R.J., Loyeland, D.H. & Cable, C. Natural concepts in pigeons. *J. Exp. Psychol: Animal Behaviour Processes,*2: 285-302 (1976).

Hilgartner, C.A. and Randolph, J.F. Psycho-logics. *J.Theoret.Biol.*,23, 285-338 (1969).

Hintzman, D.L., O'Dell, C.S. and Arndt, D.R, Orientation in Cognitive Maps. *Cognitive Psychology,* 13,149-206 (1981).

Hochberg, J. In the Mind's Eye. In Huber, R.N. (ed.) *Contemporary Theory and Research in Perception.* New York: Holt, Reinhard & Winston, 1968.

Hochberg, J. Levels of perceptual organization. In Kubovy M. and Pomenatz, J.R. (eds.) *Perceptual Organization.* Hillsdale, N.J.: Erlbaum, 1981.

Hubel, D.H. and Wiesel T.N., Receptive fields and functional architecture in two nonstriate visual areas (18 and 19) of the cat. *J.Neurophysiol.* 28, 229-89, 1965.

Hull, C.L. *Principles of Behaviour.* New York: Appleton, Century- Croft, 1943.

Humphrey, N. *Consciousness Regained,* London: Oxford U.P., 1980.

Huxley, J. *The Humanist Frame.* London: Allen & Unwin, 1961.

Irving, F.W. *Intentional Behaviour and Motivation..* New York: Lippincott, 1971.

James, W. *Varieties of Religious Experience.* London: Longman, Green & Co., 1902.

Jasper, H.H. and Penfield, W. Electrocorticograms in man: effect of the voluntary movement upon electrical activity of the precentral gyrus. *Arch. Psychiat. Z. Neurol.*, 83, 163-174 (1949).

Jaynes, J. *The Origin of Consciousness in the Breakdown of the Bicameral Mind.* London: Allen Lane, 1979.

John, E.R. *Mechanisms of Memory.* New York: Academic Press, 1967.

Johnson-Laird, P.N. *Mental Models.* London: Cambridge U.P.,1983.

Kant, I. *Kritik der Praktischen Vernunft,* 1788.

Kemp, J.M. and Powell T.P. The cortico-striate projection in the monkey. *Brain,* 93,, 525-46 (1970).

Kilmer W.L., McCulloch, W.S. and Blum, J. Some mechanisms for a theory of the reticular formation. In Mesarovic, M. (ed) *Systems Theory and Biology.* New York: Springer, 1968.

Kohler, I. The formation and transformation of the perceptual world.(transl. by H. Fiss). *Psychological Issues,* 3, 1-173, (1964).

Kornheiser, A.S. Adaptation to laterally displaced vision: a review. *Psychological Bulletin,*33, 783-816 (1976).

Kornhuber, H.H. and Deecke, L. Hirnpotentialänderungen beim Menschen vor und nach Willkürbewegungen, dargestellt mit Magnetbandspeicherung und Rückwärtsanalyse. Pflügers *Arch. Ges. Physiol.* 261:52 (1964).

Kosslyn, S.M. *Images and Mind.* Cambridge (Mass.): Harvard U.P., 1980.

Kosslyn, S.M. Seeing and imagining in the cerebral hemisphere: a computational approach. *Psych. Rev.*, 94,2, 148-175 (1987).

Krushinsky, L.V. Solution of elementary problems by animals on the basis of extrapolation. In Wiener, N. and Schade, J.P. (eds), *Cybernetics and the Nervous System*. Amsterdam: Elsevier, 1965.

Lacquaniti, F. and Maioli, C. The role of preparation in tuning anticipatory and reflex responses during catching. *J. of Neuroscience*, 9, 134- 148 (1989).

Lee, A. *Goundwork of the Philosophy of Religion*. London: Duckworth 1946.

Lenneberg, E.H. *The Biological Foundation of Language*. New York: Wiley, 1967.

Levin, M. *Metaphysics and the Mind-Body* Problem. London: Clarendon Press,1980.

Levine, D.N., Warach, J., and Farah, M.J. Two visual systems in mental imagery. Dissociation of "what" and "where" in imagery disorders due to bilateral posterior lesions. *Science*, 35, 1010-1018 (1985).

Libet, B. Neuronal vs. Subjective Timing. In Buser, P.A and Rougeul-Buser, A. (1978).

Loftus, E.G. Leading questions and the eye-witness report. *Cognitive Psychology*, 7,560-572 (1975).

Lorenz, K. *Civilized Man's Eight Deadly Sins*. London: Methuen, 1974.

MacKay, D.M., Cerebral organization and the conscious control of action. In Eccles J.C. (ed), *Brain and Conscious Experience*. New York: Springer, 1969.

Mackay, D.M. What determines my choice? In Buser, P.A. and Rougeul-Buser, A. (1978).

Maddi,S.R. *Personality Theories*. Homewood (Illinois): Dorsey, 1968.

Marr, D. *Vision*. San Francisco: Freeman, 1982.

Maslow, A.H. Self-actualizing and beyond. In Lindsey, G., Hall C.S. and Manosevitz, M. *Theories of Personality*. New York: Wiley, 1973.

Masterson, P.*Atheism and Alienation*. London: Macmillan, 1971.

Miller, N.E. Some reflections on the Law of Effect produce a new alternative to drive reduction. In Jones, R.J. (ed), *Nebraska Symposium on Motivation*. Lincoln: Univ. of Nebraska Press, 1963.

Minsky, M. A framework for representing knowlege. In Winston, P.H. (ed), *The Psychology of Computer Vision*. New York: MacGraw-Hill, 1975.

Moruzzi, G. Synchronizing influences of the brain stem.Moscow Colloquium on Electroencephalography of Higer Nervous Activity (eds. Jasper, H.H. and Smirnov, G.D.), *EEG and Clin. Neurophysiology,* Suppl. 13 (1960).

Mountcastle, V.B., Anderson, R.A. and Motter, B.C: The influence of attentive fixation upon the excitability of light- sensitive neurons of the posterior parietal cortex. *J. of Neuroscience*, 1, 1218-1235 (1981).

Mountcastle, V.B., Motter, B.C.,Steinmetz, M.A. and Duffy, C.J. Looking and seeing: the visual functions of the parietal lobe. In Edelman et. al.(eds) *Dynamic Aspects of Neocortical Function.* New York: Wiley, 1984.
Murray, H.A. *Explorations in Personality.* London: Oxford U.P., 1938.

Nagel, E. A formalization of functionalism. In Emery, F.E.(ed) *Systems Thinking.* London, Penguin, 1969.
Neisser, U. *Cognitive Psychology.* New York: Meredith, 1967.
Neisser, U. *Cognition and Reality.* San Francisco: Freeman, 1976.
Noda, H., Freeman, R.B. and Creutzfeldt, O., cited in 'Nerve cells that keep the world steady.' *New Scientist,* 24 Feb. 1972.
Norton, T., Frommer, G. and Galambos, R. Effects of partial lesions of optic tract on visual discrimination in cats. *Federation Proc.,* 25, 2168 (1966).
Nowell Smith P.H. *Ethics.* London: Penguin, 1954.

Olton, D.S. Mazes, maps and memory. *Amer. Psychol.* 34:588-96. (1979).

Pandya, D.N. and Yeteran E.H. Proposed neural circuitry for spatial memory in primate brain. *Neuropsychologia,* 22, No. 2, 109-122 (1984).
Peacocke, A. *God and the New Biology.* London: Dent, 1986.
Perret, D.I., Amanda, J. Chitty, M.& A. Visual neurons responsive to faces.*Trends in Neurosciences,* Vol.10, No.9 ,1987.
Pfurtscheller, G. and Berghold, A. Patterns of cortical activation during planning of voluntary movements. *EEG and clinical Neurophysiology,* 72, 250-258 (1989).
Phillips, C.G., Zeki, S. & Harlow H.B. Localization of function in Cerebral Cortex. *Brain,* 107: 338-61, 1984.
Piaget, J. and Inhelder, B. *Memoire et Intelligence.* Paris: Presses Universitaires de France, 1968.
Piaget, J. *The Grasp of Consciousness* (transl. Susan Wedgwood), London: Routledge and Kegan Paul, 1977.
Popper, K. *Objective Knowledge.* Oxford: Clarendon Press, 1972.
Popper, K. and Eccles J.C. *The Self and its Brain.* London: Springer, 1977.
Premack, D. & Woodruff, G. Does the chimpanzee have a theory of mind? *The Behavioural and Brain Sciences,* 4: 515-26 (1978).
Pribram, K.H. The intrinsic system of the forebrain.In File, J., Magoun, H.W. and Hall, V.E. (eds) *Handbook of Physiology,* Vol.II, Section I. Washington: Amer. Physiol. Soc., 1960.
Pribram, K.H. *Languages of the Brain.* Monterey (Calif.): Brooks- Cole, 1971.

Pucetti, R. Brain bisection and unity of consciousness *Br. J. Phil. Sc.*,24, 339-355 (1973).
Pylyshyn, Z.W. Computation and Cognition: issues in the foundation of cognitive science. *The Behavioural and Brain Sciences,* 3, 111-169 (1980).
Pylyshyn, Z.W. What the mind's eye tells the mind's brain: a critique of mental imagery. *Psychological Bulletin,* 80, 1-24 (1973).
Pylyshyn, Z.W. The imagery debate: analogue media versus tacit knowledge. *Psychological Review,* 88, 1645 (1981).

Quine, W.V. *Word and Object,* Cambridge (Mass.): MIT Press, 1960.

Rapoport, J.L. The biology of obsessions and compulsions. *Scientific American,* 3, 63 - 69 (1989).
Richmond, B.J., Wurtz, R.H. and Sato, T. Visual responses of infero-temporal neurons in awake rhesus monkey. *J. of Neurophysiology,* 50, 6, 1415-1432 (1983).
Rock, I. *The Logic of Perception.* Cambridge (Mass.): MIT Press, 1983.
Roland, P.E. Sensory feedback to the cerebral cortex during voluntary movement in man. *The Behavioural and Brain Sciences,* 1, 129-71 (1978).
Rose, S. *The Conscious Brain.* London: Weidenfeld & Nicholson, 1973.
Rosenblueth, A. and Wiener, N. Purposeful and non-purposeful behaviour. *Philosophy of Science,* 17,, 318-26 (1950).
Roszak, T. *Where the Wasteland Ends.* London: Faber & Faber, 1973.
Ruse, M. and Wilson E.O. The Evolution of Ethics. *New Scientist,* 17.10.85.
Russell, Bertrand. *The Analysis of Mind.* London: Allen & Unwin, 1921.
Russell, Bertrand. *Religion and Science.* London: Oxford U.P., 1935.
Ryle, G.. *The Concept of Mind.* London: Hutchinson, 1949.

Sakata, H., Shibatani, H. and Kawane, K. Functional properties of visual tracking neurons in posterior parietal association cortex of the monkey. *J. of Neurophysiology,* 49,, No. 6, 1983.
Schilder, P. The Image and Appearance of the Human Body. *Psychol. Monographs,* No.4. London: Kegan, Trench & Trubner, 1935.
Schlag-Rey, M. and Schlag, J. Visuomotor functions of central thalamus in monkey. *J. of Neurophysiology,* 50, 1149-95 (1984).
Segal, S,J. and Fusella, V. Influence of imaged pictures and sounds on detection of auditory and visual signals. *J. of Exp. Psych.,* 83, 458-64 (1970).
Shallice, T. Dual functions of consciousness. *Psych. Rev.,* 79, 383-93 (1972).
Shannon, C.E. and Weaver, W. *The Mathematical Theory of Communication.* Urbana: Univ. of Illinois Press, 1949.

Shephard, N.R. and Metzler, J. Mental rotation of three- dimensional objects. *Science*, 171, 701-03 (1971).

Shepard, N.R. Kinematics of perceiving, imagining, thinking and dreaming. *Psych. Rev.*, 91, 417 - 447 (1984).

Skinner, B.F. *About Behaviourism*. London: Cape, 1974.

Skinner, B.F. London: ITV Broadcast, 7th March, 1978.

Sokolov, E.N. *Perception and the Conditioned Reflex*. London: Pergamon, 1963.

Sokolov, E.N. The modelling properties of the nervous system. In *Handbook of Contemporary Soviet Psychology*, ed. M.Cole & I. Maltzman. London: Basic Books, 1969.

Sommerhoff, G. The abstract characteristics of living systems. In Emery, F.E.(ed) *Systems Thinking*. London: Penguin, 1969.

Sommerhoff, G. *Logic of the Living Brain*. London: Wiley, 1974.

Sorabji, R. *Aristotle on Memory*. London: Duckworth, 1972.

Sperry, R.W. Lateral specialization in the surgically separated hemispheres. In McGuigan, F.J. and Schoonover, R.A.(eds) *The Psychophysiology of Thinking*. New York: Academic Press, 1973.

Sperry, R.W. Consciousness, personal identity and the divided brain. *Neuropsychologica*, 22, 661-673, 1984.

Szekely, E and Montgomery, D.L. *Mol. Cell. Biology*, 4, 939 (1984)

Tanji, J. and Evarts, E.V. Anticipatory activity of motor cortex neurons in relation to direction of intended movement. *J. of Neurophysiology*, 42, 1062-68 (1976).

Thorndyke, P.W. and Hayes-Roth, B. Differences in spatial knowledge acquired from maps and navigation. *Cognitive Psychology*, 14, 560-89 (1982).

Thorpe, W.H. *Animal Nature and Human Nature*. London: Methuen, 1974.

Tinbergen, N. *The Study of Instinct*. Oxford: Clarendon Press, 1951.

Tolman, E.C. Cognitive maps in rats and men. *Psych. Review*, 55, 189-208 (1948).

Toyama, K., Komatso, Y. and Shibuki, K. Integration of retinal and motor signals in eye movement in striate cortex cells of the alert cat. *J. of Neurophysiol.*, 47, 649-665 (1984).

Treisman, A. and Gormican, S. Feature analysis in early vision: evidence from search asymmetries. *Psych. Rev.*, 95, 1, 15-48 (1988).

Turing, A.M. Computing machinery and intelligence, *Mind*, 59, 433- 60 (1950).

Unger, S.M. Habituation of the vasoconstrictive orienting reaction. *J. of Exp. Psych.*, 67, 11-18 (1964).

Verleger, R. Event-related potentials and cognition: A critique of the context updating hypothesis and an alternative interpretation of P3. *Behavioural and Brain Sciences*, 11, 343- 427 (1988).

Vernon, M.D. The function of schemata in perceiving. *Psych. Review*, 62, 180-192, 1955.

Wallach, H., Frey, K.J. and Bode, K.A. The nature of adaptation in distance perception based on oculomotor cues. *Perception and Psychophysics*, 11, 110-16 (1972).

Walter, W.Grey, Slow potential waves in the human brain associated with expectancy, attention and decision. *Arch. Psych. Zeitschrift f.d. ges. Neurologie*, 206, 309-22, 1964.

Warnocke, Mary. *Imagination*. London: Faber and Faber, 1976.

Whitehead, A.N. *Science and the Modern World*. London: Collins, 1925.

Wiener, N. *Cybernetics*. New York: Wiley 1948.

Wilson, E.O. *Sociobiology*. Cambridge (Mass.): Harvard U.P.,1975.

Winograd, T. *Understanding natural language*. New York: Academic Press, 1972.

Winograd, T. and Flores F. *Understanding Computers and Cognition*. Norwood, N.J.: Ablex, 1986.

Wittgenstein, L. *Philosophical Investigations*. Oxford: Blackwell, 1953.

Woodfield, A. *Teleology*, London: Cambridge U.P., 1976.

Woodger J.H. *Biological Principles*. London: Kegan Paul,Trench & Trubner, 1929.

Wurtz, R.H., Goldberg, M.E. and Robinson,D.L. Behavioural modulation of visual responses in the monkey: stimulus selection for attention and movement. *Progress in Psychobiology and Physiological Psychology*, 9, 44-83 (1980)

Wurtz, R.H., Goldberg, M.E. and Robinson D.L. Brain mechanisms of visual attention. *Scientific American*, 246, No.6, 100-7 (1982).

Wurtz, R.H., Richmond B.J. and Newsome, W.T. Modulation of cortical visual processing by attention, perception and movement. In Edelman et al. (eds) *Dynamic Aspects of Neocortical Function*. New York: Wiley, 1984.

Yntema, D.B. and Trask, F.P. Recall as a search process, *J.of Verbal Learning and Verbal Behaviour*, 2, 65-74 (1963).

Zaporozhets, A.V. The developmnt of perception in the pre-school child. In Mussen, P.H. (ed) *European Research in Cognitive Development*, Vol. 30, No.2. Chicago: Univ. of Chicago Press, 1965.

Zeki, S.M. Functional specialisation in the visual cortex of the rhesus monkey. *Nature*, 274, No. 5670, 423-428, (1978).

INDEX

Selection: in the interest of relevance only the names of those authors have been indexed whose views are discussed in the text.

Definitions: The page on which a term is defined is shown in *italic*.

action strategies 122, 126, 169
    formation 151 – 159
adaptation (adaptiveness) 24, 40, 43, *64*
aesthetic sensibilities 31, 243
    see also meta-sensibilities
    genetic cost-effectiveness not a cause 248
    roots 259 – 266
    and art 266
altruism, human
    see also empathy
    contrast with animal 'altruism' 244, 248
Anderson, J.R. 196, 216
anxiety 32, 252, 253, 257, 272
approaches to brain theory:
    top-down vs. bottom-up 2, 3, 169
appropriateness, concept of *63*
Armstrong, D.M. 11
artificial intelligence (AI)
    and mind-modelling 4, 7, 228 – 234
arousal 153
attention 153
    shifts of ~ 98, 120, 227
automation, in motor activities 158
Ayer, A.J. x, 270

basal ganglia 153
Battista, J.R. 74
Beauty, sense of 31, 243, 259 – 266
    see also meta-sensibilites

Behaviourism ix, 6, 8, 72, 73, 132, 133
belief
    concept of 227
    religious ~ 276 – 281
Birkhoff, G.D. 262
Blakemore, C. 250
Boden, M. 229
body-schema (body-knowledge) 18, 119, 172
    see also self-awareness, first-order
brain, as integral system 7
    ~ lesions: effects on consciousness 159 – 168
Bruner, J.S. 135

Cannon, W.W. 41
Catell, R.B. 251
Causal relationships
    concept of ~ 54, *316*
    ~ in directive activities 53 – 58
classes
    internal representation of ~ 203, 206
    logical relations between ~ 204, 292
    properties as ~ 203, 291, 293–5
    relations as ~ 203, 291, 293–5
    calculus of ~ 217, 301
class-membership relation 201
    internal representation of ~ 205
cognitive consonance 260, 262
    – dissonance 32, 253, 262

- resolution 202, 207 – 209, *210*
- theories 2, 4 – 8, 131 – 137
completion, visual 111
conditional expectancies
  see expectancies
Conditional Expectancy Hypothesis
  discussed 27, 96, 99 – 104, 181
  formulated 100
  ~ and neural responses 148
confidence, degrees of 141
conflict 32, 252, 253, 256, 272
  intrapsychic vs. psychosocial 252, 261
connectivism 231
consciousness *91*
  importance of problem 3, 4
  lack of systematic treatment 5 – 7
  definitions discussed 5, 10, 21
  various descriptions 6
  meanings of the term 9 – 15, 20, 75 – 78
  criteria for ~ 14, 87, 91
  conditions of entry into 20, 91
  seat of ~ 159
  ~ in robots 234
  more questions answered 283
coordination 24, 40, 42, *287*
Craik, K. 83
Crook, J.H. 5, 74
Croce, B. 260
cultural evolution 32
  contrasts with Darwinian evolution 246, 248 – 251, 268
  some determining factors 274
cultural norms 31, 248 – 251, 268
  see also cultural evolution
Cannon, W.B. 41
Catell R.B. 251

Dawkins, R. 175
deep structure of sentence *307*

definitions, different kinds 9
Dennett, D.C. 5, 238
Descartes 10
determinism (causal) *317*
deterministic systems 54, 296
Deutsch, J.A. 187
Directive Correlation 3, 24, 49, *56*
  comments on definition 58 – 60
  three main categories of 59
  formal place in bridge between mind and matter 221
directiveness of vital activities 23, 37 – 43
  misconceptions 43 – 46
  general characteristics 47 – 51
  common pattern of causal relationships 51 – 58
disturbance 52
Dual Aspect Theory of mind 239
Dualism 168, 238

Eccles, J.C. 168
empathy 33, 247, 254
  see also altruism, imagination
  ~ in aesthetic experience 263
enhancement, neural *144*, 145
equilibrium-seeking
  not a directive correlation 61
expectancy, states of *67*
  role of ~ 27
  formation of ~ 27
  conditional ~ 28, 98
  what-leads-to-what ~ 96, 80, 159
  act-outcome ~ 96
  categories of 104 – 109
  neural correlates of 149 – 151
  reactions to violoation of ~ 149
    see also orienting reactions
  formal position in bridge between mind and matter 221
extended standard theory 213

Index **331**

extra-pyramidal motor system 153

feedback *23*
   in servomechanisms 23, 39, 48, 156
   in directive activities generally 55
   ~ from sensory inputs 95
   scope for ~ in motor system 154
   instability through positive ~ 176
figure from ground separation 111
   see also unity of figure or shape
Fodor J.A. 135
formal logic, basic concepts 299 – 303
freedom of the will 224 – 226
Freud, Sigmund 3, 252
frontal cortex 153
frontal eye-fields 145, 147
frustration 32, 252, 253, 255, 272
function, organic 223, 38, 42, *65*
Functionalism 42, 239

Galileo 280
Gestalt theorists 260
Gibson, J. & E. 136
global world-model
   see internal representations (category 1)
Goodman, N. 135
grammars
   generative vs. transformational 305
Gregory, R.L. 281
Grossberg, S. 116

habituation 98, 115
   ~ of orienting reactions 150
   as dynamic inhibition 115
harmony (organic) 42, 273, *289*
Heisenberg Indeterminacy 238
Helmholtz, H. von 135
Hobbes, Thomas 271
Hochberg, J. 104, 135
homeostasis 41

Hull, C.L. 73
humanist movements
   cultural influence of 245, 248
Hume, David 177, 271
Humphrey, N. 5
Huxley, J. 280

Identity Theory 235
independent variables
   technical meaning 53
imagination
   see also internal representations (category 2)
   a representation of possibilities 16,18,177
   role in moral and aesthetic responses 32, 253
   ~ as internal representation 88
   ~ properties of the ~ 177 – 180
   neural correlates of the ~ 180 – 185
   ~ and memory recall 88, 194
information (concept) 24, 60, *311*
   and learning 186
   in aesthetic experience 261, 262
   in the Gibson theory of perception 136
information theory 6, 311 – 313
integration, organic 24, 40 – 43
intentionality (concept) 237, 238
intentions and their execution 151 – 158
internalization of moral values 267
internal representations
   meaning of the phrase 13
   diverse views 83
   three main categories 16 – 20, 86 – 90
   functional definition of ~
      discussed 17, 25, 82-86
      adopted *85*
   neural correlates 25 – 29, 139 – 151
   physical structure of ~ 17, 26, 97 – 99, 221

propositional form of ~ 29, 196, 202,
   205, 214 – 216
~ of past events 16, 191
introspection 15, 73, 89, 92, 93, 174

James, William 3
Johnson-Laird, P.N. 74, 82, 175, 217
Jung, C.G. 92, 252

Kant, Immanuel 177, 263, 268
knowledge (concept)
   suggested definition 226
Kohler, I. 110
Kornhuber, H.H. 6
Korsakoff syndrome 195
Kosslyn, S.M. 184, 215

language 29
   diverse aspects 199, 200
   internal representations 203 – 214
Lashley K.S. 188
learning 59
   mechanisms in the formation of internal
      representations 121
   teleological aspects of ~ 185
   ~ and information 186
   physiology of learning changes 187
learning to anticipate 188
Lee, A. 280
Lenneberg, E.H. 109
limbic system 155
Linguistic Philosophy 11, 236
Locke, John 121, 240
Logical Positivism 271
Lorenz K. 250

MacKay, D.M. 225
Maddi, S.R. 252
Malcolm N. 11 – 13
mappings 298

Marr, D. 83, 133
Maslow, A.H. 246
Materialism (philosophical) 235
Mechanists (in biological theory) 45
memory (*see also* memory recall)
   aspects of 185 – 190
   ~ storage and retrieval 95
   ~ recall, episodic 16, 190 – 197
      – and imagination 88, 194
mental events
   concept of ~ 3
   main categories of ~ 219
mental images (*see also* imagination)
   properties 178 – 183
   rotation of ~ 179, 215
   ~ in propositional form 214
   ~ in reasoning 217
meta-sensibilities and -responses *246*, 248
   roots 254
mind (concept)
   see also mental events
   relation of ~ to matter 221, 234 – 240
Minsky, M. 83, 135, 230
modular analyzers 130, 133, 229
Moore, G.E. 272
moral norms, *see* cultural norms
moral sensibilities 31, 243,
   special features 245
   genetic cost-effectiveness not a cause
      248 – 251
   roots 32, 266 – 272,
metaphors, use and abuse of 235, 240
motor area of cortex 153, 154, 166
   see also pyramidal tract
Murray, H.A. 251

nameless entities 201, 207 – 210
Neisser, U. 135, 183
neural correlates of mental events
   see internal representations

Index

Objectivism in moral theory 272
operational mode (or *set*) of the brain 99
  see also readiness sets

order in the visual field 261, 262, 313
organic integration *288*
Organicism 45
organic order 40, *289*
  its dynamic nature 39, 41
  its directiveness 40
  diverse manifestations 42
  ~ and harmony 273
orienting reactions *18*, 95, 98, 115, 149
  inhibition through habituation 150

parietal cortex 119, 144, 162 – 165
Pavlov I.P. 72, 115
Peacocke, A. 46
Penfield, W. 165
personality theories 251
phrase-markers
  logical *203*, 210 – 212
  generative *203*, 306 – 308
physical,
  modern meaning of the term 1
  ~ systems 1, 135
Piaget, J. 134, 177
Plato 271
pleasure centres 156
Popper, K. 169
predicate logic 299
premotor cortex 166
prestriate cortex 130, 144, 162
progress, biological 276
proposition (linguistic) *205*
propositional (form of) representations
  see internal representations
P3 waves 114
purpose, concept of 23, 62

Pylyshyn, Z.W. 184, 215
pyramidal tract 154

Quine, W.V. 225

rational thought
  foundations of ~, 216 – 218
  roots in the nature of classification 217
  use of imagination in ~, 217
readiness sets 125
reality testing 17, 157, 180 – 184
recognition 189
redundancy 111, *312*
  role in aesthetic experience 261, 262
reference states *156*
regeneration 41
regulation, concept of, 24, *287*
Relativism (in moral theory) 272
religion
  see also next two entries
  regulative function 279
  conflict with science 281
religious movements
  influences of ~, 245, 248, 276
religious needs 277 – 279
reticular formation 153
Rock, I. 135
Roszak, T. 265
Russell, Bertrand 270
Ryle, G. x, 74, 177, 236

schemata 134
science
  term used throughout as short for the *physical* or *natural* sciences 1
  restraints on language of ~, 1, 235
  scientists' attitude to mind/matter problem 10
searching 4, 43, *288*

self, the 19
  meaning in different contexts 172
  unity in split-brain subjects 222 – 224
self-awareness
  first-order ~ (body knowledge) 19, *89*, 119, 172, 173
  second-order ~ (self-consciousness) 19, 89, 174 – 176
self-consciousness *See* preceding entry
  see also internal representations (category 3)
self-reports 15, 73, 89
sensorimotor cortex 147, 154, 167
sentence (in descriptive statements):
  what its structure conveys 210 – 214
  see also phrase markers
Shallice, T. 74
Shannon, C.E. 6
SHRUDU 232/3
Skinner, B.F. 6, 72
split-brain subjects 197, 222 – 224
sociobiology 244
Sokolov, E.N. 116, 149
S–R theories, *see* Behaviourism
stochastic processes 318
strategies, *see* action strategies
striate cortex 130, 148, 161
subconscious, the 92
subjectivity (privacy) of experience 92, 93, 266, 283
  ~ and AI 232
substantia nigra 144
superior colliculus 144
surface structure (of sentence) 307
surprise reactions 95
  see also orienting reactions
syllogism, figures of 300
system, physical 1, 315 –318
systems theory
  nature of 2

applications 22 – 24
target situation *152*
teleological aspects of vital activities
  see directiveness of vital activities
teleological concepts *23*
  range and importance 42 – 44
  neglect 45
  need for definitions in non-teleological terms 32, 47 – 62
  specific definitions 63 – 68, 287 – 290
teleological explanations 290
  mistaken beliefs 44, 45
temporal cortex 145, 195
thalamus 14, 45
Tolman E.C. 103, 104
truth table 302
trying (concept) 208
Turing machine 230

unconscious, the, 3, 92
unconscious inferences 135
unity of figure or shape 261, 262

variables 296
  see also independent variable
Venn diagrams 217, 293, 301
Virgil 44
visual perception
  role of movement in ~, 117
  constancies 128
  role of expectancies in ~, 127 – 131
  theories of ~, 131 – 137
vital activities: general characteristics 47
Vitalists 44
voluntary activity *93*
  ~ and freedom of the will 225

Watson, J.B. ix, 6, 72
Weaver, W. 6
what-where separation 164

Whitehead, A.N. 280
will, freedom of: *see* freedom of the will
Winograd, T. 7, 233

Wittgenstein L. 81, 213
Woodfied, A. 42
Woodger, J.H. 45

www.ingramcontent.com/pod-product-compliance
Ingram Content Group UK Ltd.
Pitfield, Milton Keynes, MK11 3LW, UK
UKHW020657050526
12271UKWH00003B/11